RADIOWAVE PROPAGATION

RADIOWAVE PROPAGATION

Physics and Applications

CURT A. LEVIS
JOEL T. JOHNSON
FERNANDO L. TEIXEIRA

The cover illustration is part of a figure from R.C. Kirby, "Introduction," Lecture 1 in *NBS Course in Radio Propagation*, Ionospheric Propagation, Central Radio Propagation Laboratory, National Bureau of Standards, U.S. Department of Commerce, Boulder, CO 1961. See Figure 1.5 for the original.

WILEY
A JOHN WILEY & SONS, INC., PUBLICATION

Published by John Wiley & Sons, Inc., Hoboken, New Jersey
Published simultaneously in Canada

Library of Congress Cataloging-in-Publication Data:

Levis, C. A., 1926-
Radiowave propagation : physics and applications / C.A. Levis, J.T. Johnson, and F.L. Teixeira.
p. cm.
Includes bibliographical references.
ISBN 978-0-470-54295-8 (cloth)
1. Radio–Antennas. 2. Radio wave propagation. 3. Radio–Transmitters and transmission. I. Johnson, J. T., 1969- II. Teixeira, F. L., 1969- III. Title.
TK6565.A6L46 2010
621.384′11–dc22

2009045884

CONTENTS

PREFACE

This textbook on radiowave propagation is designed for use at the college senior and first-year graduate school level. It should also be useful to practicing engineers, particularly antenna and telecommunication systems engineers, who need a good understanding of propagation but may not have had sufficient exposure to this subject. The number of antenna courses taught in American universities seems to exceed greatly the number of propagation courses; this text is intended to encourage more radiowave propagation course offerings.

We have tried to achieve a balance between the theoretical developments, that is, the physics of radiowave propagation, and the many applications. The theory is necessary for a thorough understanding of the propagation phenomena and, especially, the limitations of various methods for predicting system performance. Although the discussions of applications are intended to be useful in themselves, they also contribute to the general understanding. A broad spectrum of applications is considered, as opposed to focusing only on a specific technology, such as cellular systems. Of course, a text of this size cannot cover all applications in complete detail. Fortunately, the recommendations of the Radiocommunication Sector of the International Telecommunication Union (ITU-R) are a repository of the currently recommended practices, and the student is often referred to these recommendations for details beyond the scope of this book.

In writing this book, we have reviewed some basic material that is all too often slighted in courses on electromagnetic theory and antennas. For example, in treating the behavior of dielectrics, electromagnetic theory courses often proceed very quickly to free space or very simple media to lay the groundwork for antennas and transmission lines. However, radiowave propagation media often are not simple: they can be anisotropic, inhomogeneous, lossy, dispersive, and time varying. Therefore,

Chapter 2 treats dielectrics with such properties in mind. Similarly, Chapters 3 and 4 review plane wave theory and antenna and noise concepts from the propagation point of view. Chapter 6 reviews basic refraction and reflection concepts before applying them to propagation over terrain. In the same spirit, Chapter 8 includes a brief introduction to the probability theory concepts necessary to characterize fading.

Since the late 1950s, courses in Radiowave Propagation have been given regularly at The Ohio State University (OSU). An initial set of course notes appeared in 1961 after one of the authors attended an intensive and well-documented three-week course given by the staff of the Central Radio Propagation Laboratory of the National Bureau of Standards in Boulder, CO. Needless to say, there have been many additions and changes over the years as new applications (e.g., satellite and cellular communications) and new tools (e.g., computer programs for ionospheric predictions) became available and others waned in importance and as the instructors for the course changed. Substantial changes were also made in converting the course notes into a book. Each of the authors has had experience in teaching the course. We are grateful to many generations of students for their comments and suggestions, and we look forward to the inputs of future generations.

We wish to express our appreciation to those who were especially helpful in the making of this book. A sabbatical year at the National Center for Atmospheric Research as Senior Post-doctoral Fellow helped Prof. Levis broaden his understanding of atmospheric effects and research; he is indebted to many members of the staff, but especially to Dr. Ray Roble for insights into the dynamics of the upper atmosphere and to Dr. Pauline Middleton for general guidance and friendship. Later, most of a year as Invited Guest Worker at the Institute for Telecommunication Sciences (ITS) of the National Telecommunications and Information Administration (NTIA) was similarly stimulating, with access also to the staff of the National Geophysical Data Center (NGDC) of the National Oceanic and Atmospheric Administration (NOAA). Many staff members were helpful, but special thanks are due to Dr. James R. Wait, Dr. George Hufford, Dr. Hans Liebe, and Dr. Kenneth Allen at ITS and to Dr. Kenneth Davies and Mr. Ray O. Conkright at NOAA. Prof. Levis also acknowledges help from Reference Librarian Jean Bankhead. He is also grateful for the support and friendship of Alan and June Roberts who opened their home and their hearts to him.

Prof. Teixeira is grateful to the Department of Electrical and Computer Engineering of OSU for providing him with a sabbatical leave in 2009 for the completion of this book. We thank Ms. Lisa Stover of the OSU ElectroScience Laboratory (ESL) for editorial assistance and Ms. Mengyuan Guo, also of ESL, for help in the preparation of many of the figures. We also thank the staff at John Wiley & Sons, Inc., including our editor, George Telecki and his assistant, Lucy Hitz; production manager Danielle Lacourciere, and project manager Nitin Vashisht.

We are ever so grateful for the support of our wives and our families, and for their patience while we were preoccupied with teaching and writing.

Curt A. Levis
Joel T. Johnson
Fernando L. Teixeira

1

INTRODUCTION

1.1 DEFINITION OF PROPAGATION

Information can be transmitted in many ways. The use of electromagnetic waves for this purpose is attractive, in part, because direct physical connections such as wires or cables are not required. This advantage gave rise to the terms "wireless telegraphy" and "wireless telephony" that were commonly used for radio in the early part of the past century and have returned to popular usage with the widespread development of "wireless" systems for personal communications in recent decades. Electromagnetic waves are utilized in many engineering systems: long-range point-to-point communications, cellular communications, radio and television broadcasting, radar, global navigation satellite systems such as the Global Positioning System (GPS), and so on. The same considerations make electromagnetic energy useful in "sensors", that is, systems that obtain information about regions from which transmitted energy is reflected. Electromagnetic sensors can be used for detecting hidden objects and people, oil and gas exploration, aircraft control, anticollision detection and warning systems, for measuring electron concentrations in the Earth's upper atmosphere (and in planetary atmospheres in general), the wave state of the sea, the moisture content of the lower atmosphere, soils, and vegetation, and in many other applications.

In most cases, it is possible to divide the complete system, at least conceptually, into three parts. The first is the transmitter, which generates the electromagnetic wave in an appropriate frequency range and launches it toward the receiver(s) or the region

Radiowave Propagation: Physics and Applications. By Curt A. Levis, Joel T. Johnson, and Fernando L. Teixeira

to be sensed. The last is the receiver, which captures some fraction of the energy that has been transmitted (or scattered from the medium being sensed) for extracting the desired information. Propagation is the intervening process whereby the information-bearing wave, or signal, is conveyed from one location to another. In communications, propagation is the link between the transmitter and the receiver, while for sensors, propagation occurs between the transmitter and the target to be sensed and between the target and the receiver.

1.2 PROPAGATION AND SYSTEMS DESIGN

Propagation considerations can, and usually do, have a profound influence on communication systems design. They are therefore of great importance to the system engineer as well as to the propagation specialist. For example, consider the case of the "White Alice" system, a communication system implemented in Alaska and northern Canada during the late 1950s, before the advent of satellite communications. It was designed partly for general communications needs and partly to convey information from the Distant Early Warning (DEW) Line to Command Centers of the U.S. Defense forces. The establishment and maintenance of communications centers in an inhospitable environment in the Arctic was a difficult and expensive task. In the high-frequency (HF) band (3–30 MHz), it is possible to transmit signals for very great distances with very modest equipment and antennas—a fact well known to radio amateurs. Thus, an HF system might seem to have been the best solution in this case. There are, however, several drawbacks to this solution. The ready propagation of HF signals would make an HF system very susceptible to interference from signals arriving from other parts of the Earth. Also, HF propagation strongly depends on the ionosphere, an ionized atmospheric region that is significantly influenced by the Sun. At times, the Sun ejects huge streams of charged particles that severely upset the ionosphere and make HF communication in the Arctic and sub-Arctic regions particularly difficult. Thus, an HF system might have been cheap, but it would have been unreliable, and unreliability was unacceptable for this application.

The method chosen was based on the "tropospheric scatter" mechanism, using a frequency of operation of about 900 MHz. This propagation mechanism uses the reflection of signals by minor irregularities that are always present in the lower atmosphere. In contrast with HF propagation links, the ranges that can be achieved by tropospheric scatter links are only on the order of 200 miles, necessitating intermediate communications (repeater) stations between the DEW Line and the more populated areas. Also, very large antennas and high-powered transmitters are required. Figure 1.1 is a picture of a typical "White Alice" site—it should be apparent that its establishment and maintenance in the far North was neither easy nor cheap. Nevertheless, the high reliability and relative freedom from interference associated with tropospheric scatter propagation outweighed the cost and other considerations, so this system was implemented. The White Alice system was eventually superseded by satellite-based systems. This is, however, an interesting historical example that illustrates how propagation considerations can play a dominant role in communication systems design.

FIGURE 1.1 "White Alice" tropospheric communications site. (Courtesy of Western Electric.)

1.3 HISTORICAL PERSPECTIVE

The idea that electromagnetic signals might propagate over considerable distances at the velocity of light was first proposed in 1865 by James Clerk Maxwell. Having added the "displacement current" term to the set of equations governing electromagnetic events (now termed Maxwell's equations), he deduced that among their possible solutions wave solutions would exist. This prediction was verified experimentally by Heinrich Hertz in a series of experiments conducted in the late 1880s. Many of his experiments utilized waves of approximately 1 m wavelength, in what is now termed the ultra high frequency (UHF) range, and transmission distances were usually on the order of only a few feet.

Such an orderly progression from theory to experimental verification has by no means been characteristic of the field of radiowave propagation in general, however. When Guglielmo Marconi attempted his first trans-Atlantic transmissions in 1901, using waves of approximately 300 m length in what is now called the medium-frequency (MF) band, there was no clear theoretical understanding of whether signal transmission over such great distances might be possible. After the experiment was a success, there remained considerable controversy about the actual propagation mechanism until the now accepted explanation—ionospheric propagation—was sufficiently well understood to be generally accepted. Such a lag of theory behind experiment was quite characteristic of the early days of radio propagation. As more powerful transmitters and more sensitive receivers became available over increasing frequency ranges, the body of knowledge regarding electromagnetic wave propagation has grown enormously and has become increasingly complex.

1.4 THE INFLUENCE OF SIGNAL FREQUENCY AND ENVIRONMENT

In part, this complexity is due to the extraordinary range of frequencies (or wavelengths) that are useful for signal propagation. The lowest of these are in the vicinity of 10 kHz (30 km wavelength), although even lower frequencies (longer wavelengths) are useful for underwater communications and for observing some geomagnetic phenomena. For optical systems, the highest frequencies of interest for communicating information over considerable distances are on the order of 10^{15} Hz, corresponding to a wavelength of a few tenths of a micron. Thus, a frequency (or wavelength) range of 11 orders of magnitude is spanned! A corresponding range in the case of material structures would span from the lengths of the largest bridges to those of the smallest viruses!

The variety and complexity of the observed electromagnetic signal propagation phenomena are further increased by the variety of intervening environments. For example, the depth to which an electromagnetic wave penetrates in seawater varies by a factor of more than 50 over the frequency range 100 kHz to 300 MHz. There are also geographic variations, since salinity varies geographically. In the very low frequency (VLF) band, all wave structures in the ocean are small compared to the wavelength of the signal, and the ocean can be approximated as a smooth conducting surface. At higher frequencies, the signal wavelength decreases until it may be of the same order of magnitude as the large ocean swells. The ocean thus behaves as a rough lossy dielectric in the very high frequency (VHF) range; but in this frequency range, the small capillary wavelets due to wind can still be ignored. As the frequency is increased further, so that the signal wavelength becomes a few millimeters or less, the swells represent randomly tilted flat plates; it is now the capillary wavelets that represent roughness. Clearly, the sea surface is a complicated propagation medium boundary. Land exhibits similar variations except that in most cases there are no important short temporal scale changes.[1] The atmosphere, like the sea, is highly variable both temporally and geographically since its lower levels are strongly influenced by the weather, and its upper ones by solar activity.

It is not surprising, then, that different propagation calculation techniques are appropriate to the various frequency and environmental regimes, and that in many cases the resulting predictions are only of an approximate and statistical nature. This book does not attempt to develop any of these techniques exhaustively, but rather gives a survey of the most common calculations and phenomena.

Since signal frequency is such a very important parameter, a rough indication of which range is being considered is often necessary. The frequency band designations recommended by the IEEE as given in Table 1.1 will be used for this purpose. These are frequently used in the literature as well. They are generally not meant to be taken too literally, as propagation phenomena do not fall so neatly into decade frequency regions. Nevertheless, this nomenclature is useful for giving a quick, rough indication of the frequency range under discussion. The UHF and SHF bands, commonly used in radar applications, have an additional set of frequency band designations used for radar systems, as indicated in Table 1.2.

[1] An exception would be those produced by some land vegetation under strong winds.

TABLE 1.1 IEEE Frequency Band Designations

Band Name	Abbreviation	Frequencies	Wavelengths
Very low frequency	VLF	3–30 kHz	10–100 km
Low frequency	LF	30–300 kHz	1–10 km
Medium frequency	MF	300 kHz to 3 MHz	100 m to 1 km
High frequency	HF	3–30 MHz	10–100 m
Very high frequency	VHF	30–300 MHz	1–10 m
Ultra high frequency	UHF	300 MHz to 3 GHz	0.1–1 m
Super high frequency	SHF	3–30 GHz	1–10 cm
Extremely high frequency	EHF	30–300 GHz	1–10 mm

The wide frequency range employed and the variety of natural environments give rise to a surprising number of propagation mechanisms for electromagnetic signals. By "propagation mechanism" we mean a physically distinct process by which the signal may travel from the transmitter to the receiver or to and from the region being sensed. Depending both on frequency and on the environmental conditions, one or only a few mechanisms usually produce much higher signal strengths at the receiver than the others. The former are said to be the dominant mechanisms for the conditions considered (some authors use the word "modes"), and the others can often be neglected. One of the aims of this book is to develop an understanding of which mechanisms are likely to be dominant for particular frequency ranges and environmental conditions.

As an example, consider ordinary ionospheric reflection. Under many conditions, the ionosphere is very effective in guiding signals in the range VLF to HF. Signals propagated by this mechanism are likely to be much stronger than those received in any other way over the same distance, and other mechanisms may be neglected when this is true. But this is not always true. For example, the ionospheric reflection process is not useful when distances between the transmitter and the receiver are relatively short. In that case, signals arriving at the ionosphere at steep angles pass right through it, while those going up at shallow angles end up at too great distances; the ionospheric signals are then said to "skip" over the receiver. Also, in the LF range during the

TABLE 1.2 Microwave Frequency Band Designations

Band Name	Frequencies (GHz)	Wavelengths (cm)
L	1.0–2.0	15–30
S	2.0–4.0	7.5–15
C	4.0–8.0	3.75–7.5
X	8.0–12.0	2.5–3.75
K_u	12.0–18.0	1.67–2.5
K	18.0–27.0	1.11–1.67
K_a	27.0–40.0	0.75–1.11
V	40.0–75.0	0.40–0.75
W	75.0–110	0.27–0.40

daytime, certain regions of the ionosphere absorb signals very effectively so that the signal strength is attenuated. The same is true of HF at medium and high latitudes when the Sun is highly disturbed. Under any of these conditions, other propagation mechanisms may become dominant—otherwise, no *efficient* propagation mechanism may exist, so that terrestrial communications in certain frequency ranges becomes very difficult or impossible.

One example of this variability of propagation mechanisms is easy to observe: the daytime/nighttime effect in U. S. standard (AM) radio broadcasts, which fall into the MF band. During the daytime, only stations within a 200 or 300 mile radius are received, and the reception is likely to be interference free. At sunset, the picture changes, sometimes with dramatic abruptness. Now distant stations can be received with ease. Unfortunately, on those frequencies that are shared by several stations, many are likely to be received simultaneously, much to the dissatisfaction of the listener! The nighttime propagation mechanism is dominated by ionospheric reflection, the so-called "skywave" mode, but since (as noted above) the ionosphere absorbs signals well in this frequency range in the daytime, the daytime mechanism is a different one, namely, the so-called "groundwave" mode. Thus, a change in the dominant propagation mechanism is responsible for the difference in reception conditions. Note that the groundwave mode is present at night as well, but its effect is not noticed then at large distances from the transmitter because of the much higher signal strength of the skywave ionospheric signal.

1.5 PROPAGATION MECHANISMS

Before proceeding further, it will be useful to first develop some cursory acquaintance with the propagation modes or mechanisms to be discussed later in this book in more detail. Propagation mechanisms are separated below into "typical" and "unusual" categories, depending on the degree to which they are used in practice. We begin with the "typical" mechanisms, which are treated in later chapters.

Direct Propagation The simplest mechanism is perhaps *direct propagation*, involving the travel of the signal directly from the transmitter to the receiver quite unaffected by the intervening medium. It assumes the form of a spherical wave emanating from the transmitter. Since the receiver is often sufficiently far away from the transmitter, this wave can be approximated as a plane wave at the receiver location. Direct propagation may seem to be a highly idealized situation, but it has important applications. For frequencies in the UHF and higher frequency bands, the ionosphere has little influence, essentially because the electrons responsible for its conductivity at lower frequencies are unable to follow the rapid variations at such high frequencies. Also, at higher frequencies it is possible to build very directive antennas, so that the signal beam may be kept isolated from ground effects (except perhaps at the end point of its intended path, if this is very close to the ground). Under these conditions, the signal propagation is mostly unaffected by ground or sky effects: propagation is essentially direct. Since most radars and satellite communications systems operate in this fashion, and because a narrow beam is also advantageous for separating a particular radar target from its surroundings, direct propagation is the dominant mechanism, and sometimes the only one to be considered, for many microwave radar and satellite communications calculations. The frequency spectrum for direct propagation is not open ended at the higher frequency end, however, because then a band of frequencies (in the upper SHF range and above) is reached in which atmospheric constituents are able to absorb energy efficiently. In this range, the direct propagation assumption must be modified to account for this absorption by the inclusion of an additional attenuation term. As frequency is increased even further, the wavelength decreases until it becomes of the order of magnitude of atmospheric dust and water droplet particles. As a result, these particles can scatter or absorb the signal quite strongly, which requires further modification of the propagation model. In short, direct propagation is the appropriate mechanism to consider only when all other mechanisms are inoperative, a situation most frequently encountered in the atmosphere at UHF and SHF with systems utilizing highly directive antennas and when the transmitter and receiver are in plain view (line of sight) with respect to one another at elevation angles that preclude ground effects.

The effect of gravity causes the atmosphere to be generally more dense and moist at lower altitudes than at higher ones. Though the effect is small, it causes a significant bending of the propagated signal path under many conditions. For example, in the design of microwave links that are sometimes used for long-distance telephone voice and data communications, care must be taken that the link will perform adequately for a variety of atmospheric conditions that may cause the beam to bend upward or downward. This bending is known as *tropospheric refraction.* The atmospheric effects that cause tropospheric refraction also cause time delays as signals propagate through the atmosphere; such time delays have a significant impact on systems used for global navigation such as the Global Positioning System (GPS).

Ducting The bending effect of tropospheric refraction may be strong enough to cause signals to follow closely along the curvature of the Earth, so that they are in effect guided along the Earth. Such behavior is called ducting. Ducts are most frequently observed at VHF and UHF; they also exist at higher frequencies but the more directive antennas employed at these frequencies are less likely to couple efficiently into a duct. Ducting is much more common at some localities than others since it is closely related to meteorological phenomena. In most areas of the world, it is a source of potential interference rather than a means of reliable communication.

Earth Reflections If the antennas used are not very directional, or if they are located near the ground, signals may travel from the transmitter to the receiver by reflection from the ground. In this case, both the directly propagated signal and the ground-reflected signal must be considered in evaluating the propagation performance of a system. A typical case is ground-to-air or air-to-air communication at UHF. The size limitation of aircraft antennas makes it impossible to use highly directive antennas in this frequency regime, so it is not possible to keep signals from reaching the ground. The ground reflected signals can be added to or subtracted from the directly propagated

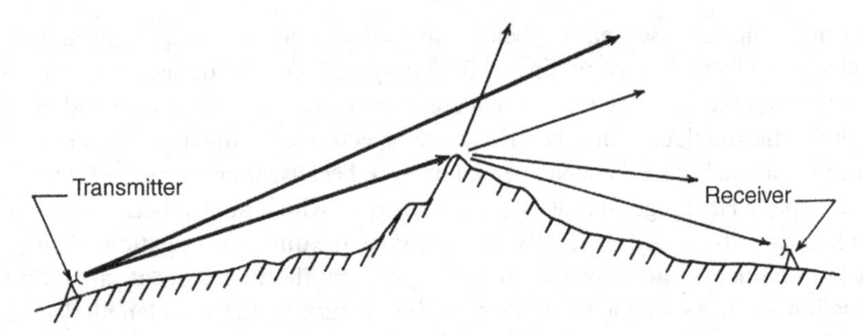

FIGURE 1.2 Diffraction by terrain.

signal (constructive or destructive interference, respectively), so both terms must be considered.

Terrain Diffraction All the mechanisms considered thus far can be described using the concept of rays, that is energy traveling along straight or nearly straight paths. Therefore, these propagation mechanisms would allow no signal transmission when the transmitter and receiver are not within the line of sight. However, such transmission is still possible, and the reason becomes apparent when diffraction is taken into account. Diffraction by the Earth's curvature itself is important, but even more pronounced is the effect of sharper obstacles, such as mountains. These obstacles scatter energy out of the incident beam, some of it toward the receiver as shown in Figure 1.2.

Multipath Environments In many cases, it is possible for transmitted signals to reach a receiver by multiple reflection or diffraction paths instead of a single reflection from the Earth's surface or a single diffraction point from the terrain. When many time-delayed and/or distorted copies of a transmitted signal are received, the term "multipath environment" is used to describe the propagation mechanism. Multipath is usually an important factor for ground-based point-to-point links, especially in urban environments, and must be considered, for example, in the design of wireless cellular communications and data networks. Because the consideration of multiple paths between the source and the receiver can become very complex, it is common to use a statistical methodology (in terms of statistical "fading models") to describe the average properties and variability of propagation links in multipath environments. Recent years have seen extensive efforts in developing communications modulation and signal processing strategies to combat the impact of multipath fading effects. Our discussion will focus on empirical models and fading statistics used to model such propagation channels.

Groundwave When both the transmitting and receiving antennas operate near or on the ground, it is found that the direct and reflected waves cancel almost completely. In this case, however, one also finds that a wave can be excited that travels along the ground surface, one of several types of waves denoted as "surface waves." Since efficient transmitting antennas at MF and lower frequencies are necessarily large in size (since the wavelength is long), they are generally positioned close to the ground, and groundwave propagation becomes important at these lower frequencies. Groundwave propagation is the dominant mechanism for standard daytime (AM) radio broadcast transmissions in the United States; it is usually not an important mechanism at frequencies above the HF band.

Ionospheric Reflections In the MF and HF bands, electromagnetic signals can be well described in terms of rays that are reflected by the ionosphere and the ground—actually the rays are bent rather than sharply reflected in the ionosphere, but the net effect is essentially the same as shown in Figure 1.3. Signal transmission by this means can be very efficient, and great distances can be spanned with modest power and equipment. For this reason, "short-wave" bands, as these frequencies are often called popularly, are utilized for broadcasting, point-to-point communications, and amateur (ham) radio. Depending on the signal frequency, the reflection can occur in different regions (also called layers) of the ionosphere.

In the VLF and LF parts of the spectrum, the ionosphere and the Earth may be considered, respectively, as the top and bottom of a waveguide that guides energy around the Earth. This point of view is particularly useful at the lower end of these frequencies, because then the wavelength is so long that the spacing of the "waveguide walls", that is, the Earth's surface and the effective ionospheric region, is on

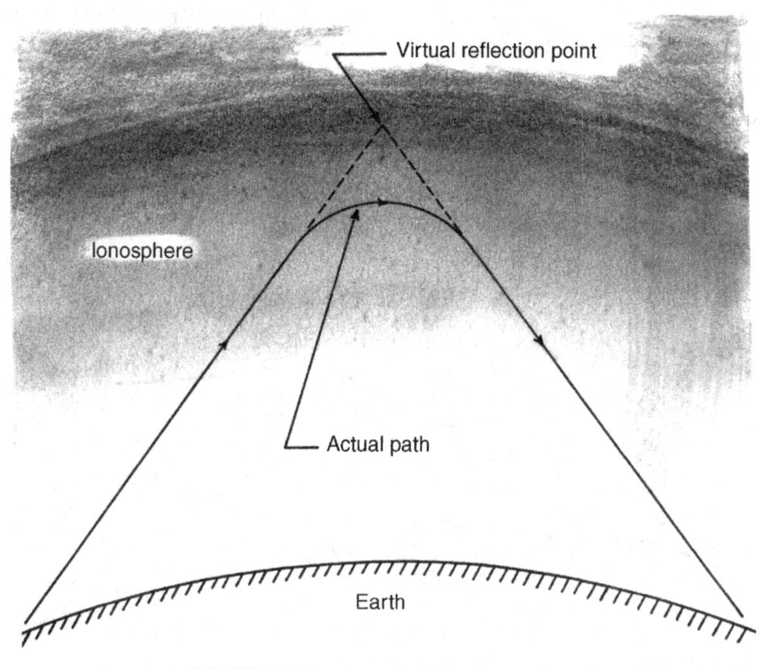

FIGURE 1.3 Ionospheric "reflection."

the order of a wavelength, and the mode description becomes relatively simple for calculation. Strictly speaking, it is not quite correct to distinguish between ionospheric reflection and waveguide modes as different physical mechanisms: in both cases, the signal is guided between the Earth and the ionosphere. However, if the wavelength is sufficiently long, it is more convenient to treat the problem in terms of waveguide modes. If the wavelength is very short compared to the Earth–ionosphere spacing, the ray picture becomes more convenient, and one treats the problem as a series of reflections. In between, in the LF region, computations by either technique become difficult. The distinction between ionospheric hops and waveguide modes is based more on the mathematical description employed than on the actual physical process itself. This book focuses on the ray model of ionospheric propagation.

In addition to apparent reflections or guiding of signals, the ionosphere can also cause important effects on higher frequency systems (up to about 3 GHz) on Earth-to-space paths. Electromagnetic waves propagating through the ionosphere can experience time delays (as in tropospheric propagation), polarization rotation, and scintillation effects.

Unusual Propagation Mechanisms More "unusual" mechanisms that can still be quite effective methods for communication in certain circumstances (as in the "White Alice" example) are discussed below. Brief summaries of some of these mechanisms are provided in Chapter 12.

Tropospheric Scatter: The troposphere is never truly homogeneous, as common experience with wind gusts and other meteorological phenomena suggests. Tropospheric irregularities may be used to advantage when communications are needed over a path of several hundred miles, so that the transmitter and receiver are not within the line of sight of each other. If very strong signals are beamed at a region of the atmosphere that is within the line of sight of both stations as in Figure 1.4, the relatively small signal scattered out of the beam may

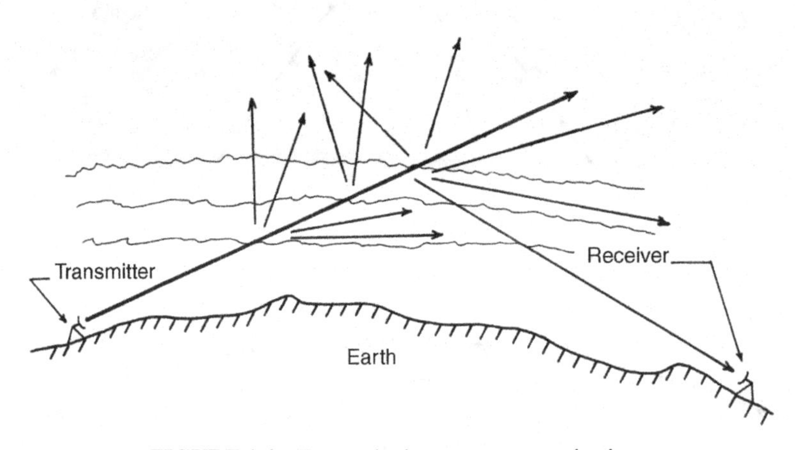

FIGURE 1.4 Tropospheric scatter communications.

be sufficient to allow significant information transfer between the terminals. This is the mechanism employed in the "White Alice" system mentioned at the beginning of this chapter.

Ionospheric Scatter: Signals of a frequency too high to be reflected from the ionosphere may nevertheless still be slightly affected by it. One of these effects is the scattering out of the beam of a small amount of the energy by ionospheric irregularities, quite analogously to the scattering by tropospheric irregularities discussed above. Ionospheric scattering is most noticeable in the frequency regime immediately above the end point of ionospheric reflection. Ionospheric scatter communication systems have been operated successfully in the VHF band.

Meteor Scatter: Many persons think of meteors as rare phenomena, perhaps because our daytime habits and increasingly urban existence lead us to see them rarely. Actually, visually observable meteors are not rare by any means. Most importantly, those too small to be observed visually are even vastly more abundant. As meteors enter the atmosphere, trails of ionized gas are formed that are capable of reflecting electromagnetic signals. Since the ionization is more intense than that of the ionosphere, signals of a high enough frequency to be relatively unaffected by the ionosphere may be returned from meteor trails. Systems in the VHF band have been built based on this mechanism and operated successfully.

Whistlers: Electromagnetic signals in the audio range of frequencies can propagate through the ionosphere in a peculiar mode, in which they follow closely the lines of the Earth's magnetic field. It is not easy to launch man-made signals of such very long wavelengths, but lightning strokes generate and launch energy in this frequency range quite effectively. The lightning-generated signals travel along the Earth's magnetic field lines, often going out a distance of several Earth radii, and are guided by the line to the point on the opposite hemisphere where the field line terminates. This point is called the antipode; at the antipode, the signal may be detected. Part of the signal may be reflected at the antipode to travel back along the same magnetic field line to the point of origin, and so on, back and forth. The peculiar sound of the signals has led to the name "whistlers". This mode of propagation has not been utilized for information transmission because of the very small bandwidth available at such low frequencies and the very restricted area of reception, but it has been a means of obtaining information about the upper ionosphere.

Non-Atmospheric Propagation The propagation phenomena discussed so far have dealt with electromagnetic signal propagation through space or through the Earth's atmosphere over considerable distances. Of course, a totally different environment prevails when electromagnetic signals are propagated through the ocean, the Earth's crust, or other planetary atmospheres, and this can have many applications. Electromagnetic wave propagation into the Earth's crust, for example, can aid in the detection of underground objects and tunnels and in the search for oil and gas fields. In this book, we will be concerned only with propagation in the Earth's atmosphere.

1.6 SUMMARY

Propagation is the process whereby a signal is conveyed between the transmitter and the receiver. An advantage of signal transmission based on electromagnetic waves is that no material link, such as a wire or a cable, is required between the transmitter and the receiver. Propagation considerations can have profound influence on systems design. The signal frequency and the environment determine which propagation mechanisms are dominant. Although these mechanisms generally appear to involve distinct physical processes, in some cases what is different is not the physical process but the mathematical model used to represent it.

A pictorial summary of many of these mechanisms is provided in Figure 1.5. The use of natural and artificial satellites to reflect or retransmit signals is also indicated in this figure.

Table 1.3 lists the most common mechanisms with some applications to which they are most appropriate. The list of applications is, of course, far from inclusive. The notation used for the ionosphere (D-layer, E-layer, and F-layer) is standard, the D-layer being the lowest region useful for reflecting signals back to the Earth, and the F region being the highest (see Chapter 10, section 10.4).

TABLE 1.3 Examples of the Application of Various Propagation Mechanisms

Propagation Mechanism	Applications
Direct	Most radar systems. SHF ground-to-satellite links.
Direct plus Earth reflections	UHF broadcast TV with high-gain antennas. Ground-to-air and air-to-air communications.
Multipath environments	VHF and higher ground-based point-to-point links (especially in urban areas).
Groundwave	Local standard broadcast (AM). Local HF links.
Ionospheric waveguide (D layer)	VLF and LF systems for long-range communications and navigation.
Ionospheric skywave (E and F layers)	MF and HF broadcast and communications (including long-range amateur radio).
Tropospheric scatter	UHF medium-range communications.
Ionospheric scatter	Experimental medium-range communications in lower VHF band.
Meteor scatter	VHF narrow-band long-range communications.

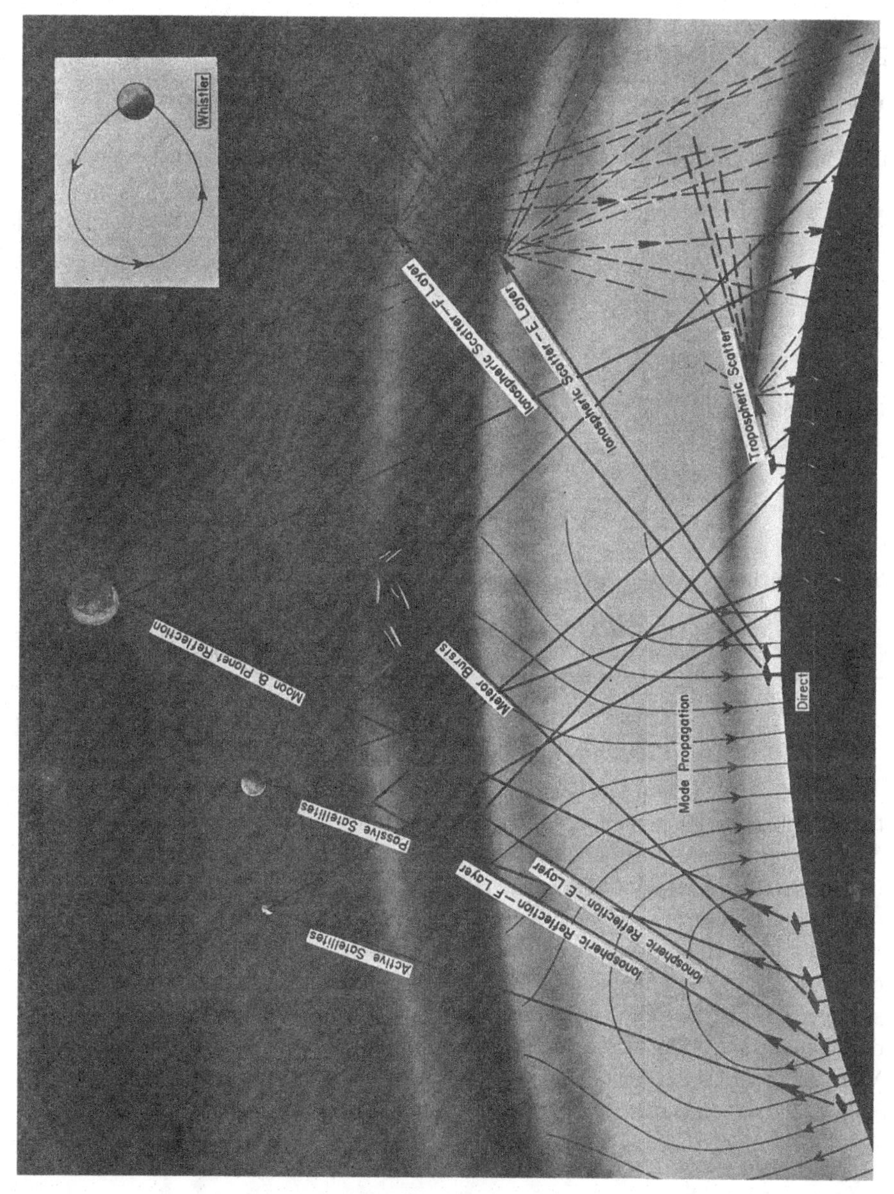

FIGURE 1.5 Summary of propagation mechanisms. (*Source*: R. C. Kirby, "Introduction," Lecture 1 in *NBS Course in Radio Propagation,* Ionospheric Propagation, Central Radio Propagation Laboratory, National Bureau of Standards, U.S. Deptartment of Commerce, Boulder, CO, 1961.)

TABLE 1.4 The Most Likely Propagation Mechanisms for Each Frequency Band

Frequency Band	Propagation Mechanism
VLF to LF (10–200 kHz)	Waveguide mode between Earth and D-layer. Groundwave at short distances.
LF to MF (200 kHz to 2 MHz)	Transition between groundwave and waveguide mode predominance to skywave (ionospheric hops). Skywave especially pronounced at night.
HF (2–30 MHz)	Ionospheric hops. Very long-range communications with low power and simple antennas. The "short-wave" band.
VHF (30–100 MHz)	Low power and small antennas. Primarily for short range using direct or direct-plus-Earth-reflected propagation (ducting can greatly increase the propagation range).
UHF (80–500 MHz)	Direct: early-warning radars, air-to-satellite and satellite-to-satellite communications. Direct-plus-Earth-reflected: air-to-ground communications, local television. Tropospheric scatter: when large highly directional antennas and high power are used.
SHF (500 MHz to 10 GHz)	Direct: most radars, satellite communications. Tropospheric refraction, terrain diffraction, and multipath become important in ground-to-ground links and in satellite communications at low elevation angles.

In Table 1.4, much of the same information is displayed by frequency bands. The frequency ranges given in parentheses emphasize that the propagation phenomena do not group themselves neatly by the frequency bands as defined by the IEEE and ITU. Of course, the frequency limits in the table are only approximate, since the status of the ionosphere exerts a strong influence for HF and lower frequencies.

It should be emphasized again that the information in this introductory chapter is only cursory and that the tables are meant only to convey a general sense of the importance of both the frequency (wavelength) and the environment in determining the dominant propagation mechanism(s). For example, the entry for "waveguide mode" VLF and LF in Table 1.4 would clearly be inappropriate to a radio-astronomy satellite stationed in space to receive LF signals from distant stars!

1.7 SOURCES OF FURTHER INFORMATION

This book is intended only as an introduction to the vast subject of radiowave propagation. Several organizations, both nationally and internationally, support research in the area, and readers are encouraged to become familiar with the publications of these organizations for more detailed information. Probably the most used source for propagation information is the International Telecommunication Union (ITU), based in

Geneva, Switzerland, which regulates telecommunications internationally and compiles international research that affects these regulations. The Radio Communication sector of the ITU, called the ITU-R, issues a set of reports and recommendations that often serve as useful guidelines for propagation predictions. ITU-R publications can be downloaded via the Internet for a modest charge. Radio regulations of the ITU are also available, but consist mostly of frequency assignments with little propagation information per se. It is interesting to note that these regulations have the force of U.S. law when ratified by the Senate.

Within the United States, the National Telecommunications and Information Administration, Institute for Telecommunications Sciences (NTIA/ITS) of the U.S. Department of Commerce provides reports and expert advice. Computer programs for propagation predictions are also available, usually for a fee. Several Department of Defense agencies, particularly the U.S. Navy, also maintain propagation-focused divisions that often produce computer codes that are publicly available. The U.S. Federal Communications Commission (FCC) is responsible in the United States for regulating electromagnetic communications and can provide publications of its rules and regulations. Information about atmospheric conditions is provided by the National Oceanic and Atmospheric Administration (NOAA) of the U.S. Department of Commerce and also by the National Aeronautics and Space Administration (NASA). Many of these U.S. agencies have counterparts in other countries.

Of course, the convenience of the Internet for obtaining up-to-date information is of note. Many of the organizations listed above maintain web sites that undergo regular updates.

1.8 OVERVIEW OF TEXT

The remaining book is divided into 11 chapters, each of which discusses a different aspect or mechanism of radiowave propagation. The next three chapters serve primarily as a review and introduction to the basic electromagnetic theory that underlies propagation studies, with discussions of propagation media properties, plane wave propagation, and antenna and noise concepts. Chapter 5 then begins the study of direct line-of-sight propagation through the atmosphere, with consideration of atmospheric absorption, rain attenuation, and site diversity improvements. Chapter 6 reviews the basic theory of reflection and refraction at material interfaces such as the ground and discusses the effective Earth radius model for refraction in an inhomogeneous atmosphere along with the ducting effects that can result. Chapter 7 introduces procedures for analysis of reflection, refraction, and diffraction in microwave link design for a given (known) terrain profile. Chapter 8 extends this discussion to include empirical path loss models for point-to-point ground links, as well as a discussion of statistical models that are commonly used to describe fading effects in multipath environments prevalent, for example, in wireless cellular communications. Chapter 9 discusses standard techniques for prediction of surface or ground wave propagation.

Chapter 10 discusses the basic physical properties of the ionosphere, while Chapter 11 considers ionospheric propagation with emphasis on the skywave mechanism at MF and HF and on ionospheric perturbations for Earth–space links at VHF and higher frequencies. Finally, Chapter 12 provides a brief description of other more unusual propagation mechanisms, including tropospheric scatter, as well as other applications, such as radar, involving propagation effects.

2

CHARACTERIZATION OF PROPAGATION MEDIA

2.1 INTRODUCTION

It is assumed that the reader has some familiarity with electromagnetic field theory and has encountered plane waves before. Nevertheless, a brief review here seems appropriate because introductory treatments, particularly for engineers, often steer quickly toward the simplest dielectric and magnetic materials. This simplifies the equations and allows an early introduction to waveguides, antennas, and other applications. Propagation media are not necessarily simple, however, and departure from the simple model can strongly impact signal propagation.

2.2 MAXWELL'S EQUATIONS, BOUNDARY CONDITIONS, AND CONTINUITY

The starting point for electromagnetic theory [1–5] is the set of four equations named after James Clerk Maxwell:

$$\nabla \times \overline{H} = \frac{\partial \overline{D}}{\partial t} + \overline{J}, \tag{2.1}$$

$$\nabla \times \overline{E} = -\frac{\partial \overline{B}}{\partial t}, \tag{2.2}$$

Radiowave Propagation: Physics and Applications. By Curt A. Levis, Joel T. Johnson, and Fernando L. Teixeira

$$\nabla \cdot \overline{D} = \rho, \tag{2.3}$$

$$\nabla \cdot \overline{B} = 0. \tag{2.4}$$

In these equations, the overbars denote vectors, t represents time, and the usual notation of vector calculus is used. $\overline{H}$ denotes the magnetic field intensity (A/m), $\overline{B}$ the magnetic flux density or induction (T), $\overline{E}$ the electric field intensity (V/m), $\overline{D}$ the electric flux density or induction (C/m^2), ρ the volume density of free charge (C/m^3), and $\overline{J}$ the current flow density due to free charges (A/m^2).

From these equations some auxiliary relations can be deduced. Among these are the boundary conditions, which hold at the interface of any two regions of space:

$$\hat{n} \times \left(\overline{H}_2 - \overline{H}_1\right) = \overline{J}_S, \tag{2.5}$$

$$\hat{n} \times \left(\overline{E}_2 - \overline{E}_1\right) = 0, \tag{2.6}$$

$$\hat{n} \cdot \left(\overline{D}_2 - \overline{D}_1\right) = \rho_S, \tag{2.7}$$

$$\hat{n} \cdot \left(\overline{B}_2 - \overline{B}_1\right) = 0. \tag{2.8}$$

Here $\hat{n}$ denotes a unit vector normal to the interface and pointing out of region 1 into region 2 (see Figure 2.1), the subscripts indicate the regions in which the fields are to be evaluated at the boundary, $\overline{J}_S$ denotes the surface current density, and ρ_S denotes surface charge density on the boundary between the regions. These boundary conditions are very useful when problems involving several materials are considered: Maxwell's equations can then be solved in each homogeneous region and the resulting solutions matched by applying the proper boundary conditions across the interfaces. This usefulness should not obscure the fact that the above boundary conditions add no real information to the Maxwell's equations: by assuming finite amplitude fields, equation (2.5) can be deduced from (2.1), (2.6) from (2.2), and so on. A similarly

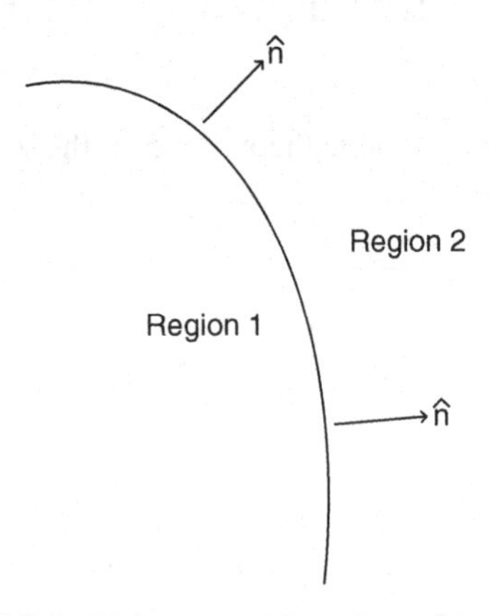

FIGURE 2.1 Unit vector at boundary of two regions.

useful, but also dependent set of equations are the charge continuity equations

$$\nabla \cdot \overline{J} = -\frac{\partial \rho}{\partial t}, \tag{2.9}$$

$$\nabla \cdot \overline{J}_S = -\frac{\partial \rho_S}{\partial t}. \tag{2.10}$$

Equation (2.9) results when the divergence of both sides of equation (2.1) is taken; the left-hand side can then be shown to be zero by a vector identity, and by application of equation (2.3) to the right-hand side, equation (2.9) can be obtained. Equation (2.10) can be obtained from equation (2.9) by considering current flow parallel to a boundary surface in a boundary region of vanishingly small thickness. The continuity equations are very useful because they express an important physical concept that a net outflow of current from a region depletes the free charge within that region (charge conservation); in short, that current *is* the flow of charge. Again, the importance of that concept should not obscure the fact that the continuity equations follow from Maxwell's equations.

2.3 CONSTITUTIVE RELATIONS

Maxwell's equations by themselves are not sufficient to specify a problem. From the mathematical point of view, the Helmholtz theorem states that it is necessary to specify both the divergence and curl of a vector to determine its departure from a constant vector. Maxwell's equations deal with two electric vectors, $\overline{D}$ and $\overline{E}$, but specify only the curl of one and the divergence of the other. Without a relationship between the two, there is insufficient information about either. The same is true for the magnetic field vectors $\overline{B}$ and $\overline{H}$. From a physical point of view, Maxwell's equations by themselves *must* be incomplete because they take no account of the material medium in which they are considered. Experimentally, one finds that the material has a very pronounced effect on the fields. In fact, practical problems are often specified by giving (a) the material properties and (b) the field values at the boundary of the region of interest.

The equations that, mathematically, link the various field vectors and, physically, account for the effects of a material medium are called *constitutive relations*. For particularly simple materials, the link is very convenient and direct:

$$\overline{D} = \epsilon \overline{E}, \tag{2.11}$$

$$\overline{B} = \mu \overline{H}, \tag{2.12}$$

where ϵ is the *permittivity* and μ the *permeability* of the material. Equations (2.11) and (2.12) are, however, based on a host of assumptions that are often not satisfied in propagation media. It is therefore worthwhile to go back to a more basic set of relationships to examine the nature of the constitutive equations.

2.4 DIELECTRIC BEHAVIOR OF MATERIALS: MATERIAL POLARIZATION

According to the atomic theory, matter is composed of atoms consisting of positively charged nuclei and orbiting electrons. The atoms combine to form molecules; in liquids and solids, these may, in turn, be arranged into more complex structures. The separation between each orbiting electron and the corresponding positive nuclear charge causes these charges to constitute an electric dipole (an electric dipole is an equal amount of positive and negative charge separated by a small distance). Thermal motions cause a constant reorientation of these dipoles, and under many conditions the orientation of the dipoles is completely random; the average dipole moment per unit volume is then zero. This average dipole moment per unit volume is called the polarization vector $\overline{P}(x, y, z)$, where the average is to be taken over a volume surrounding the point (x, y, z) small enough to be associated with a point as far as field calculations are concerned, but large enough to contain many molecules, or other ordered structures (such as crystals). No specific coordinate system is implied, and we will write $\overline{P}(\overline{r})$ instead of $\overline{P}(x, y, z)$ to emphasize this point. In the presence of an externally applied electric field (and even without such a field in the case of ferroelectric materials), some order is injected into the randomness of the dipole orientation and in that case the polarization vector, $\overline{P}$, which represents the average dipole moment per unit volume, assumes nonzero values. The electric flux density is then related to the actually existing electric field intensity by the relation

$$\overline{D}(\overline{r}, t) = \epsilon_0 \overline{E}(\overline{r}, t) + \overline{P}(\overline{r}, t), \tag{2.13}$$

where the first term on the right-hand side represents the flux corresponding to the same electric field intensity in free space (vacuum, with permittivity $\epsilon_0 = 8.854 \times 10^{-12}$ F/m), and the second term represents the contribution of the polarization, that is, the effect of the material. Of course, the electric field due to a given set of sources (charges) in the presence of the dielectric is different from what it would be in free space. Equation (2.13) is the generalization from which equation (2.11) can be derived, but only for very simple materials. Two words of caution may be appropriate at this point. First, the electric field intensity $\overline{E}$ in equation (2.13) is the actual macroscopic field that exists in the material, not the applied field that would exist in the absence of the material. This is unfortunate because the actual field is not always easy to calculate. Second, the polarization of a *material*, which is under discussion here, should not be confused with the polarization of a *wave*, which will be discussed later. It is unfortunate that the same word is used for two different concepts, and we need to be careful to distinguish between *material polarization* and *wave polarization* in contexts where confusion might arise.

Equation (2.13), while more general than equation (2.11), is not as effective in providing, together with Maxwell's equations, a complete set of equations for the description of electromagnetic phenomena. The reason is that we have added not only an equation but also an additional vector, $\overline{P}(\overline{r}, t)$. We must bring in now the relation between $\overline{P}$ and $\overline{E}$, which depends on the material. For particular materials the relationship can be quite simple, but for others it can be very complicated. Let us look briefly and qualitatively at the complicating properties.

2.5 MATERIAL PROPERTIES

Nonlinearity When the applied field strength is sufficiently low, as is generally true in typical propagation problems, the three components of the polarization vector $\overline{P}$ depend in some linear fashion on the three components of the electric field intensity $\overline{E}$. This characterizes *linear media*. Since linear differential equations are much easier to solve than nonlinear ones, this is a great simplification. However, if the fields are strong enough, the approximation breaks down. For example, linear optical propagation theory is not a valid theory for a high-powered focused laser propagating in air at normal pressures, because such a beam will produce significant ionization and hence nonlinear effects.

Anisotropy Most media have no directional properties of their own. In such a medium, each component of $\overline{P}$ depends only on the corresponding component of $\overline{E}$, and the dependence is the same for all components. This greatly simplifies the relationship, although it does not guarantee that $\overline{P}$ and $\overline{E}$ have the same direction at all instants of time. (For example, in a dispersive medium (see below) if the direction of $\overline{E}$ changes, the $\overline{P}$ vector cannot follow that direction change instantly.) Media that have no directional properties of their own are called *isotropic*; those that do are called *anisotropic*. The ionosphere is an example of an anisotropic medium since the Earth's magnetic field gives some directionality to the motions of charges in it. Therefore, in the ionosphere, equation (2.11) is invalid (except in a very approximate sense) and equation (2.13) will need to be used. All other media considered in this text are isotropic.

Dispersion The material polarization is the result of a statistical alignment or orientation of the charge dipoles in a medium, counteracted in part by the constant disordering influence of thermal motion. This alignment cannot occur instantaneously, and neither can the disordering after the applied field is removed. For many materials, the ordering process is much faster than the disordering process, and disordering can be characterized approximately by an exponential function. The time constant of this function is then called the *relaxation time* of the polarization. When the applied signal varies slowly with respect to the ordering and disordering processes, the polarization appears to follow the applied field almost instantaneously. In that case, to find the polarization at a given time, we need to know only the instantaneous applied field at that same time. On the other hand, if the signal varies significantly within the time required for ordering and disordering to take place, the polarization effects lag behind the applied field that causes them. Then we need to know the prior time history of the applied field to find the polarization at any given time. In this case, the rate of variation of the signal becomes important, hence different frequencies will produce different polarization responses. A medium whose response depends on the frequency is called *dispersive*, otherwise it is called *nondispersive*. From the foregoing discussion it should be apparent that most materials exhibit nondispersive properties up to some frequency regime characteristic of the polarization process involved and hence the particular material. They will act dispersively at frequencies close to that regime. At sufficiently high frequencies (i.e., those much higher than the reciprocal of all the characteristic times of the ordering–disordering process), the polarization charges will be unable to follow the field variations at all, and the material will exhibit the dielectric properties of free space.

Inhomogeneity and Time Dependence The dielectric properties of a medium may be the same at every point within the medium or they may vary from place to place. In the first case, the medium is called *homogeneous*; in the second, *inhomogeneous*. The properties may also be invariant with time or time-dependent. In natural propagation media, inhomogeneities are often associated with time variation. For example, tropospheric inhomogeneities, which make possible the tropospheric scatter mechanism, drift with winds and change in size and shape with time.

Fortunately, not all the complicating properties discussed above need to be considered simultaneously. For most problems, it is only necessary to consider one or two of them, but even then the resulting solutions of Maxwell's equations become complicated. The relationship between the electric flux density and the electric field intensity for some relatively simple media of practical relevance for propagation problems will be considered in the next several sections.

2.5.1 Simple Media

The Simplest Medium Consider first the simplest possible dielectric media, namely, those that are isotropic, linear, homogeneous, time-invariant, and nondispersive. The last property is a restriction on the rate of change of the applied signal as well as on the material properties, as discussed previously. For such a medium, the relationship between $\overline{P}$ and $\overline{E}$ can be written as

$$\overline{P}(\overline{r}, t) = \chi \epsilon_0 \overline{E}(\overline{r}, t). \tag{2.14}$$

The constant χ is called the *electric susceptibility* of the medium. Clearly, equation (2.14) is linear. Furthermore, each component of $\overline{P}$ depends only on the corresponding component of $\overline{E}$ and all have the same proportionality constant, satisfying the isotropic assumption. At any time, the polarization depends only on the field intensity at that same moment satisfying the nondispersive assumption. Since χ is independent of position or time, the medium is also homogeneous and time-invariant. This is the simplest relationship between $\overline{P}$ and $\overline{E}$ that a material can exhibit. Use of equation (2.14) in equation (2.13) gives

$$\overline{D}(\overline{r}, t) = \epsilon_0 (1 + \chi) \overline{E}(\overline{r}, t) = \epsilon_0 \epsilon_r \overline{E}(\overline{r}, t) = \epsilon \overline{E}(\overline{r}, t), \tag{2.15}$$

where

$$\epsilon_r = 1 + \chi \tag{2.16}$$

and

$$\epsilon = \epsilon_0 \epsilon_r. \tag{2.17}$$

The constant ϵ_r is called the *relative dielectric constant* or the *relative permittivity*.

Simple Inhomogeneous and Time-Varying Media The word "simple" will be used to describe a medium that has properties identical to the simplest medium just discussed, with the exception of those specifically mentioned in the name. Thus, "simple inhomogeneous and time-varying media" are isotropic, linear, and dispersionless, by definition. Allowing inhomogeneity and time variation has, indeed, a very simple effect on the constitutive relations, which become

$$\overline{P}(\overline{r}, t) = \epsilon_0 \chi(\overline{r}, t) \overline{E}(\overline{r}, t), \tag{2.18}$$

so

$$\overline{D}(\overline{r}, t) = \epsilon(\overline{r}, t) \overline{E}(\overline{r}, t). \tag{2.19}$$

The "simplicity" ceases at this point, however, because solutions of Maxwell's equations in conjunction with (2.19) are generally much more difficult than when ϵ is a constant.

Simple Dispersive Media As mentioned before, a medium acts dispersively when the characteristic relaxation times are comparable to the rate of change of the impressed electric field. Under these conditions, the polarization is unable to follow the field instantaneously.

The polarization vector arises from an average displacement of charges from their statistically neutral positions. Each charge individually obeys an equation of motion with respect to the electric field intensity. If one assumes that all forces except those due to the field are local (on a macroscopic scale), these equations are ordinary differential equations; then the relationship between the polarization and the electric field intensity is also an ordinary differential equation with time as the independent variable. Assuming the material is isotropic, we can consider each of the field components separately. For example, we have for the x components

$$\mathcal{L}[P_x(t)] = E_x(t), \tag{2.20}$$

where $\mathcal{L}$ is a differential operator involving differentiation with respect to t only. Since the material is to be simple except for dispersion, we have ruled out nonlinear or time-varying constitutive effects. As a result, we can assume that the operator $\mathcal{L}$ is linear with constant coefficients. The solution for $P_x(t)$ can be written as a superposition integral

$$P_x(t) = \epsilon_0 \int_{-\infty}^{t} E_x(t') f_{\mathrm{I}}(t - t') dt', \tag{2.21}$$

where $f_{\mathrm{I}}(t)$ is the *impulse response* of the medium. Since the "simple" dispersive medium under discussion is assumed isotropic, similar equations may be written for the y and z components, so a more general equation is

$$\overline{P}(\overline{r}, t) = \epsilon_0 \int_{-\infty}^{t} \overline{E}(\overline{r}, t') f_{\mathrm{I}}(t - t') dt'. \tag{2.22}$$

The limits in the above integral imply causality: the electric field for $-\infty < t' < t$ influences the behavior of $\overline{P}$ at time t, while the electric field for $t' > t$ does not. In other words, only the *past* history of the electric field determines the polarization at any instant of time.

The above considerations have been mostly heuristic. Moreover, a classical model has been assumed, whereas one might argue that in the discussion of microscopic phenomena, a quantum model would have been more accurate. Actually, when the problem is formulated quantum mechanically, one finds that summing over all allowed quantized orientations is equivalent, as far as the final results are concerned, to the classical model integration over a random orientation of orbits.

In many cases, the polarization function $f_{\mathrm{I}}(t)$ can be approximated for $t > 0$ by a sum of the form

$$f_{\mathrm{I}}(t) = \sum_{i=1}^{N} k_i e^{-(t/\tau_i)}, \tag{2.23}$$

where the time constants τ_i are the relaxation times of the individual polarization processes. Since $f_{\mathrm{I}}(t - t')$ characterizes the polarization at time t due to an electric field impulse at t', and since random thermal motion operates to disorder the polarized molecules, $f_{\mathrm{I}}(t - t')$ approaches zero for $t \gg t'$. If the electric field is static, or if the electric field is so slowly varying that $\overline{E}(\overline{r}, t')$ remains essentially constant over the time interval over which $f_{\mathrm{I}}(t - t')$ is much different from zero, then $\overline{E}(\overline{r}, t')$ may be taken out of the integral to yield

$$\overline{P}(\overline{r}, t) = \epsilon_0 \overline{E}(\overline{r}, t) \int_{-\infty}^{t} f_{\mathrm{I}}(t - t')dt'. \tag{2.24}$$

This leads to the electrostatic or quasi-static representation

$$\chi_{\mathrm{es}} = \int_{-\infty}^{t} f_{\mathrm{I}}(t - t')dt' = \int_{0}^{\infty} f_{\mathrm{I}}(\zeta)d\zeta, \tag{2.25}$$

where the last equality results from letting $t - t' = \zeta$.

The dielectric constant for the electrostatic or quasi-static case then becomes

$$\epsilon_{\mathrm{es}} = (1 + \chi_{\mathrm{es}})\epsilon_0. \tag{2.26}$$

This representation is valid for all materials that are polarized entirely by electronic and atomic alignments up to the submillimeter wave region; however, for materials that exhibit molecular (Debye) or interfacial polarization effects, equation (2.22) must be used for radio waves.

So far we have considered polarization as related to an arbitrary time-varying electric field. The assumption of sinusoidal (time-harmonic) time dependence greatly simplifies the analysis, and this assumption will be very useful in much of the material to be covered in this book. If we express a sinusoidal field in equation (2.22) by the usual phasor representation

$$\overline{E}(\overline{r}, t) = \mathrm{Re}\left[\overline{E}_0(\overline{r})e^{j\omega t}\right], \tag{2.27}$$

where $j = \sqrt{-1}$, ω is the radian frequency of the sinusoidal field, and the underbar denotes a complex number, one gets from equation (2.22)

$$\overline{P}(\overline{r}, t) = \text{Re}\left[\epsilon_0 \underline{\overline{E}}_0(\overline{r}) \int_{-\infty}^{t} e^{j\omega t'} f_{\text{I}}(t - t')dt'\right]. \tag{2.28}$$

Again letting $t - t' = \zeta$ gives

$$\overline{P}(\overline{r}, t) = \text{Re}\left[\epsilon_0 \underline{\overline{E}}_0(\overline{r}) e^{j\omega t} \int_{0}^{\infty} e^{-j\omega\zeta} f_{\text{I}}(\zeta)d\zeta\right]. \tag{2.29}$$

Letting

$$\overline{P}(\overline{r}, t) = \text{Re}\left[\underline{\overline{P}}_0(\overline{r}) e^{j\omega t}\right], \tag{2.30}$$

one gets

$$\underline{\overline{P}}_0(\overline{r}) = \underline{\chi}(\omega)\epsilon_0 \underline{\overline{E}}_0(\overline{r}), \tag{2.31}$$

where

$$\underline{\chi}(\omega) = \int_{0}^{\infty} e^{-j\omega\zeta} f_{\text{I}}(\zeta)d\zeta. \tag{2.32}$$

The last equation defines the *complex susceptibility* $\underline{\chi}(\omega)$.

It can be shown easily that equations (2.32) and (2.13) lead to a complex dielectric constant, or complex permittivity,

$$\underline{\epsilon}(\omega) = \epsilon_0 \left(1 + \underline{\chi}(\omega)\right), \tag{2.33}$$

which can be related to the fields by

$$\underline{\overline{D}}_0(\overline{r}) = \underline{\epsilon}(\omega) \underline{\overline{E}}_0(\overline{r}), \tag{2.34}$$

where

$$\overline{D}(\overline{r}, t) = \text{Re}\left[\underline{\overline{D}}_0(\overline{r}) e^{j\omega t}\right] \tag{2.35}$$

is implied. It is also easy to show that the complex susceptibility and complex dielectric constant approach the quasi-static values as ω approaches zero.

The appearance of equation (2.34) is deceptively similar to that of equation (2.15),

$$\overline{D}(\overline{r}, t) = \epsilon \overline{E}(\overline{r}, t),$$

that applies to the quasi-static case (and hence to the simplest media, since a simple dispersive medium reduces to that when the quasi-static assumption is satisfied). However, they are really quite different in meaning. In equation (2.15), ϵ is a real constant relating the arbitrarily time-varying vectors $\overline{D}$ and $\overline{E}$. In contrast, $\underline{\epsilon}(\omega)$ in equation (2.34) is a complex function of ω that relates the *phasor representations* of $\overline{D}$ and $\overline{E}$, for sinusoidal time variations.

The imaginary part of $\underline{\epsilon}(\omega)$ can be shown to result in a loss of energy in the material. This phenomenon occurs because $\overline{D}(\overline{r}, t)$ lags $\overline{E}(\overline{r}, t)$, and it is called *dielectric hysteresis* loss. The most general behavior of the complex dielectric constant of a

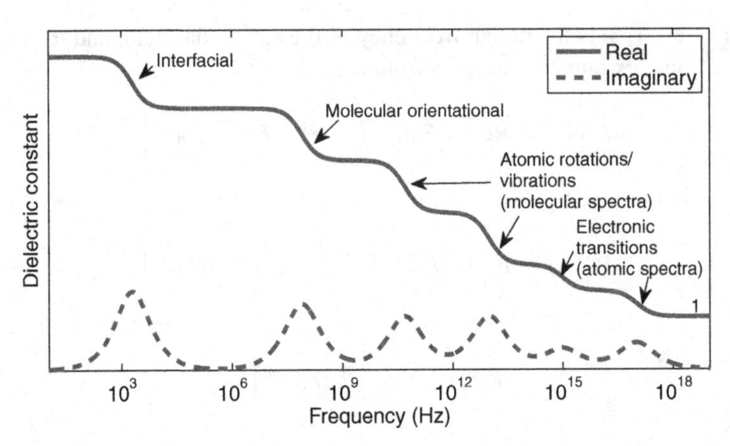

FIGURE 2.2 Schematic representation of complex dielectric constant variation.

material is shown in Figure 2.2. In this figure, the legends refer to four different polarization mechanisms corresponding to progressively longer relaxation times. At the lowest frequencies, all are effective and the complex dielectric constant is essentially real and large. As the frequency ω is raised to a value such that the period $T = 2\pi/\omega$ is on the order of the relaxation time τ of the interfacial polarization process, the dipoles corresponding to that process can no longer follow and the dielectric constant due to that process begins to disappear. In this region, the lagging dipole moments cause an increase in dielectric hysteresis loss that shows up as a hump in the imaginary part of $\underline{\epsilon}$. The same behavior is observed for the other polarization mechanisms. Not all materials exhibit all the polarization mechanisms: the dispersion curves for water, which exhibits molecular polarization, and carbon tetrachloride, which has only atomic and electronic polarization, are shown in Figures 2.3. One can notice that there is a correlation between variations in the real part of the permittivity ϵ' and the value of the imaginary part of the permittivity ϵ''. Indeed, ϵ' and ϵ'' are related by the so-called Kramers–Kronig relations. The details of Figures 2.2 and 2.3 are not important to our discussion here. The important concepts to remember are that the dielectric constant can vary considerably with frequency, and so can the dielectric loss.

At this point, let us summarize the important points about the dispersive characteristics of materials otherwise simple. Simple relations of the type

$$\overline{P}(\overline{r}, t) = \chi\epsilon_0\overline{E}(\overline{r}, t),$$
$$\overline{D}(\overline{r}, t) = \epsilon\overline{E}(\overline{r}, t)$$

are not valid unless the applied electric field is slowly varying compared to the ordering–disordering processes responsible for polarization, a condition we have called quasi-static. The more general relation takes the form

$$\overline{P}(\overline{r}, t) = \epsilon_0 \int_{-\infty}^{t} \overline{E}(\overline{r}, t') f_1(t - t') dt', \tag{2.36}$$

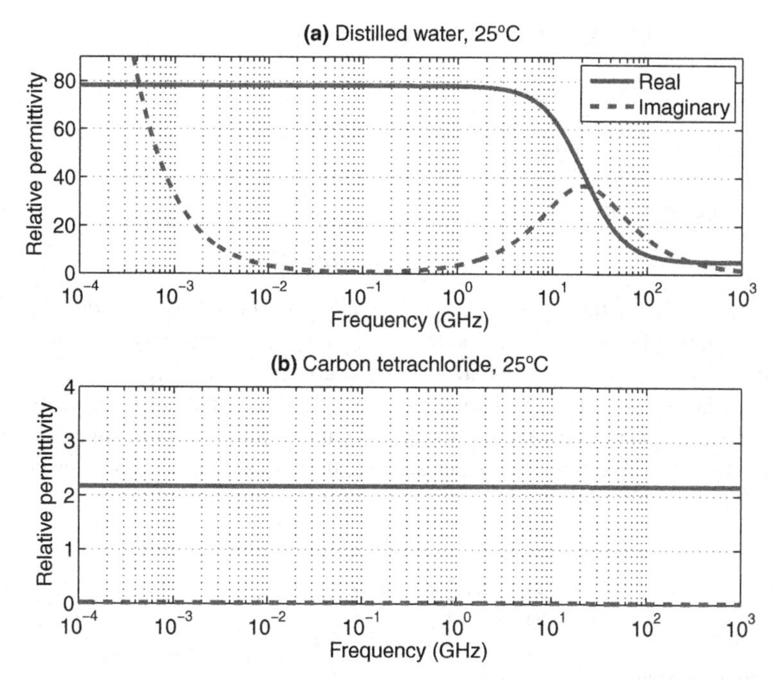

FIGURE 2.3 Dielectric behavior at 25°C (a) distilled water and (b) carbon tetrachloride.

and susceptibility and dielectric constant cannot be defined for arbitrary time variation. When the excitation is sinusoidal, however, equation (2.22) can be converted to a relation between the phasors that characterize $\overline{P}$ and $\overline{E}$, namely, equation (2.31), and hence equation (2.34) follows. The complex susceptibility and complex dielectric constant that appear in these equations are functions of frequency, and their imaginary part corresponds to a dielectric hysteresis loss.

A distinction should also be made here between the dispersive properties of a dielectric *material*, which have been discussed here, and the dispersive properties of an entire signal path of which the material may be only a part. Consider, for example, a slab of the "simplest material", which in itself has no dispersion, and let a wave be normally incident upon it. The surrounding medium is free space. It will be found that there is no reflection if the frequency is chosen so that the thickness of the slab is a precise multiple of half-wavelengths, but at other frequencies there will be reflections. Clearly, the path is dispersive, since it does not treat all frequencies alike, even though the material in the slab is not a dispersive material.

Simple Anisotropic Media An anisotropic material exhibits certain preferential polarization directions. As a result, the polarization does not necessarily align itself with the electric field. Thus, it is no longer true that the x component of $\overline{P}$ responds only to the x component of $\overline{E}$ and so forth, although such simple relations may still hold in special coordinate systems. In general, we must write

$$\begin{bmatrix} P_x \\ P_y \\ P_z \end{bmatrix} = \epsilon_0 \begin{bmatrix} \chi_{xx} & \chi_{xy} & \chi_{xz} \\ \chi_{yx} & \chi_{yy} & \chi_{yz} \\ \chi_{zx} & \chi_{zy} & \chi_{zz} \end{bmatrix} \begin{bmatrix} E_x \\ E_y \\ E_z \end{bmatrix}, \tag{2.37}$$

which can be abbreviated symbolically as

$$\overline{P}(\overline{r}, t) = \epsilon_0 \, [\chi] \cdot \overline{E}(\overline{r}, t). \tag{2.38}$$

The elements χ_{ij} of the susceptibility matrix $[\chi]$ are not all independent, and the matrix can often be simplified considerably by taking advantage of the symmetries that occur when coordinate axes are chosen to coincide with symmetry axes of the material. These details need not concern us here; for the moment we wish to stress only that the simple relation of equation (2.14) must be replaced by the matrix relation (2.37) in the case of anisotropic materials. The corresponding relationships between $\overline{D}$ and $\overline{E}$ follow from equation (2.13),

$$\begin{bmatrix} D_x \\ D_y \\ D_z \end{bmatrix} = \begin{bmatrix} \epsilon_{xx} & \epsilon_{xy} & \epsilon_{xz} \\ \epsilon_{yx} & \epsilon_{yy} & \epsilon_{yz} \\ \epsilon_{zx} & \epsilon_{zy} & \epsilon_{zz} \end{bmatrix} \begin{bmatrix} E_x \\ E_y \\ E_z \end{bmatrix}, \tag{2.39}$$

which may be written as

$$\overline{D}(\overline{r}, t) = [\epsilon] \cdot \overline{E}(\overline{r}, t), \tag{2.40}$$

where

$$[\epsilon] = \epsilon_0 \left([I] + [\chi] \right), \tag{2.41}$$

in which $[I]$ represents the identity matrix,

$$[I] = \begin{bmatrix} 1 & 0 & 0 \\ 0 & 1 & 0 \\ 0 & 0 & 1 \end{bmatrix}. \tag{2.42}$$

Simple Nonlinear Media In nonlinear media, the polarization is no longer linearly related to the electric field intensity. In an otherwise simple nonlinear medium, the most general relationship is

$$\hat{a}_P(\overline{r}, t) = \hat{a}_E(\overline{r}, t), \tag{2.43}$$

$$P(\overline{r}, t) = f_N \left(E(\overline{r}, t) \right). \tag{2.44}$$

The first of these equations states that the directions of $\overline{P}$ and $\overline{E}$ are same, the $\hat{a}$'s denoting unit vectors of the respective field directions. The second equation leaves the relationship between their magnitudes quite general. Often, it is possible to expand the magnitude function f_N in a Taylor series, to obtain

$$P(\overline{r}, t) = \epsilon_0 \chi_1 E(\overline{r}, t) + \epsilon_0 \chi_2 E^2(\overline{r}, t) + \cdots, \tag{2.45}$$

and to neglect higher order terms. The electric field strength that must be exceeded before a given material becomes noticeably nonlinear is a property of the material and varies greatly from one medium to another. For the media prevalent in radiowave propagation, the required field strengths are usually quite high, and it is unusual to consider nonlinear effects. An exception is the propagation of very high powered waves in the ionosphere, in which case the nonlinearity can produce an intermodulation of signals, the so-called "Luxembourg effect". Another exception is the parametric generation of coherent light in a nonlinear propagation path. In general, nonlinear phenomena are likely to become more important in optical propagation because of the very high local field strengths that can be achieved with lasers. Nonlinear effects are not considered further in this book.

More Complicated Media Any number of the properties seen above — nonlinearity, anisotropy, dispersion, inhomogeneity, time variability — can, of course, sometimes coexist in a propagation medium. In many cases, the extension from the simple media described above is conceptually straightforward, but the solution of Maxwell's equations becomes increasingly difficult in more complicated media. For example, the ionosphere must be treated in some calculations as an anisotropic, dispersive, and inhomogeneous medium. From the discussion of simple dispersive media, we know that we should not expect to derive real susceptibilities or real dielectric constants, valid for arbitrary time dependence, but should look for complex quantities relating the phasors corresponding to $\overline{P}$ and $\overline{E}$ or $\overline{D}$ and $\overline{E}$. From the discussion of simple anisotropic media, we seek not single constants but a matrix to relate the components. From the discussion of simple inhomogeneous media, we expect the matrix elements to be functions of position. Indeed, relationships of the form

$$\underline{\overline{P}}(\overline{r}) = \epsilon_0 \left[\underline{\chi}(\overline{r}, \omega)\right] \cdot \underline{\overline{E}}_0(\overline{r}), \tag{2.46}$$

$$\underline{\overline{D}}_0(\overline{r}) = \left[\underline{\epsilon}(\overline{r}, \omega)\right] \cdot \underline{\overline{E}}_0(\overline{r}) \tag{2.47}$$

turn out to be appropriate for the ionosphere, where

$$\left[\underline{\epsilon}(\overline{r}, \omega)\right] = \epsilon_0 \left([I] + \left[\underline{\chi}(\overline{r}, \omega)\right]\right). \tag{2.48}$$

Time variations of the form of equation (2.27) are understood, and the $[\underline{\chi}]$ and $[\underline{\epsilon}]$ matrices differ from those in equations (2.37) and (2.39) in that each element is complex and a function of position and frequency.

There are some cases, however, where the treatment of complicated media is difficult even at a conceptual level. These arise primarily when nonlinearity occurs in conjunction with dispersion or anisotropy. Linearity played an important role in the discussion of the simple dispersive and simple anisotropic medium, but there is no simple way to generalize these treatments to the nonlinear case. We note that there is great current interest in the production of engineered "metamaterials" to achieve greater control over constitutive relations for fabricating new electromagnetic devices.

2.6 MAGNETIC AND CONDUCTIVE BEHAVIOR OF MATERIALS

Just as the dielectric behavior of materials can be very simple or very complicated, so can their magnetic and conductive behavior. For propagation problems, it is seldom necessary to consider materials that exhibit strong magnetic effects. For all the propagation mechanisms considered in this book, the simple, free-space relation

$$\overline{B}(\overline{r}, t) = \mu_0 \overline{H}(\overline{r}, t) \tag{2.49}$$

is valid with sufficient accuracy, where $\mu_0 = 4\pi \times 10^{-7}$ H/m is the permeability of free space.

Conductive behavior does occur in propagation problems, for example in the ionosphere and in the ground. The same complicating characteristics (nonlinearity, anisotropy, etc.) that came up in connection with dielectrics occur in regard to conductivity as well. The analogy is so direct that it would be repetitious to examine conductive behavior in great detail. For example, in direct analogy with equation (2.14), the simplest conductive behavior is characterized by

$$\overline{J}(\overline{r}, t) = \sigma \overline{E}(\overline{r}, t), \tag{2.50}$$

where $\overline{J}$ is the current density and σ (S/m or mho/m) is the conductivity of the material.

For a simple anisotropic conduction process, one has in analogy with equation (2.39)

$$\overline{J}(\overline{r}, t) = [\sigma] \cdot \overline{E}(\overline{r}, t), \tag{2.51}$$

where

$$[\sigma] = \begin{bmatrix} \sigma_{xx} & \sigma_{xy} & \sigma_{xz} \\ \sigma_{yx} & \sigma_{yy} & \sigma_{yz} \\ \sigma_{zx} & \sigma_{zy} & \sigma_{zz} \end{bmatrix}. \tag{2.52}$$

The other analogies should be obvious.

2.6.1 Equivalence of Ohmic and Polarization Losses

Consider a medium in which conduction by free charges can take place; the medium also exhibits simple dispersive polarization. Because of the dispersive nature of the polarization, it is easiest to deal with the phasor representations of the fields. Letting

$$\overline{H}(\overline{r}, t) = \text{Re}\left[\underline{H}_0(\overline{r}) e^{j\omega t}\right], \tag{2.53}$$

and similarly for $\overline{D}$ and $\overline{J}$, one gets for equation (2.1),

$$\nabla \times \underline{H}_0(\overline{r}) = \underline{J}_0(\overline{r}) + j\omega \underline{D}_0(\overline{r}), \tag{2.54}$$

and by equations (2.34) and (2.50)

$$\nabla \times \underline{\overline{H}}_0(\overline{r}) = (\sigma + j\omega\underline{\epsilon})\,\underline{\overline{E}}_0(\overline{r}). \tag{2.55}$$

If one separates out the real and imaginary parts of the dielectric constant,

$$\underline{\epsilon} = \epsilon' - j\epsilon'', \tag{2.56}$$

and rearranges the terms slightly, the result is

$$\nabla \times \underline{\overline{H}}_0(\overline{r}) = \left[(\sigma + \omega\epsilon'') + j\omega\epsilon'\right] \underline{\overline{E}}_0(\overline{r}). \tag{2.57}$$

The σ term in this last equation represents conduction current due to free charges and results in ohmic loss. The $\omega\epsilon''$ term represents lagging polarization, producing an effective current in phase with the conduction current. It is responsible for the polarization hysteresis loss. The effects of these two losses on a signal propagating through a medium are same; indeed, it is not easy to design an experiment for measuring them separately. At a first glance, from equation (2.57) it might appear that the frequency dependence of the polarization term would make it easy to separate the two, but this is not so since in the dispersive material σ and ϵ'' themselves are frequency dependent. In any case, from the propagation point of view there is no need to separate the two effects, and generally they are lumped together.

For this purpose, two representations are commonly employed. The first, commonly used to specify the ground properties at MF (e.g., the U.S. AM broadcast band), is based upon lumping the two terms into an effective conductivity,

$$\sigma_e = \sigma + \omega\epsilon''. \tag{2.58}$$

Defining the relative dielectric constant

$$\epsilon_r = \epsilon'/\epsilon_0 \tag{2.59}$$

and a loss factor

$$x = \frac{\sigma_e}{\omega\epsilon_0}, \tag{2.60}$$

Ampere's law becomes

$$\nabla \times \underline{\overline{H}}_0 = j\omega\epsilon_0\,(\epsilon_r - jx)\,\underline{\overline{E}}_0. \tag{2.61}$$

The quantity $(\epsilon_r - jx)$ appears frequently in calculations. A handy formula for computing x is

$$x \approx 18 \times 10^3 \sigma_e/f, \tag{2.62}$$

where σ_e is measured in mho/m and f in MHz. The subscript "e" on σ_e is often omitted; whenever a real dielectric constant and conductivity are specified for a lossy dielectric, such as ground, it is safe to assume that the effective conductivity is meant.

A second convention, used especially at microwave frequencies, is based upon representing the material as though polarization accounted for all the losses. In this case, an effective imaginary dielectric constant is introduced as

$$\epsilon_{\mathrm{e}}'' = \epsilon'' + \sigma/\omega \tag{2.63}$$

and a loss tangent as

$$\tan\delta = \epsilon_{\mathrm{e}}''/\epsilon', \tag{2.64}$$

so Ampere's law becomes

$$\nabla \times \overline{H}_0 = j\omega\epsilon' (1 - j\tan\delta)\,\overline{E}_0. \tag{2.65}$$

The term "loss tangent" is derived from representing the dielectric constant as a complex number

$$\underline{\epsilon}^{\mathrm{e}} = \epsilon' - j\epsilon_{\mathrm{e}}'', \tag{2.66}$$

as in Figure 2.4. Under this convention, too, it is usual to omit the subscript denoting "effective"; whenever the dielectric behavior is specified by ϵ' and ϵ'' or $\tan\delta$, it is safe to assume that conductivity effects are also included.

It is sometimes necessary to convert between the two representations. By equating the right-hand sides of equations (2.61) and (2.65), one finds

$$\epsilon_{\mathrm{r}} = \epsilon'/\epsilon_0, \tag{2.67}$$

$$x = \left(\epsilon'/\epsilon_0\right)\tan\delta. \tag{2.68}$$

Figure 2.5 provides a plot of dielectric constant and conductivity values for various soils and ocean water, from ITU-R Recommendation 527-3 [6]. A map of the ground conductivity of the continental United States (in millimhos per meter) from ITU-R Recommendation 832-2 "World Atlas of Ground Conductivities" [7] is shown in Figure 2.6. It was compiled for use in the U.S. AM broadcast band, 540–1600 kHz. The real part of the relative dielectric constant ϵ_{r} also varies with soil type, but not over as wide a range.

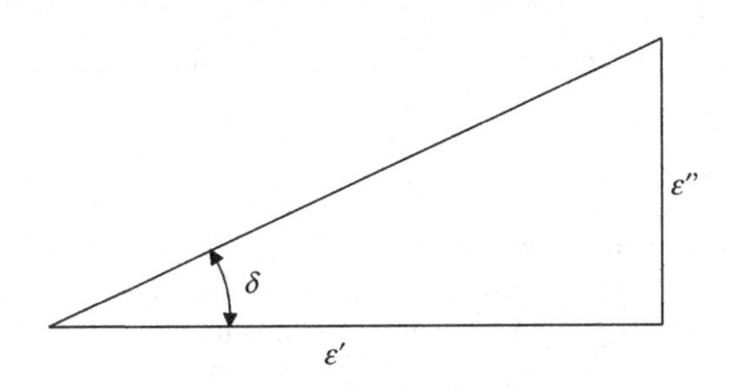

FIGURE 2.4 Loss tangent diagram.

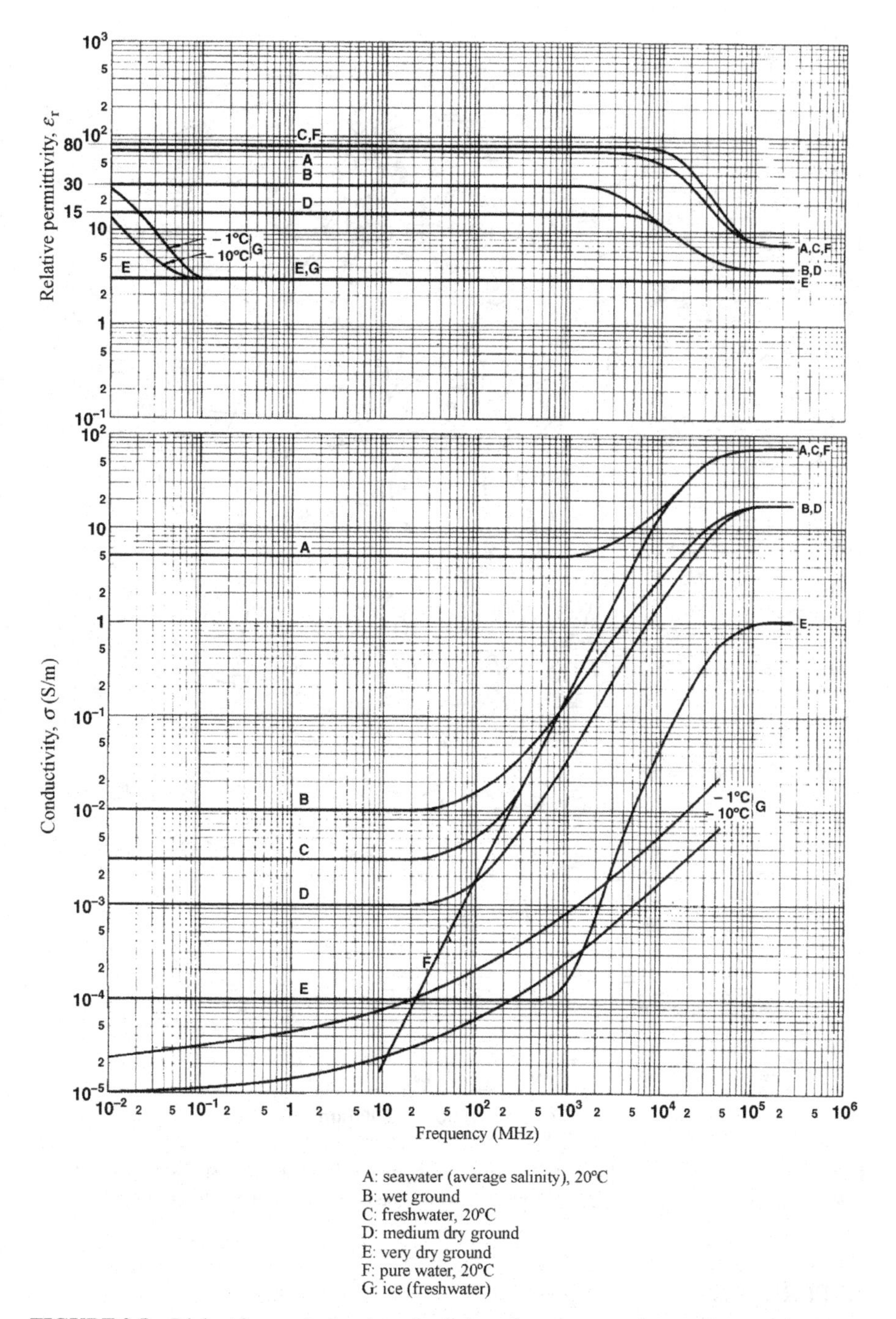

FIGURE 2.5 Dielectric constant and conductivity values for several natural materials versus frequency. (*Source:* ITU-R Recommendation 527-3, used with permission). The convention of (2.58) and (2.59) is implied.

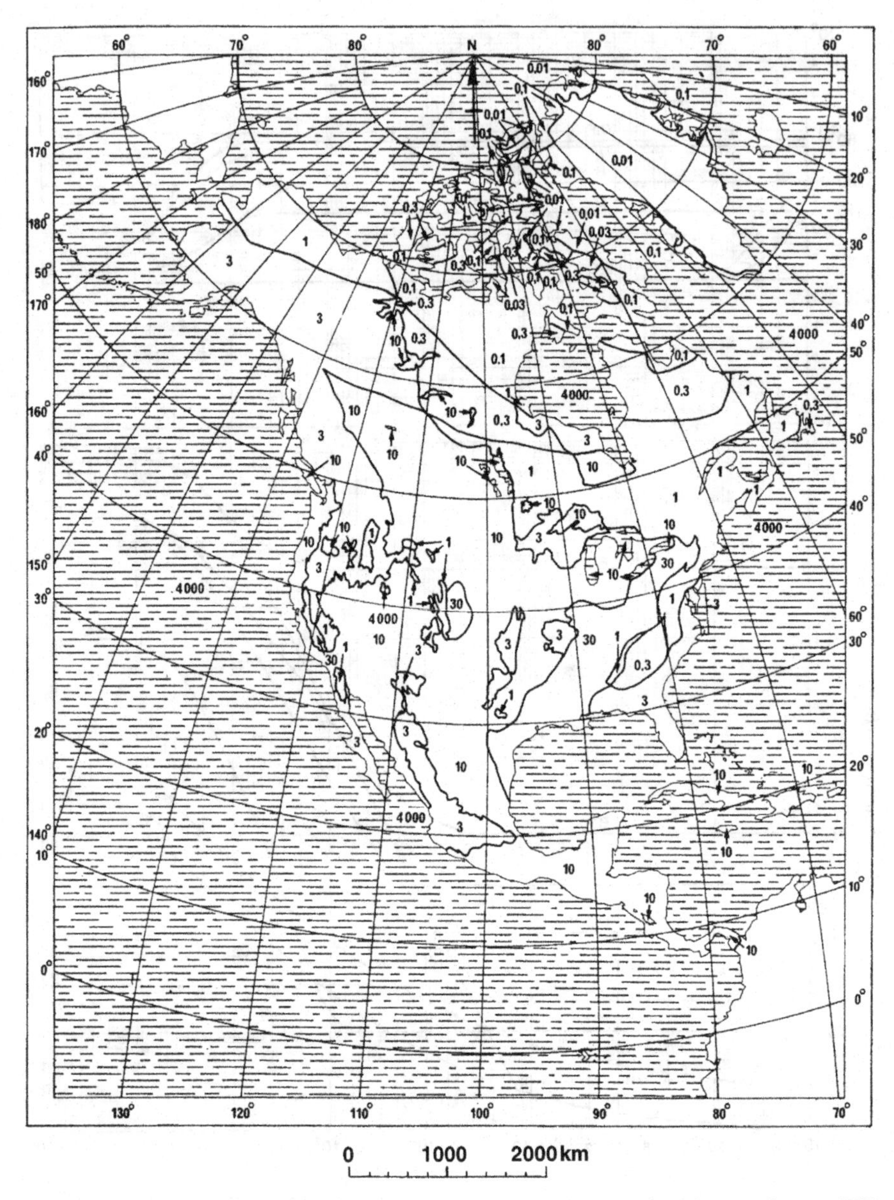

FIGURE 2.6 Ground conductivity (millimhos per meter) in the United States. (*Source:* ITU-R Recommendation 832-2, used with permission). The convention of (2.58) is implied.

REFERENCES

1. Stratton, J. A., *Electromagnetic Theory*, Wiley–IEEE Press, 2007.
2. Kong, J. A., *Electromagnetic Wave Theory*, EMW Publishing, 2008.

3. Harrington, R. F., *Time-Harmonic Electromagnetic Fields*, McGraw-Hill, 1961.
4. Chew, W. C., *Waves and Fields in Inhomogeneous Media*, IEEE Press, 1999.
5. Landau, L. D., L. P. Pitaevskii, and E. M. Lifshitz, *Electrodynamics of Continuous Media*, second edition, Butterworth-Heinemann, 1984.
6. ITU-R Recommendation P.527-3, "Electrical characteristics of the surface of the Earth," International Telecommunication Union, 1992.
7. ITU-R Recommendation P.832-2, "World Atlas of Ground Conductivities," International Telecommunication Union, 1999.

3

PLANE WAVES

3.1 INTRODUCTION

Maxwell's equations (equations (2.1)–(2.4)) are very elegant and concise, but elegance should not be confused with simplicity. There is nothing simple about four coupled vector partial differential equations! Nor should we expect simplicity since these equations are to describe all propagation modes in all possible media. The complexity of the mathematics reflects the complexity of the physical phenomena it serves to describe.

Since the general case is too complex to solve analytically, one is forced to make simplifying assumptions to find some simpler analytical solutions. Plane waves turn out to be the simplest solutions of Maxwell's equations [1–4]. Despite their analytical simplicity, plane waves find physical applications in a wide range of scenarios. More complex solutions are generally required to describe electromagnetic fields in the vicinity of sources, or close to material discontinuities and/or inhomogeneities in the propagation medium. Far from such regions, plane waves are in general a very good description for the local electromagnetic field behavior. Furthermore, in more complex cases, the total solution can often be represented as the superposition of a set of plane waves with varying amplitudes and propagation directions, in a manner analogous to a Fourier (or spectral) decomposition. Therefore, plane wave solutions are worthy of special attention.

Radiowave Propagation: Physics and Applications. By Curt A. Levis, Joel T. Johnson, and Fernando L. Teixeira

3.2 D'ALEMBERT'S SOLUTION

To simplify the derivations below, we consider a charge-free and nonconducting medium of the simplest kind discussed in Chapter 2, that is, isotropic, linear, dispersionless, homogeneous, and time invariant. Maxwell's equations then reduce to

$$\nabla \times \overline{H} = \epsilon \frac{\partial \overline{E}}{\partial t}, \tag{3.1}$$

$$\nabla \times \overline{E} = -\mu \frac{\partial \overline{H}}{\partial t}, \tag{3.2}$$

$$\nabla \cdot \overline{E} = 0, \tag{3.3}$$

$$\nabla \cdot \overline{H} = 0. \tag{3.4}$$

If one takes the curl of both sides of the first equation, makes use of the fact that ϵ is a constant for a "simplest" material, that time and space are independent variables so that order of differentiation is immaterial, and finally substitutes from equation (3.2), one gets

$$\nabla \times \nabla \times \overline{H} = \nabla \times \left(\epsilon \frac{\partial \overline{E}}{\partial t} \right) = \epsilon \nabla \times \frac{\partial \overline{E}}{\partial t} = \epsilon \frac{\partial}{\partial t} \nabla \times \overline{E} = \epsilon \frac{\partial}{\partial t} \left(-\mu \frac{\partial \overline{H}}{\partial t} \right) \tag{3.5}$$

Noting that μ is constant (for the same reasons as those applied to ϵ), there results

$$\nabla \times \nabla \times \overline{H} = -\epsilon\mu \frac{\partial^2}{\partial t^2} \overline{H}. \tag{3.6}$$

Similarly, if one takes the curl of equation (3.2) and uses equation (3.1) as the auxiliary relation, one obtains

$$\nabla \times \nabla \times \overline{E} = -\epsilon\mu \frac{\partial^2}{\partial t^2} \overline{E}. \tag{3.7}$$

Equations (3.6) and (3.7) are known as the vector Helmholtz (or wave) equations. Their advantage over Maxwell's equations is that their independent variables appear uncoupled, equation (3.7) involving only the electric field and equation (3.6) only the magnetic field. By use of the vector identity

$$\nabla \times \nabla \times \overline{a} = \nabla \nabla \cdot \overline{a} - \nabla^2 \overline{a}, \tag{3.8}$$

and equations (3.3) and (3.4), one obtains the vector wave equations

$$\nabla^2 \overline{E} = \epsilon\mu \frac{\partial^2}{\partial t^2} \overline{E}, \tag{3.9}$$

$$\nabla^2 \overline{H} = \epsilon\mu \frac{\partial^2}{\partial t^2} \overline{H}, \tag{3.10}$$

where the symbol ∇^2 operating on a vector denotes the Laplacian operator applied to each Cartesian component; for example, equation (3.9) stands for

$$\hat{x}\left(\frac{\partial^2 E_x}{\partial x^2}+\frac{\partial^2 E_x}{\partial y^2}+\frac{\partial^2 E_x}{\partial z^2}\right)+\hat{y}\left(\frac{\partial^2 E_y}{\partial x^2}+\frac{\partial^2 E_y}{\partial y^2}+\frac{\partial^2 E_y}{\partial z^2}\right)$$
$$+\hat{z}\left(\frac{\partial^2 E_z}{\partial x^2}+\frac{\partial^2 E_z}{\partial y^2}+\frac{\partial^2 E_z}{\partial z^2}\right)=\mu\epsilon\left(\hat{x}\frac{\partial^2 E_x}{\partial t^2}+\hat{y}\frac{\partial^2 E_y}{\partial t^2}+\hat{z}\frac{\partial^2 E_z}{\partial t^2}\right). \tag{3.11}$$

Again, we must distinguish between elegance and simplicity: equation (3.9) is elegantly concise, but its equivalent equation (3.11) shows that it is far from simple. Nor should this surprise us. We have assumed the medium to be a simple one locally, but at any distance there might be obstacles or inhomogeneities, and nothing whatever has been said about sources, so that the field may be quite a complicated one.

To simplify equation (3.11) so that a solution may be obtained, let us assume that all derivatives with respect to y and z vanish. Physically, this is equivalent to requiring the problem to be invariant along those directions. Taking the three components of equation (3.11) separately, we now have three relatively simple equations

$$\frac{\partial^2 E_x}{\partial x^2}=\mu\epsilon\frac{\partial^2 E_x}{\partial t^2}, \tag{3.12}$$

$$\frac{\partial^2 E_y}{\partial x^2}=\mu\epsilon\frac{\partial^2 E_y}{\partial t^2}, \tag{3.13}$$

$$\frac{\partial^2 E_z}{\partial x^2}=\mu\epsilon\frac{\partial^2 E_z}{\partial t^2}. \tag{3.14}$$

It can be shown now that equation (3.12) does not lead to a useful solution. Equation (3.3) and the assumption we have made that all derivatives with respect to y and z vanish require

$$\frac{\partial E_x}{\partial x}=0. \tag{3.15}$$

Using the fact that derivatives with respect to y and z vanish in equation (3.1) results in

$$\frac{\partial E_x}{\partial t}=0, \tag{3.16}$$

and hence E_x is constant in time (a static or DC field). Since *information* is carried only by signal *variations*, this field plays no part in the information transmission process. Therefore, we will neglect it, while keeping in mind that static fields may always exist superimposed on the signals we are treating.

Equations (3.13) and (3.14) are of the form

$$\frac{\partial^2 f}{\partial x^2}=\frac{1}{v^2}\frac{\partial^2 f}{\partial t^2}, \tag{3.17}$$

which was first shown by d'Alembert to admit solutions of the form

$$f = f_1(x - vt) + f_2(x + vt), \tag{3.18}$$

where f_1 and f_2 are arbitrary twice-differentiable functions. Accordingly, we find

$$E_y = f_1(x - \frac{1}{\sqrt{\mu\epsilon}}t) + f_2(x + \frac{1}{\sqrt{\mu\epsilon}}t), \tag{3.19}$$

$$E_z = f_3(x - \frac{1}{\sqrt{\mu\epsilon}}t) + f_4(x + \frac{1}{\sqrt{\mu\epsilon}}t). \tag{3.20}$$

Here, the $f_n, n = 1, 2, 3, 4$ are arbitrary twice-differentiable functions. The time variation of the source will determine the particular functions that are appropriate to a given problem. In vector form,

$$\overline{E} = \hat{y}\left[f_1\left(x - \frac{1}{\sqrt{\mu\epsilon}}t\right) + f_2\left(x + \frac{1}{\sqrt{\mu\epsilon}}t\right)\right] + \hat{z}\left[f_3\left(x - \frac{1}{\sqrt{\mu\epsilon}}t\right) + f_4\left(x + \frac{1}{\sqrt{\mu\epsilon}}t\right)\right]. \tag{3.21}$$

Taking the curl of both sides of this equation and recalling that only x derivatives are nonvanishing, we get

$$\nabla \times \overline{E} = \hat{z}\left(f_1' + f_2'\right) - \hat{y}\left(f_3' + f_4'\right), \tag{3.22}$$

where the prime denotes a derivative with respect to the argument. Now we can apply equation (3.2) to get

$$-\mu\frac{\partial \overline{H}}{\partial t} = \hat{z}\left(f_1' + f_2'\right) - \hat{y}\left(f_3' + f_4'\right). \tag{3.23}$$

Noting that

$$\int dt f'\left(x \pm \frac{1}{\sqrt{\mu\epsilon}}t\right) = \pm\sqrt{\mu\epsilon} f\left(x \pm \frac{1}{\sqrt{\mu\epsilon}}t\right) + C \tag{3.24}$$

and that the constant C represents a static or DC field, which can be neglected as before, we have

$$\overline{H} = \sqrt{\frac{\epsilon}{\mu}}\left[-\hat{y}(f_3 - f_4) + \hat{z}(f_1 - f_2)\right], \tag{3.25}$$

with the argument of each f_n in (3.27) the same as in (3.21).

3.3 PURE TRAVELING WAVES

The arguments of the f_n functions appearing in equations (3.19)–(3.25) are of two kinds. The functions f_1 and f_3 always appear with the argument $\left(x - \frac{t}{\sqrt{\mu\epsilon}}\right)$, while f_2 and f_4 always appear with argument $\left(x + \frac{t}{\sqrt{\mu\epsilon}}\right)$. Let us assume for the moment that f_2 and f_4 are zero and that f_1 at $t = 0$ has the space dependence as shown in Figure 3.1a. From this we infer that $f_1(u)$, where $u = x - \frac{t}{\sqrt{\mu\epsilon}}$, has the form of

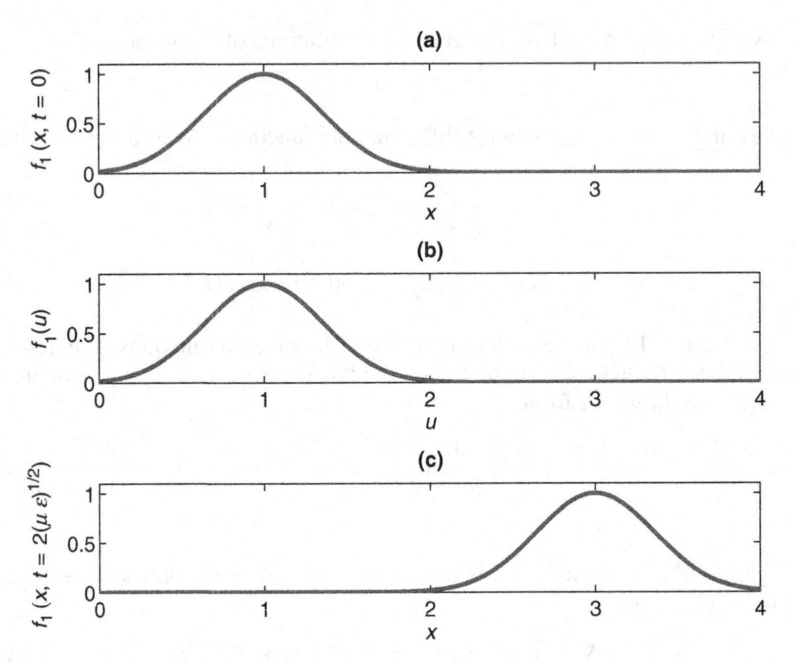

FIGURE 3.1 (a) Space dependence of f_1 at $t = 0$. (b) Dependence of f_1 on u as found from (a). (c) Space dependence of f_1 at $t = 2/v$, as found from (b).

Figure 3.1b. Hence, we can find the space dependence of f_1 at any other time; the dependence at time $t = 2\sqrt{\mu\epsilon}$ is shown in Figure 3.1c. We see that the form of the wave has not changed at all, being determined entirely by the shape of f_1. What has happened is that between $t = 0$ and $t = 2\sqrt{\mu\epsilon}$ the wave has traveled 2 units in the x direction. Thus, f_1 represents a wave traveling, without change of form or amplitude, in the x direction. Its velocity, called the wave velocity, is $1/\sqrt{\mu\epsilon}$. The same is true of f_3. Any wave for which $f_2 = 0 = f_4$ is a pure traveling wave in the $+x$ direction. Similarly, f_2 and f_4, which have the argument $x + \frac{t}{\sqrt{\mu\epsilon}}$, can be shown to represent a wave traveling without change of amplitude or shape in the $-x$ direction; thus, any wave for which $f_1 = f_3 = 0$ represents a pure traveling wave in the $-x$ direction. From equations (3.21) and (3.25), it follows that for either set of traveling waves the relations

$$\overline{H} = \left(\hat{k} \times \overline{E}\right)/\eta, \tag{3.26}$$

$$\overline{E} = -\eta\left(\hat{k} \times \overline{H}\right) \tag{3.27}$$

are obeyed, where the *wave impedance* η (also known as the *characteristic impedance* or the *intrinsic impedance* of the medium) is given by

$$\eta = \frac{\left|\hat{k} \times \overline{E}\right|}{\left|\overline{H}\right|} = \sqrt{\frac{\mu}{\epsilon}}, \tag{3.28}$$

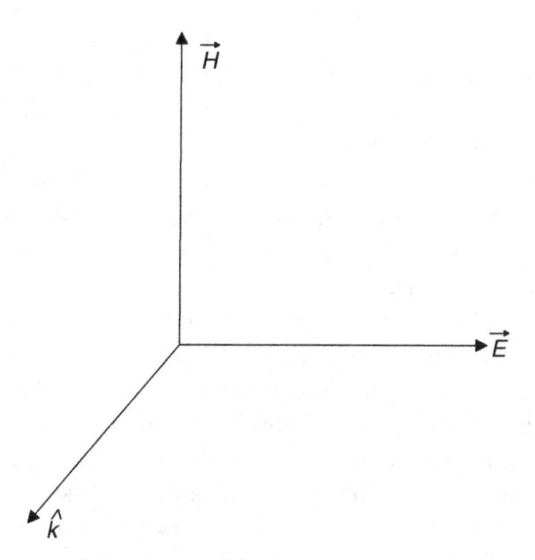

FIGURE 3.2 Space relationship for TEM wave vectors; $\hat{k}$ denotes the propagation direction.

and $\hat{k}$ is the unit vector in the propagation direction; that is, $\hat{k} = \hat{x}$ for f_1 and f_3, and $\hat{k} = -\hat{x}$ for f_2 and f_4. Equations (3.26) and (3.27) are by far the easiest way for finding one of the fields (electric or magnetic) of a pure traveling wave when the other is known, and they will be used so often that it is worthwhile to know them by rote. They show that $\overline{E}$, $\overline{H}$, and $\hat{k}$ are three mutually perpendicular vectors. Such waves are called *transverse electromagnetic* (TEM) *waves* because both the electric and magnetic vectors are transverse to the direction of propagation, as shown in Figure 3.2. They are also referred to as *plane waves* because the $\overline{E}$ and $\overline{H}$ fields do not vary in any plane perpendicular to the direction of propagation.

3.4 INFORMATION TRANSMISSION

D'Alembert's solution is of fundamental importance to the concept of transmitting information by electromagnetic waves. It guarantees that, to the extent that a medium satisfies the approximation of a nonconducting "simplest" medium, a plane wave propagating in this medium with a waveform impressed upon it will emerge at the receiver unaltered in shape, with the information intact. The process of shaping the waveform at the transmitter according to the information to be transmitted is called *modulation* and that of recovering the information from the wave at the receiver is called *demodulation* or *detection*.

The conditions that were stipulated for the medium, that it be nonconducting and otherwise simple, were necessary, as can be shown by substituting equations (3.21) and (3.25) into Maxwell's equations: when these conditions are not satisfied it is found that they are no longer solutions. Thus, departure from the ideal properties causes the signal to be distorted during propagation; the greater the departure from the ideal and the longer the path, the greater the distortion. The concept of bandwidth comes in handy here. In general, media that are nearly ideal have wide bandwidth, while those that depart greatly from the "simplest" case have narrow bandwidths. The troposphere departs from an ideal medium primarily by having only slight inhomogeneities; its bandwidth for most propagation processes is many megahertz. The ionosphere is a plasma that has a very strongly dispersive as well as anisotropic and conductive character; its bandwidth for most propagation modes is on the order of only a fraction of a kilohertz to a few kilohertz.

In principle, it is not necessary for a medium to be nondistorting to propagate information satisfactorily, as long as the signal information can be reconstructed at the receiver. In practice, of course, such reconstruction can be difficult if the distortion is strong. For certain digital signals, distortion in the medium can cause intersymbol interference (ISI), necessitating the use of channel equalizers, for example. For digital signal transmission, the reliability of data transmission can be improved by adding redundant data in the signal by means of forward error correcting codes. One complicating factor is that for natural media such as the troposphere and ionosphere, the medium properties vary in time and are generally not predictable in sufficient detail for a deterministic model to apply. In this case, probabilistic (stochastic) models need to be assumed for the propagation medium (channel), as we will see later in this chapter. In a practical sense, the close relation between the simplicity of the medium and the simplicity of systems for information transmission is valid and a good concept for system designers to have in mind.

3.5 SINUSOIDAL TIME DEPENDENCE IN AN IDEAL MEDIUM

Sinusoidal (time-harmonic) time dependence is of special importance in the treatment of propagation problems for many reasons. First, it is the limiting case of a narrowband signal, so the behavior of such signals is often adequately characterized by that of the (sinusoid) carrier. Second, for linear media the general case may be obtained by Fourier analysis as a superposition of sinusoids. Third, time-harmonic waves simplify cases involving complicated media. In the ideal medium (nonconducting, isotropic, homogeneous, linear, and time invariant) discussed above, sinusoidal time dependence emerges from d'Alembert's solution as a special case. For example, for the y component of the electric field intensity of a pure traveling wave in the $+x$ direction, we have from equation (3.21)

$$E_y = f_1\left(x - \frac{1}{\sqrt{\mu\epsilon}}t\right). \tag{3.29}$$

If this is to represent a time-harmonic function, we can also require

$$E_y = \text{Re}\left[\underline{f}(x)e^{j\omega t}\right]. \tag{3.30}$$

This can only be satisfied if f_1 is an exponential function

$$f_1 = \mathrm{Re}\left[\underline{A}e^{-jk\left(x-\frac{1}{\sqrt{\mu\epsilon}}t\right)}\right], \tag{3.31}$$

where k is chosen so that

$$k/\sqrt{\mu\epsilon} = \omega, \tag{3.32}$$

for then we have

$$f_1 = \mathrm{Re}\left[\underline{A}e^{j(\omega t-kx)}\right] = \mathrm{Re}\left[\underline{A}e^{j\omega t}e^{-jkx}\right]. \tag{3.33}$$

This last set of equations shows that the choice of f_1 as in equation (3.31) combines both the traveling waveform of equation (3.29) as required by d'Alembert's solution and the time-exponential form of equation (3.30) required for a time-harmonic signal. Thus, for a monochromatic plane wave traveling in the $+x$ direction, we obtain the phasor representation

$$\underline{\overline{E}} = (\hat{y}\underline{A} + \hat{z}\underline{B})\,e^{-jkx}, \tag{3.34}$$

$$\underline{\overline{H}} = \frac{1}{\eta}\,(-\hat{y}\underline{B} + \hat{z}\underline{A})\,e^{-jkx}, \tag{3.35}$$

and for a similar plane wave propagating in the $-x$ direction

$$\underline{\overline{E}} = (\hat{y}\underline{C} + \hat{z}\underline{D})\,e^{jkx}, \tag{3.36}$$

$$\underline{\overline{H}} = \frac{1}{\eta}\,(\hat{y}\underline{D} - \hat{z}\underline{C})\,e^{jkx}. \tag{3.37}$$

Note that both time and space enter these equations through complex exponentials, since equation (3.30) is implied. A consequence of the symmetrical way in which time and space coordinates enter into these equations is the fact that in an ideal medium a time-harmonic plane wave is also harmonic in space. The separation between equal phase points in time is called a *period*. That separation in space is called a *wavelength*. Both will be discussed in more detail later in connection with more general media. It should also be pointed out that the ideal medium considered so far in this chapter must be lossless. This is true because a nonmagnetic medium can have only two loss mechanisms: Joule heating by conducting currents and dielectric hysteresis. The former was ruled out by specifying a nonconducting medium, and the latter by specifying a nondispersive one. It is the lossless nature of the medium that allows the wave to continue unattenuated, as required by d'Alembert's solution, since it is giving up no energy to the medium.

The assumption of a lossless medium is sometimes justified, at other times it is not, depending on the frequency range and the environment considered. For example, below about 3 GHz the troposphere exhibits very low losses, and above several hundred MHz ionospheric effects become very small. Thus, the assumption of a lossless medium would be a good one for a signal at 2 GHz directively propagating to a satellite. The same would be true at 10 GHz in clear weather, but in this frequency range the polarization hysteresis of water has a significant effect, so losses cannot be neglected during heavy rainfalls.

At first sight, the condition that the fields vary only in the x direction, that is, that the plane wave travel in the $\pm x$ direction, might seem to be overrestrictive. After all, there is no physical reason why plane waves could not be excited to travel in other directions. The more general case can, of course, be obtained from equations (3.34) and (3.35) by rotating the coordinate system. The general expressions for a sinusoidal traveling wave progressing in the direction of an arbitrary unit vector $\hat{k}$ are found to be

$$\underline{\overline{E}} = \underline{\overline{E}}_0 e^{-j(kx\cos\alpha + ky\cos\delta + kz\cos\gamma)} = \underline{\overline{E}}_0 e^{-j\overline{k}\cdot\overline{r}}, \tag{3.38}$$

$$\underline{\overline{H}} = \frac{1}{\eta}\hat{k} \times \underline{\overline{E}}, \tag{3.39}$$

where $\underline{\overline{E}}_0$ is a constant complex vector perpendicular to $\hat{k}$, such that

$$\underline{\overline{E}}_0 \cdot \hat{k} = 0, \tag{3.40}$$

and $\eta = \sqrt{\frac{\mu}{\epsilon}}$ is the wave impedance of the medium. The cosines appearing in equation (3.38) are the direction cosines of the propagation direction, illustrated in Figure 3.3, and $\overline{r}$ denotes the position vector $\overline{r} = \hat{x}x + \hat{y}y + \hat{z}z$. The $\overline{k}$ vector is defined as $\overline{k} = k\hat{k}$, where $k = \omega\sqrt{\mu\epsilon}$ as discussed previously. For a pure propagating wave, the $\overline{k}$ vector is purely real, so the amplitude and the phase of the plane wave are constant along planar surfaces where $\overline{k} \cdot \overline{r}$ is constant. We will later learn that nonuniform plane waves having a similar form, but where the $\overline{k}$ vector has some complex-valued components, are also possible.

By manipulating Maxwell's equations, it is also possible to develop descriptions of the power flow in electromagnetic fields. The basic quantity that describes the time-averaged power density (watts per unit area) carried by a plane wave is the *time-averaged Poynting vector*, given by

$$\overline{S} = \frac{1}{2}\text{Re}\left[\underline{\overline{E}} \times \underline{\overline{H}}^*\right] = \hat{k}\frac{|\underline{\overline{E}}_0|^2}{2\eta}, \tag{3.41}$$

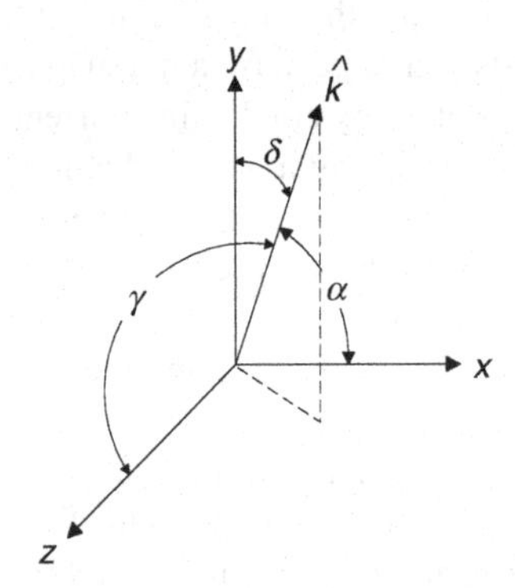

FIGURE 3.3 Direction angles for the propagation vector $\hat{k}$.

where the star denotes complex conjugation and the last equality is valid only for uniform plane waves in lossless media. The amplitude of $\overline{S}$ indicates the power density (W/m^2) carried by a propagating field, and the direction of $\overline{S}$ indicates the direction of power flow, which for a plane wave in a simple medium coincides with the direction of propagation $\overline{k}$.

In equation (3.41), the amplitude of the complex vectors $\underline{\overline{E}}$ and $\underline{\overline{H}}$ is equal to the *peak* amplitude of the respective time-domain fields $\overline{E}$ and $\overline{H}$; see, for example, equation (2.27). Sometimes, the amplitude of a complex vector is chosen instead to be equal to the *root mean square* (rms) amplitude of the respective time-harmonic field. Because the peak amplitude is $\sqrt{2}$ times the rms value for a time-harmonic field, equation (2.27) would be replaced by

$$\overline{E}(\overline{r}, t) = \text{Re}\left[\sqrt{2}\,\underline{\overline{E}}_0(\overline{r})e^{j\omega t}\right], \tag{3.42}$$

and the $1/2$ factor in equation (3.41) is not present. Instead, we have

$$\overline{S} = \text{Re}\left[\underline{\overline{E}} \times \underline{\overline{H}}^*\right] = \hat{k}\frac{|\underline{\overline{E}}_0|^2}{\eta}. \tag{3.43}$$

The reader should be aware that, in general, equations that relate power quantities to complex vectors will differ by a $1/2$ factor depending on whether the rms or the peak amplitude convention is used to represent the complex vectors. The peak amplitude is used throughout this text.

The following examples are presented to help the reader build skills in identifying properties of uniform plane waves. Our goal is first to assess whether the example is an acceptable plane wave, and then to identify the plane wave's direction of propagation and its frequency. All examples are assumed to be propagating in free space ($\mu = \mu_0$, $\epsilon = \epsilon_0$, $\sigma = 0$).

- $\underline{\overline{E}} = \hat{x}e^{-j2\pi y}$: By examining the exponent, we can see that this is a plane wave propagating in the $\hat{y}$ direction since here $\overline{k} = 2\pi\hat{y}$. The electric field direction is perpendicular to the direction of propagation, so this is an acceptable plane wave. The frequency of the plane wave can be determined from $k = 2\pi = \omega\sqrt{\mu_0\epsilon_0}$, so that when y is measured in meters, $f = \omega/(2\pi) = 1/\sqrt{\mu_0\epsilon_0}$ or 300 MHz.
- $\underline{\overline{H}} = \hat{z}e^{j\pi z}$: By examining the exponent, this is a plane wave propagating in the $-\hat{z}$ direction. Here, the magnetic field direction is parallel to the direction of propagation, which is not allowed since we must have $\nabla \cdot \underline{\overline{H}} = 0$. The example is erroneous.
- $\underline{\overline{E}} = \hat{x}e^{-j8\pi(y+z)}$: By examining the exponent, we have

$$\overline{k} = 8\pi\,(\hat{y} + \hat{z}) = 8\pi\sqrt{2}\left[(\hat{y} + \hat{z})\,/\sqrt{2}\right],$$

so this plane wave propagates in the $\left[(\hat{y} + \hat{z})\,/\sqrt{2}\right]$ direction. The electric field direction is perpendicular to the direction of propagation; therefore this is a plane wave. The frequency is $f = \omega/(2\pi) = 4\sqrt{2}/\sqrt{\mu_0\epsilon_0}$ or 1.7 GHz.

- $\underline{\overline{E}} = (2\hat{x} + \hat{y} - 3\hat{z})\, e^{-j2\pi(x+y+z)}$: By examining the exponent, we find that

$$\hat{k} = (\hat{x} + \hat{y} + \hat{z})\,/\sqrt{3}$$

and $k = 2\pi\sqrt{3}$. Now check $\underline{\overline{E}}_0 \cdot \hat{k}$ to find that this is indeed zero, so that the electric field direction is perpendicular to the direction of propagation, as required for a plane wave. Finally, the frequency is $300\sqrt{3}$ MHz.

3.6 PLANE WAVES IN LOSSY AND DISPERSIVE MEDIA

The plane wave solutions obtained so far are applicable only to media that are "simplest" in the sense of Chapter 2 (linear, isotropic, dispersionless, homogeneous, and time invariant) and also nonconducting. As mentioned above, this is not a satisfactory description when loss is an important consideration, for example, the satellite communications example at 10 GHz in heavy rain. We now wish to obtain solutions that are valid in the presence of conduction and dispersion. We shall restrict ourselves to the time-harmonic regime. The medium is still restricted to be simple except for conduction and dispersion, and therefore it is linear, so that the general case can be obtained via Fourier analysis as a superposition of sinusoidal waves.

Using again phasor representation for sinusoidal quantities, for example,

$$\overline{E}(\overline{r}, t) = \text{Re}\left[\underline{\overline{E}}(\overline{r})e^{j\omega t}\right], \tag{3.44}$$

Maxwell's curl equations (equations (2.1) and (2.2)) may be rewritten as

$$\nabla \times \underline{\overline{H}} = (\sigma_{\text{e}} + j\omega\epsilon)\,\underline{\overline{E}} = j\omega\underline{\epsilon}^{\text{e}}\underline{\overline{E}}, \tag{3.45}$$

$$\nabla \times \underline{\overline{E}} = -j\omega\mu\underline{\overline{H}}, \tag{3.46}$$

where σ_{e} is the effective conductivity (see equation (2.58)) and ϵ is the real part of the complex permittivity $\underline{\epsilon}^{\text{e}}$, so that σ and ϵ are functions of ω. If we take the divergence of both sides of equation (3.45) and recall that in a homogeneous medium σ and ϵ do not vary in space, we get

$$\nabla \cdot \underline{\overline{E}} = 0, \tag{3.47}$$

and similarly from equation (3.46) follows

$$\nabla \cdot \underline{\overline{H}} = 0. \tag{3.48}$$

It is possible to separate the variables by taking the curl of one of the curl equations, say equation (3.45), and substituting from the other in the result, obtaining

$$\nabla \times \nabla \times \underline{\overline{H}} = \omega^2\mu\underline{\epsilon}^{\text{e}}\underline{\overline{H}}. \tag{3.49}$$

Again, one can use the vector identity of equation (3.8) to arrive at the analog of the vector wave equation for the monochromatic case

$$\nabla^2\underline{\overline{H}} = -\omega^2\mu\underline{\epsilon}^{\text{e}}\underline{\overline{H}}. \tag{3.50}$$

To obtain a simple solution, we again assume no variation in the y and z directions and end up with the three equations

$$\frac{\partial^2 \underline{H}_x}{\partial x^2} = -\omega^2 \mu \underline{\epsilon}^{\mathrm{e}} \underline{H}_x , \tag{3.51}$$

$$\frac{\partial^2 \underline{H}_y}{\partial x^2} = -\omega^2 \mu \underline{\epsilon}^{\mathrm{e}} \underline{H}_y , \tag{3.52}$$

$$\frac{\partial^2 \underline{H}_z}{\partial x^2} = -\omega^2 \mu \underline{\epsilon}^{\mathrm{e}} \underline{H}_z . \tag{3.53}$$

From equation (3.48) and the vanishing of the y and z derivatives follows

$$\frac{\partial}{\partial x} \underline{H}_x = 0 \tag{3.54}$$

and using this in the left half of equation (3.51) yields

$$\underline{H}_x = 0, \tag{3.55}$$

indicating that the wave is transverse magnetic (TM).

Solutions of equations (3.52) and (3.53) are of the form $\underline{M} e^{\pm j\underline{k}x}$, where $\underline{M}$ is a complex constant and

$$\underline{k} = \omega \sqrt{\mu \underline{\epsilon}^{\mathrm{e}}} , \tag{3.56}$$

so we obtain for the phasor representation of the magnetic field

$$\overline{\underline{H}} = (\hat{y}\underline{M} + \hat{z}\underline{N})\, e^{j\underline{k}x} + (\hat{y}\underline{P} + \hat{z}\underline{R})\, e^{-j\underline{k}x} . \tag{3.57}$$

The corresponding electric field follows from equation (3.45)

$$\overline{\underline{E}} = \frac{\underline{k}}{\omega \underline{\epsilon}^{\mathrm{e}}} \left[(-\hat{y}\underline{N} + \hat{z}\underline{M})\, e^{j\underline{k}x} + (\hat{y}\underline{R} - \hat{z}\underline{P})\, e^{-j\underline{k}x} \right] . \tag{3.58}$$

A word needs to be said about $\underline{k}$ in these equations, since only $\underline{k}^2 = \omega^2 \mu \underline{\epsilon}^{\mathrm{e}} = \omega^2 \mu (\epsilon - j\sigma_{\mathrm{e}}/\omega)$ enters into the original equation (3.50). From this definition, it can be seen that $\underline{k}^2$ lies always in the fourth quadrant of the complex plane. Hence, one of its square roots lies in the second quadrant and the other in the fourth. We choose for $\underline{k}$ the root in the fourth quadrant, that is,

$$\underline{k} = k_{\mathrm{R}} - jk_{\mathrm{I}}, \tag{3.59}$$

where k_{R} and k_{I} are both positive. Note that k_{R} in this section plays the same role as the purely real k for lossless media and becomes identical to it when $\sigma_{\mathrm{e}} = 0$.

With this definition of $\underline{k}$ in mind, it is clear that the factor $e^{j\underline{k}x}$ in equations (3.59) and (3.60) represents an increase in amplitude and phase as x increases, whereas the factor $e^{-j\underline{k}x}$ represents a decrease in amplitude and phase with increasing x. The former is appropriate to a wave traveling in the $-x$ direction while the latter belongs to a $+x$ directed wave. Thus, we have for the phasor representation of a monochromatic pure plane wave traveling in the x direction

$$\overline{\underline{E}} = \underline{\eta}\, (\hat{y}\underline{R} - \hat{z}\underline{P})\, e^{-k_{\mathrm{I}}x} e^{-jk_{\mathrm{R}}x}, \tag{3.60}$$

$$\underline{\overline{H}} = (\hat{y}\underline{P} + \hat{z}\underline{R})\, e^{-k_{\mathrm{I}}x} e^{-jk_{\mathrm{R}}x}, \tag{3.61}$$

and for a similar wave traveling in the negative x direction

$$\underline{\overline{E}} = \underline{\eta}(-\hat{y}\underline{N} + \hat{z}\underline{M})\, e^{k_{\mathrm{I}}x} e^{jk_{\mathrm{R}}x}, \tag{3.62}$$

$$\underline{\overline{H}} = (\hat{y}\underline{M} + \hat{z}\underline{N})\, e^{k_{\mathrm{I}}x} e^{jk_{\mathrm{R}}x}, \tag{3.63}$$

where $\underline{\eta}$ is a generalized wave impedance

$$\underline{\eta} = \sqrt{\frac{\mu}{\underline{\epsilon}^{\mathrm{e}}}} \tag{3.64}$$

that reduces to the η of the last section when $\sigma_{\mathrm{e}} = 0$. Indeed, it can be verified easily that the fields in equations (3.60)–(3.63) reduce to those of equations (3.34)–(3.37) when the medium is lossless.

Note here that the real part of the wavenumber, k_{R}, determines the phase variation (as will be discussed further in the next section) while the imaginary part of the wavenumber, k_{I}, gives rise to an exponential attenuation of the field amplitude. It is common to define the "skin depth" or "penetration depth" of an electromagnetic wave in a lossy medium as $1/k_{\mathrm{I}}$, since this is the distance within which the field amplitude is reduced by the factor e^{-1}. Generally, it is found that lower frequency fields penetrate further into lossy media, and therefore are preferred, for example, for underwater communications or deep subsurface sensing applications. It is also common to describe the signal attenuation properties of a lossy medium in terms of "decibels of power loss per meter." By computing the Poynting vector for a plane wave in a lossy medium by the first equality of equation (3.41), we can find that the decibels of power loss in 1 m is given by $10 \log_{10} e^{-2k_{\mathrm{I}}} = 8.68k_{\mathrm{I}}$.

Analogously to equations (3.38) and (3.39), we can generalize equations (3.60)–(3.63) to an arbitrary propagation direction

$$\underline{\overline{E}} = \underline{\overline{E}}_0 e^{-j\underline{k}(x\cos\alpha + y\cos\delta + z\cos\gamma)} = \underline{\overline{E}}_0 e^{-j\underline{\overline{k}}\cdot\overline{r}}, \tag{3.65}$$

$$\underline{\overline{H}} = \frac{\underline{\overline{k}} \times \underline{\overline{E}}}{\underline{\eta}}, \tag{3.66}$$

where again $\underline{\overline{E}}_0$ is a constant complex vector perpendicular to $\hat{k}$, so Figure 3.2 is again applicable at any instant of time. However, as discussed previously, such purely propagating waves are not the only possibility. Note that all three Cartesian components of the $\underline{\overline{k}}$ vector are in phase in equation (3.65). It can be shown by returning to the most general form of the wave equation in three dimensions, equation (3.50), that actually all that is required of the $\underline{\overline{k}}$ vector is that it satisfy

$$\underline{\overline{k}} \cdot \underline{\overline{k}} = \omega^2 \mu \underline{\epsilon}^{\mathrm{e}}, \tag{3.67}$$

allowing more types of waves to exist than those contained in equation (3.65) alone. These nonuniform waves will be studied in more detail in Chapter 6. A uniform plane wave is one for which all components of the $\underline{\overline{k}}$ vector have the same phase, so the propagation direction is constant and the field vectors do not vary in directions transverse to the propagation direction.

3.7 PHASE AND GROUP VELOCITY

The waves of equations (3.65) and (3.66) are uniform plane waves. They are not truly harmonic in space, however, since in traveling a distance d the wave attenuates by a factor $e^{-k_I d}$, so it is continuously decreasing in amplitude. Neglecting this amplitude attenuation, one has a sinusoid in space for which the phase can be identified.

The period T of the wave is the smallest displacement in time required for the wave to repeat itself; for a sinusoid this requires 2π radians of phase, so

$$\omega T = 2\pi, \tag{3.68}$$

$$T = 2\pi/\omega = \frac{1}{f}. \tag{3.69}$$

Similarly, we can define the wavelength as the smallest displacement in space required for a wave to repeat itself, or in the present case where the wave is attenuated, for the phase to repeat itself. In direct analogy to the period, we find the wavelength λ to be related to the phase constant by

$$k_R \lambda = 2\pi, \tag{3.70}$$

$$\lambda = 2\pi/k_R = \frac{v_p}{f}. \tag{3.71}$$

A concept linking the spatial and temporal domains is the *phase velocity*, v_p, which is the velocity at which an hypothetical observer would have to move in the propagation direction to remain at a point of constant phase. The time function corresponding to equation (3.61) can be written as

$$\overline{H}(\bar{r}, t) = \left[(\hat{y} P_R + \hat{z} R_R)\cos(\omega t - k_R x) - (\hat{y} P_I + \hat{z} R_I)\sin(\omega t - k_R x)\right] e^{-k_I x}, \tag{3.72}$$

where the subscripts R and I denote the real and imaginary parts of $\underline{P}$ and $\underline{R}$. Note in this expression that the wave phase propagates in time, but the envelope of the wave amplitude $e^{-k_I x}$ does not, but is a fixed function of space. Also, we see that points that have the same phase are characterized by a constant factor $(\omega t - k_R x)$. The phase velocity is found from this dependence as

$$v_p = \frac{dx}{dt}\Big|_{\omega t - k_R x \,=\, \text{const}} = \frac{\omega}{k_R}. \tag{3.73}$$

From this and equations (3.69) and (3.71), it follows that

$$\lambda = v_p T. \tag{3.74}$$

Another, more vague but still useful, concept is that of *group velocity*. This is the speed with which the information content of the wave may be considered to travel, provided the medium is not too dispersive or the bandwidth is not too great. A perfect sinusoid in time cannot convey information; it must start, stop, or at least change in phase or amplitude to convey a message. To discuss the signal velocity, we must therefore consider a varying sinusoidal signal, that is, a modulated wave. For the sake of simplicity and definiteness, consider an amplitude-modulated wave progressing in the $+x$ direction and having only a y component of the electric field. The field propagates in a medium that is lossless, but dispersive, for the frequencies of interest. In the plane $x = 0$, let this component be given by

$$E_y(0, t) = E_0 \left[1 + m \cos(\omega_m t)\right] \cos(\omega_c t). \tag{3.75}$$

In this equation, and in the remainder of this section, we shall employ parenthesis exclusively to denote function arguments and braces exclusively to denote multiplication factors to avoid any ambiguities. In equation (3.75), the bracketed term is the amplitude variation or envelope, which carries the information. By use of a trigonometric identity for the product of two cosines, this equation may be rewritten as

$$E_y(0, t) = E_0 \left\{\cos(\omega_c t) + \frac{m}{2}\left[\cos([\omega_c + \omega_m] t) + \cos([\omega_c - \omega_m] t)\right]\right\}. \tag{3.76}$$

Here, we see the total signal displayed as the sum of three sinusoids: the carrier, ω_c, the upper sideband at angular frequency $\omega_c + \omega_m$, and the lower sideband at $\omega_c - \omega_m$. Since each of these is a pure sinusoid, we can use the plane wave concepts developed above to predict the wave in space away from the $x = 0$ plane,

$$\begin{aligned} E_y(x, t) = E_0 \Big\{&\cos(\omega_c t - k_R(\omega_c) x) + \frac{m}{2}\Big[\cos([\omega_c + \omega_m] t - k_R(\omega_c + \omega_m) x) \\ &+ \cos([\omega_c - \omega_m] t - k_R(\omega_c - \omega_m) x)\Big]\Big\}. \end{aligned} \tag{3.77}$$

We return this to the form of an envelope modulating a carrier by using the trigonometric identity

$$\frac{1}{2}(\cos a + \cos b) = \cos\left(\frac{a - b}{2}\right) \cos\left(\frac{a + b}{2}\right), \tag{3.78}$$

which yields

$$\begin{aligned} E_y(x, t) = E_0 \Big\{&\cos(\omega_c t - k_R(\omega_c) x) + m \cos\Big(\omega_m t - \frac{1}{2}\left[k_R(\omega_c + \omega_m) - k_R(\omega_c - \omega_m)\right] x\Big) \\ &\times \cos\Big(\omega_c t - \frac{1}{2}\left[k_R(\omega_c + \omega_m) + k_R(\omega_c - \omega_m)\right] x\Big)\Big\}. \end{aligned} \tag{3.79}$$

This expression involves k_R evaluated not only at ω_c but also at the sideband frequencies. To obtain some commonality, use the Taylor's series expansions

$$k_R(\omega_c + \omega_m) = k_R(\omega_c) + \omega_m \frac{dk_R}{d\omega}(\omega_c) + \frac{1}{2}\omega_m^2 \frac{d^2 k_R}{d\omega^2}(\omega_c) + \cdots \tag{3.80}$$

$$k_R(\omega_c - \omega_m) = k_R(\omega_c) - \omega_m \frac{dk_R}{d\omega}(\omega_c) + \frac{1}{2}\omega_m^2 \frac{d^2 k_R}{d\omega^2}(\omega_c) + \cdots \tag{3.81}$$

to obtain the first-order approximations

$$k_R(\omega_c + \omega_m) - k_R(\omega_c - \omega_m) \approx 2\omega_m \frac{dk_R}{d\omega}(\omega_c), \tag{3.82}$$

$$k_R(\omega_c + \omega_m) + k_R(\omega_c - \omega_m) \approx 2k_R(\omega_c). \tag{3.83}$$

With these we can write

$$E_y(x,t) = E_0 \left\{ 1 + m \cos\left(\omega_m \left[t - \frac{dk_R}{d\omega}(\omega_c)x \right] \right) \right\} \cos(\omega_c t - k_R(\omega_c)x). \tag{3.84}$$

This expression is of the same form as equation (3.75) that gives the field at $x = 0$, but the carrier has been shifted in phase by $k_R(\omega_c)x$ that corresponds to a phase velocity of ω_c/k_R (not much of a surprise here), while the envelope has been shifted in time by $x\frac{dk_R}{d\omega}(\omega_c)$. The group velocity results from the time shift of the envelope:

$$v_g = \left(\frac{dk_R}{d\omega}(\omega_c) \right)^{-1}. \tag{3.85}$$

This is the actual velocity of (information carrying) signal propagation. At this point, it seems wise to check under what conditions the assumptions of equations (3.82) and (3.83) are satisfied. If k_R is linear in ω, that is, if all terms of power two or greater in equations (3.80) and (3.81) vanish, then the assumptions are valid for arbitrary ω_m. Going back to equations (3.56) and (3.59), one finds that this requires $\sigma_e = 0$ and ϵ independent of ω: in other words, a nonconducting and dispersionless medium. This brings us back to an ideal medium in which d'Alembert's solution is applicable. In such a medium, even wideband (large ω_m) signals travel undistorted, and their group velocity and phase velocity turn out to be the same and equal to the wave velocity defined before for such media. If the medium is not ideal, the group velocity can still be calculated with good accuracy from equation (3.85), but it would depend on frequency and not be equal to the phase velocity, which, by the way, would also depend on frequency. This *frequency dispersion* (the fact that each component of the signal propagates with a different phase velocity) causes the waveform to be gradually distorted as it propagates in the medium.

Now look at this from a physical point of view. Consider, for example, a pulse that becomes stretched as it propagates, as shown in Figure 3.4. The velocity of the top of the leading edge would be computed as approximately $2.5/t_1$ and that of the top of the trailing edge would be computed as approximately $1.5/t_1$. Clearly, there is no single group velocity that can describe the motion of the entire pulse. When the approximations implicit in equations (3.82) and (3.83) break down, the wave suffers distortion, and any statements about the velocity of information travel must be made with caution and sophistication.

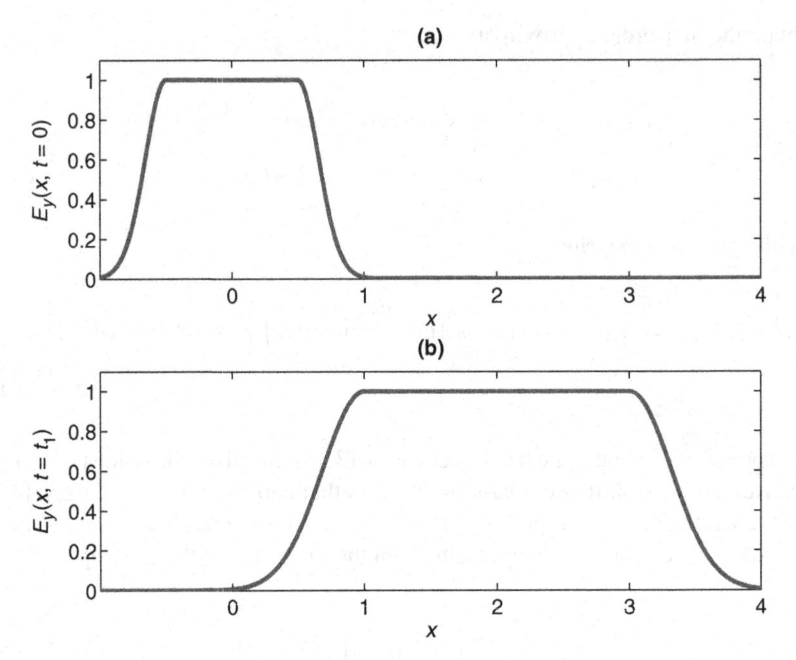

FIGURE 3.4 (a) Rectangular pulse envelope at $t = 0$. (b) Same pulse envelope at $t = t_1$.

3.8 WAVE POLARIZATION

For sinusoidal time dependence, we have found that the electric field of a wave traveling in the positive x direction is given in phasor form as

$$\underline{\overline{E}}(x) = (\hat{y}\underline{A} + \hat{z}\underline{B})\, e^{-j\underline{k}x}, \tag{3.86}$$

where $\underline{A}$ and $\underline{B}$ are complex constants and $\underline{k}$ is the complex propagation constant defined in equation (3.56). We wish to determine how the electric field behaves as a function of time at a given distance x. To do this, return from the phasor domain to the time domain by

$$\overline{E}(x, t) = \text{Re}\left[\underline{\overline{E}}(x)e^{j\omega t}\right] = \text{Re}\left[(\hat{y}\underline{A} + \hat{z}\underline{B})e^{j\omega t - j\underline{k}x}\right]. \tag{3.87}$$

Letting

$$\begin{aligned} \underline{A} &= Ae^{ja}, \\ \underline{B} &= Be^{jb}, \end{aligned} \tag{3.88}$$

where a, b, A, and B are all real, and using equation (3.59), we get

$$\begin{aligned} \overline{E}(x, t) &= \text{Re}\left[\hat{y}Ae^{-k_{\text{I}}x}e^{j(\omega t - k_{\text{R}}x + a)} + \hat{z}Be^{-k_{\text{I}}x}e^{j(\omega t - k_{\text{R}}x + b)}\right] \\ &= \hat{y}Ae^{-k_{\text{I}}x}\cos(\omega t - k_{\text{R}}x + a) + \hat{z}Be^{-k_{\text{I}}x}\cos(\omega t - k_{\text{R}}x + b). \end{aligned} \tag{3.89}$$

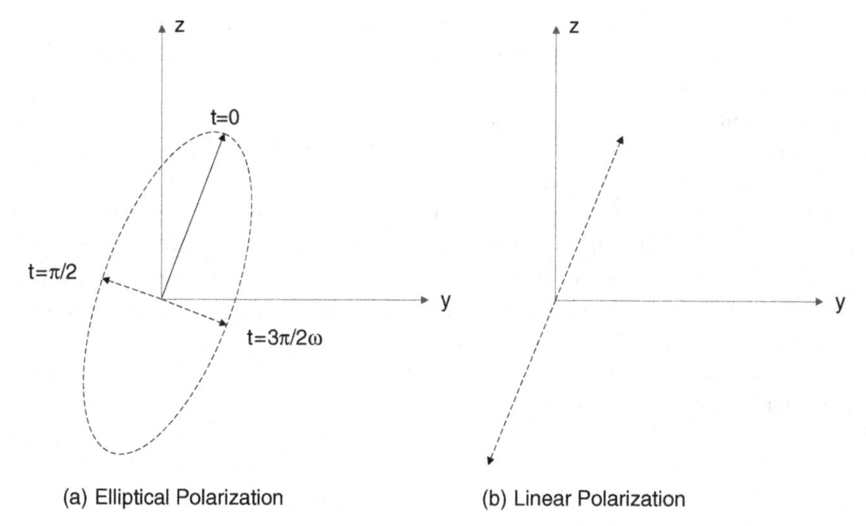

FIGURE 3.5 Polarization examples.

Setting

$$r = Ae^{-k_I x}, \tag{3.90}$$

$$s = Be^{-k_I x}, \tag{3.91}$$

and

$$u = \omega t - k_R x, \tag{3.92}$$

this simplifies to

$$E_y(x, t) = r\cos(u + a), \tag{3.93}$$

$$E_z(x, t) = s\cos(u + b). \tag{3.94}$$

These are the equations of an ellipse (or degenerate ellipse, that is, a circle or a straight line) in parametric form with $u = \omega t - k_R x$ as the parameter. Therefore, in the most general case the $\overline{E}$ vector will, during each period, trace out an ellipse in the yz plane (perpendicular to the propagation direction) as shown in Figure 3.5a. This is known as *elliptical* polarization.

When $a = b \pm n\pi$ (n an integer), that is, when $\underline{A}$ and $\underline{B}$ have the same or opposite phase angle (including the case when they are both real numbers), the ellipse degenerates into a straight line as shown in Figure 3.5b. This special case is known as *linear* polarization. Linear polarization also occurs if either $A = 0$ or $B = 0$. For example, $\underline{\overline{E}}(x) = (\hat{y}A + \hat{z}B)\,e^{-j\underline{k}x}$ and $\underline{\overline{E}}(x) = \hat{y}\underline{A}e^{-j\underline{k}x}$, both representing linearly polarized waves.

When $A = B$ and $b = a \pm \pi/2$, the ellipse degenerates into a circle. This is known as *circular* polarization. If $b - a = -\pi/2$, the rotation as time progresses will be clockwise as we look toward the propagation direction ($+x$ axis). This is termed *right-*

circular (or *right-hand circular*) polarization. If $b - a = +\pi/2$, the rotation will be counterclockwise as we look toward the propagation direction. This is termed *left-circular* (or *left-hand circular*) polarization. For example, $\overline{E}(x) = \underline{A}\,(\hat{y} - j\hat{z})\,e^{-j\underline{k}x}$ represents a right-circular wave and $\overline{E}(x) = \underline{A}\,(\hat{y} + j\hat{z})\,e^{-j\underline{k}x}$ represents a left-circular wave. Note that $\overline{E}(x) = \underline{A}\,(\hat{y} + j\hat{z})\,e^{+j\underline{k}x}$ represents a right-circular wave since the direction of propagation ($-x$ axis) has been reversed.

These concepts are important because propagation mechanisms may discriminate between waves of differing polarizations. For example, the surface of the Earth produces different effects on vertical and on horizontal linear polarization. The ionosphere may, under some conditions, attenuate waves with one sense of circular polarization more strongly than the other.

The polarization state of a wave can be specified in several ways. A useful measure is the polarization ratio $\underline{R}$ defined by

$$\underline{R} = \underline{B}/\underline{A}. \tag{3.95}$$

In particular, if $\underline{R}$ is a real number, we have a linearly polarized wave; if $\underline{R} = \pm j$, we have a circularly polarized wave.

A wave with a certain polarization can also be decomposed into two waves with a different polarization type. For example, it is obvious that an elliptically polarized wave such as $\overline{E}(x) = (\hat{y}\underline{A} + \hat{z}\underline{B})\,e^{-j\underline{k}x}$ is a superposition of two linearly polarized waves, $\overline{E}_1(x) = \hat{y}\underline{A}\,e^{-j\underline{k}x}$ and $\overline{E}_2(x) = \hat{z}\underline{B}e^{-j\underline{k}x}$. Likewise, an elliptically (and, hence, a linear) polarized wave can be decomposed into two circular polarized waves since

$$\underline{A}\hat{y} + \underline{B}\hat{z} = C_+\,(\hat{y} + j\hat{z}) + C_-\,(\hat{y} - j\hat{z}), \tag{3.96}$$

with

$$C_\pm = \frac{1}{2}\,(\mathrm{Re}[A] + j\mathrm{Im}[A]) \mp \frac{j}{2}\,(\mathrm{Re}[B] + j\mathrm{Im}[B]). \tag{3.97}$$

Such polarization decompositions can be important in treating propagation through anisotropic media, as will be shown in Chapter 11.

The basic polarization categories presented here for a wave propagating in the $\hat{x}$ direction are also applicable for waves propagating in other directions. Consider the example

$$\underline{\overline{E}} = \left[-2\hat{y} + \left(1 + j\sqrt{3}\right)\hat{x} + \left(1 - j\sqrt{3}\right)\hat{z}\right] e^{-j2\pi(x+y+z)/\sqrt{3}}. \tag{3.98}$$

Here, we have a plane wave propagating in the $\hat{k} = (\hat{x} + \hat{y} + \hat{z})\,/\sqrt{3}$ direction, and it is easy to show that $\underline{\overline{E}}_0 \cdot \hat{k} = 0$, so that the field is indeed perpendicular to the direction of propagation.

To classify the polarization of this plane wave, we need to resolve $\underline{\overline{E}}_0$ into two components perpendicular to each other and to $\hat{k}$:

$$\underline{\overline{E}} = \left(\hat{a}\underline{E}_a + \hat{b}\underline{E}_b\right) e^{-j2\pi(x+y+z)/\sqrt{3}}. \tag{3.99}$$

Once the complex amplitudes $\underline{E}_a$ and $\underline{E}_b$ are determined, the polarization can be classified in a manner similar to that for the plane wave propagating only in the $\hat{x}$ direction.

Choosing $\hat{a}$ and $\hat{b}$ is arbitrary, so to begin, any $\hat{a}$ can be chosen that is perpendicular to $\hat{k}$. Here, we choose $\hat{a} = (\hat{x} - \hat{z})/\sqrt{2}$. Now $\hat{b}$ can be determined, since we must have

$$\begin{aligned} \hat{k} \times \hat{a} &= \hat{b} \\ &= (2\hat{y} - \hat{x} - \hat{z})/\sqrt{6}. \end{aligned} \tag{3.100}$$

Now

$$\underline{E}_a = \hat{a} \cdot \underline{\overline{E}}_0 = j\sqrt{6}, \tag{3.101}$$

$$\underline{E}_b = \hat{b} \cdot \underline{\overline{E}}_0 = -\sqrt{6}. \tag{3.102}$$

so that the two polarizations components have the same amplitude and are out of phase 90°: this is circular polarization.

The handedness of the circular polarization can be classified by plotting the field vector at a fixed point in space as time elapses. This implies returning to the time domain

$$\begin{aligned} \overline{E}(t) &= -\sqrt{6}\left[\hat{b}\cos\left(\omega t - \overline{k}\cdot\overline{r}\right) + \hat{a}\sin\left(\omega t - \overline{k}\cdot\overline{r}\right)\right] e^{-j\overline{k}\cdot\overline{r}} \\ &= -\sqrt{6}\left[\hat{b}\cos(\omega t) + \hat{a}\sin(\omega t)\right] \end{aligned} \tag{3.103}$$

with the second equation at a convenient location in space, namely, the origin. We see that at time zero, the field vector is pointing in the $-\hat{b}$ direction, and then as time elapses the field traces out a circle in the direction at first of the $-\hat{a}$ direction. Noting that $\hat{a} \times \hat{b} = \hat{k}$, this is a left-hand circular polarization because one's right hand thumb points opposite the direction of propagation when the right hand fingers are circulated in the direction of the field vector rotation.

REFERENCES

1. Harrington, R. F., *Time-Harmonic Electromagnetic Fields*, McGraw-Hill, 1961.
2. Budden, K. G., *The Propagation of Radio Waves*, Cambridge University Press, 1985.
3. Chew, W. C., *Waves and Fields in Inhomogeneous Media*, IEEE Press, 1999.
4. Kong, J. A., *Electromagnetic Wave Theory*, EMW Publishing, 2008.

4

ANTENNA AND NOISE CONCEPTS

4.1 INTRODUCTION

The function of a communication link is to transmit information from one location to another. If wireless radiowave propagation is to be used, the signal must be radiated from the transmitter site by a transmitting antenna. It then propagates toward the receiver site where a receiving antenna captures the signal and delivers it to a receiver system. As a result, it is difficult to specify the propagation process without reference to the antennas that transmit and receive the signal. The ability of a signal to be properly received at the receiver will depend not only on the power carried by the incoming electromagnetic wave, but also on the sensitivity of the receiver and the presence or absence of other (interfering) signals. Consideration of noise effects is therefore very important in the design of any communication system. The aim of this chapter is to develop the basic antenna concepts necessary for a discussion of propagation effects in wireless communication systems and to introduce standard methods for describing noise in communication systems so that prediction of signal-to-noise ratios (SNR) for realistic systems becomes feasible.

4.2 ANTENNA CONCEPTS

The field radiated from an antenna is always more complicated near the antenna than at a large distance from it. If one considers free-space propagation at distances

Radiowave Propagation: Physics and Applications. By Curt A. Levis, Joel T. Johnson, and Fernando L. Teixeira

sufficiently far from the transmitting antenna, the field appears to be a spherical wave emanating from a point (the antenna location). Under these conditions, the observation point is said to be in the *far-field* or *Fraunhofer region* of the transmitter antenna. Three conditions must be satisfied for this to be the case

$$\begin{aligned} r &> 2(D+d)^2/\lambda, \\ r &> 1.6\lambda, \\ r &> 5(D+d), \end{aligned} \tag{4.1}$$

where D is the largest dimension of the transmitting antenna, d the largest dimension of the receiving antenna, r the distance between the two, and λ is the wavelength of operation.[1] In communication systems, the receiving antenna is almost always in the far field of the transmitter antenna. Therefore, we will consider only far-field properties in this chapter.

4.3 BASIC PARAMETERS OF ANTENNAS

There are many ways of specifying antenna performance [1,2], and perhaps not all antenna experts would agree as to which ones are the most basic. For the purpose of this discussion, let us consider impedance, directionality, and efficiency as the most basic. Some other useful parameters that can be derived from these three will also be discussed.

The *impedance* of an antenna at a given frequency is defined to be the impedance it presents to a generator connected to its terminals. It can be measured by connecting the antenna to the source via an impedance bridge or a slotted line, just as other load impedances are measured. Simple in principle, such measurements can nevertheless be tricky in practice. Care must be taken that the environment, including the measurement apparatus, does not reflect energy, which the antenna has radiated, back into it. Therefore, such measurements must be performed outdoors or in an anechoic chamber; that is, one that has walls made of absorbing material. Another precaution is to make sure that the antenna is not in the field of any signal that it can receive, especially if the measurement requires obtaining a null. For example, when the impedance of a broadcasting antenna is measured with an impedance bridge, a good balance cannot be obtained when a signal from another station on the same channel is present.

The antenna impedance is, in general, composed of a resistive component (real part) and a reactive component (imaginary part),

$$\underline{Z}_{\mathrm{A}} = R_{\mathrm{A}} + jX_{\mathrm{A}}. \tag{4.2}$$

[1]Consideration of the $r > 2(D+d)^2/\lambda$ criterion shows why, for example, to the human eye the moon does not appear as a point source, while a star does: at optical wavelengths and with the human eye as the receiving "antenna," stars are in the far field, but the moon is not.

The resistance indicates the time averaged power P delivered to the antenna when a sinusoidal current is applied to the terminals:

$$P = \frac{1}{2}|\underline{I}|^2 R_{\mathrm{A}}, \tag{4.3}$$

where $|\underline{I}|$ is the peak current amplitude. Part of the input power is radiated by the antenna P_{r}, and part is dissipated as heat in the antenna structure (ohmic losses) P_{l}, so

$$P = P_{\mathrm{r}} + P_{\mathrm{l}}. \tag{4.4}$$

Although the focus of the current discussion is on the antenna impedance, note that the time-averaged radiated power P_{r} can be obtained as an integral of the real part of the complex Poynting vector over a closed surface A surrounding the antenna

$$P_{\mathrm{r}} = \int\int_A \overline{S} \cdot \hat{n}\, dA = \frac{1}{2}\int\int_A \mathrm{Re}\left[\underline{\overline{E}} \times \underline{\overline{H}}^*\right] \cdot \hat{n}\, dA, \tag{4.5}$$

if the radiated fields $\underline{\overline{E}}$ and $\underline{\overline{H}}$ are known on the surface A.

In terms of P_{l} and P_{r}, the loss resistance of the antenna is defined as

$$R_{\mathrm{l}} = \frac{2P_{\mathrm{l}}}{|\underline{I}|^2}, \tag{4.6}$$

and the radiation resistance as

$$R_{\mathrm{r}} = \frac{2P_{\mathrm{r}}}{|\underline{I}|^2}. \tag{4.7}$$

Note that the factor $1/2$ in equations (4.3) and (4.5) and the factors of 2 in equations (4.6) and (4.7) are a consequence of the use of peak as opposed to rms value phasors.

It is then found from equations (4.3) and (4.4) that

$$R_{\mathrm{A}} = R_{\mathrm{r}} + R_{\mathrm{l}}. \tag{4.8}$$

The *radiation efficiency* of an antenna at a given frequency is a second basic quantity. It is defined as the ratio of the total power radiated by the antenna to the total power delivered to its terminals:

$$\upsilon = P_{\mathrm{r}}/P. \tag{4.9}$$

From (4.4)–(4.8) follow the relations

$$\upsilon = R_{\mathrm{r}}/R_{\mathrm{A}} = 1 - \frac{R_{\mathrm{l}}}{R_{\mathrm{A}}}. \tag{4.10}$$

The concept of radiation efficiency is simple, but its measurement is again by no means an easy matter. One method consists of measuring the radiated power density at a sufficient distance in all directions for which it is significant and integrating this to get the total radiated power P_{r}. The input power P is measured at the same time, and radiation efficiency is calculated from equation (4.9). In another approach, the antenna impedance is first measured with the antenna radiating into space. Next the antenna is surrounded at a sufficient distance with a conducting shell,

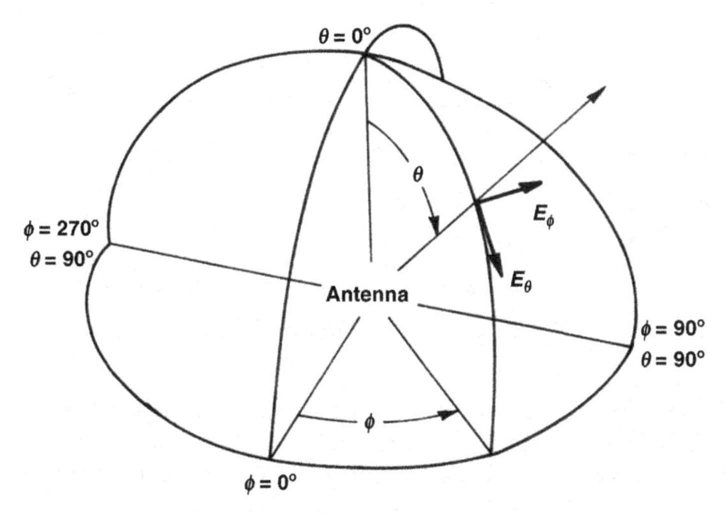

FIGURE 4.1 Antenna coordinate system.

and its impedance is remeasured. Assuming the antenna current is unaffected by the shell, and also assuming the losses in the shell are much smaller than those in the antenna, the losses will be the same for both measurements, and the resistive part of the impedance measured in the second case will be R_l while that measured in the first case is R_A. The radiation efficiency can then be computed from equation (4.10).

A third basic property of an antenna system is its directional performance at a given frequency. To specify this, a spherical coordinate system with the antenna at its center can be employed, as shown in Figure 4.1. At the given frequency, the relative field strength of the θ component amplitude and phase in a given direction can be given by a phasor $\underline{f}_{\theta}(\theta, \phi)$; similarly, the relative ϕ component can be specified by a phasor $\underline{f}_{\phi}(\theta, \phi)$. The total electric field at a distance R in the far field is then given by the complex vector function

$$\overline{\underline{E}}(R, \theta, \phi) = \left[\hat{\theta}\underline{f}_{\theta}(\theta, \phi) + \hat{\phi}\underline{f}_{\phi}(\theta, \phi)\right] \frac{e^{-jkR}}{R}, \tag{4.11}$$

where the term e^{-jkR}/R captures the range dependence of the field and describes a "spherical wave" behavior. This dependence on range is true in the far field of *any* antenna, with different antennas producing different $\underline{f}_{\theta}(\theta, \phi)$ and $\underline{f}_{\phi}(\theta, \phi)$. Unlike plane waves, fields radiated from an antenna decrease in amplitude as the reciprocal of the distance from the antenna. However, for large R, small changes in R do not cause significant amplitude variations and the field can be approximated as a plane wave in many applications.

In practice, it is more usual to measure the "relative power patterns" $P_{\theta}(\theta, \phi)$ and $P_{\phi}(\theta, \phi)$ for θ and ϕ polarizations, respectively, expressed as

$$P_{\theta}(\theta, \phi) = \mathcal{C}|\underline{f}_{\theta}(\theta, \phi)|^2, \tag{4.12}$$

$$P_{\phi}(\theta, \phi) = \mathcal{C}|\underline{f}_{\phi}(\theta, \phi)|^2, \tag{4.13}$$

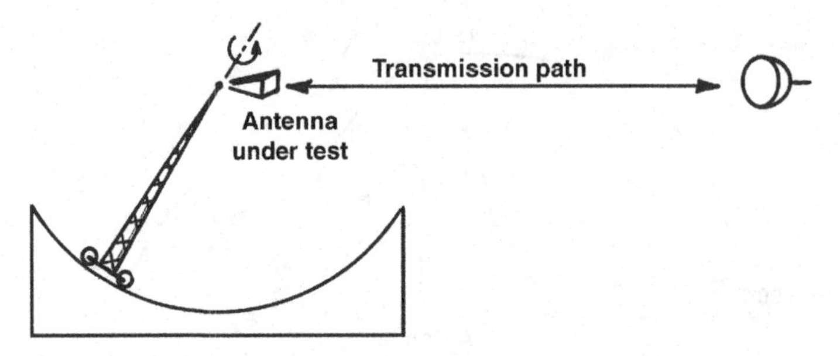

FIGURE 4.2 Directional pattern measurement.

where C is an arbitrary normalization constant. Unless one of these power patterns is zero, this is not enough information to give the antenna response to arbitrary polarization. The missing information is the phase relationship between the $\underline{f}_{\theta}$ and $\underline{f}_{\phi}$ phasor functions in a given direction, which can be obtained by measuring the power pattern of a judiciously chosen third polarization.

Relative field strength and power patterns can be measured by rotating the test antenna while a signal is being transmitted between it and another sufficiently distant antenna, as shown in Figure 4.2. The transmission path must be long enough to ensure that, if the test antenna is used as a transmitter, the field produced by it will be represented with sufficient accuracy by a plane wave over the receiving antenna aperture. In practice, transmission may be in either direction; that is, the test antenna can equally well be at the receiving end. Sometimes it is not practical to have sufficiently long links to carry out such measurements. If a measurement is done using a shorter link, lenses or reflectors can sometimes be used to generate a planar wavefront at the receiver antenna location. Such antenna pattern measurement ranges are called *compact ranges*. Care must be taken to ensure that the measured signal from the transmitter antenna reaches the receiver antenna solely via the direct transmission path, with no significant reflections from the ground or other structures (such as supporting structures) reaching the receiver antenna. Obviously, rotating an antenna structure is practical only when the antenna is reasonably small. One way of overcoming this difficulty for large antennas (meant to operate at relatively low frequencies) is based on the concept that the antenna patterns depend not on the absolute size of the antenna, but instead on the size measured in wavelengths—in other words, on the *electrical size* of the antenna. Thus, it is possible to obtain the patterns by measuring a scale model of the desired antenna. For example, the antenna may be scaled down in size by a factor of 20, provided the wavelength is also scaled down by a factor of 20, that is, provided the frequency used for measuring the model is 20 times the frequency at which the full-scale antenna is intended to operate. It is assumed in such an approach that the dielectric properties of all the antenna materials do not vary significantly between the original and scaled frequencies.

In a typical measurement, the patterns will be only relative values, but in theoretical treatments, it is sometimes useful to normalize the patterns. A particularly convenient normalization is the *directional gain* of an antenna $D(\theta, \phi)$, defined as the ratio of the power density radiated in the direction (θ, ϕ) to the power density at the same distance that would be radiated by an isotropic antenna[2] with the same polarization and radiation efficiency, and with the same input power. If a receiving system is available that is calibrated absolutely, the directional gain may be determined as

$$D(\theta, \phi) = \frac{\tilde{P}_\theta(\theta, \phi) + \tilde{P}_\phi(\theta, \phi)}{P_r/4\pi R^2} = \frac{\tilde{P}_\theta(\theta, \phi) + \tilde{P}_\phi(\theta, \phi)}{\upsilon P_{in}/4\pi R^2}, \tag{4.14}$$

where the tilde denotes an absolute measurement (in W/m^2) and R is the measurement distance. Alternatively, $D(\theta, \phi)$ can be expressed from uncalibrated relative power measurements by

$$D(\theta, \phi) = \frac{4\pi \left[P_\theta(\theta, \phi) + P_\phi(\theta, \phi)\right]}{\int_{\theta=0}^{\pi} \int_{\phi=0}^{2\pi} \left[P_\theta(\theta, \phi) + P_\phi(\theta, \phi)\right] \sin\theta d\theta d\phi}, \tag{4.15}$$

given the relative power patterns $P_\theta(\theta, \phi)$ and $P_\phi(\theta, \phi)$. It is then necessary to measure $P_\theta(\theta, \phi)$ and $P_\phi(\theta, \phi)$ for all directions (θ, ϕ) into which significant power is radiated. It is also commonplace to report the directional gain for only a given polarization. For example, for vertical polarization only the first term in the numerator of equations (4.14) or (4.15) would be used.

The *gain* of an antenna in a given direction is the product of directional gain and efficiency

$$G(\theta, \phi) = \upsilon D(\theta, \phi). \tag{4.16}$$

By the use of equation (4.14), this can be shown to be equivalent to

$$G(\theta, \phi) = \frac{\left[\tilde{P}_\theta(\theta, \phi) + \tilde{P}_\phi(\theta, \phi)\right] 4\pi R^2}{P_{in}}, \tag{4.17}$$

which at a specified distance R determines the power density radiated in a given direction relative to the input power. This is a very important measure of antenna performance.

Some examples will illustrate these concepts. Consider a vertically polarized (θ polarized) isotropic antenna; for such an antenna, we should have $P_\theta = C_1$ and $P_\phi = 0$, where C_1 is some arbitrary constant. From equation (4.15), one then finds $D(\theta, \phi) = 1$, or 0 dB. Consider now a small dipole with uniform current distribution (Hertzian dipole). For such a dipole, the radiated power density is distributed in space according to $P_\theta = C_2 \sin^2\theta$, $P_\phi = 0$, where C_2 is again an arbitrary constant (depending on the current strength and the distance) and θ is the polar angle measured from the dipole axis. Use of these P_θ and P_ϕ in equation (4.15) gives $D(\theta, \phi) = 1.5 \sin^2\theta$.

The *directivity* of an antenna is defined to be the directional gain in the direction that maximizes it; that is, it is the maximum directional gain for a given antenna. It is

[2] An isotropic antenna radiates the same power density in all directions. It turns out that an isotropic antenna cannot be realized in practice, but it is a very useful concept.

typically denoted by D (without any arguments since it is a number). For the isotropic vertically polarized antenna, $D = 1$, or 0 dB; for the Hertzian dipole, $D = 1.5$, or 1.8 dB.

The gain of an antenna can also be specified as G without any argument. In this case, it refers to the gain $G(\theta, \phi)$ in the direction (θ, ϕ) that maximizes it. For a 100% efficient vertically polarized isotropic antenna, equation (4.16) shows that $G = 1$, or 0 dB, while for a perfectly efficient Hertzian dipole, it shows $G = 1.5$, or 1.8 dB. This means that for a given input power and distance a Hertzian dipole produces 1.5 times the power density that would be produced by an isotropic antenna of the same efficiency in the direction of maximum gain. It should be apparent that gain is a very important measure of antenna performance.

4.3.1 Receiving Antennas

Our discussion so far has focused on the use of antennas as transmitters. It is also important to understand the reception of electromagnetic waves by antennas. It is possible to develop this subject through use of the "reciprocity theorem" of electromagnetics. However, a more direct approach is utilized here by considering the case of a short wire receiving antenna and generalizing the equations that result to other antenna types.

In the early days of radio, the prototypical transmitting antenna was a metal mast fed against the Earth, which was sometimes made more conductive by a buried screen of wires. A wire screen was sometimes also inserted at the top of the mast (called the "top hat") to give the vertical current along the mast (which was the main contributor to the radiation) a place to flow. The current would enter the top hat, gradually dropping to zero at its outer edges. In this fashion, the vertical current amplitude was relatively constant over the length of the vertical mast. A sketch of such an antenna is shown in Figure 4.3a. A uniform current dipole that would radiate (by the image theorem) an identical field in the region above ground is shown in Figure 4.3b. At the receiver in the plane $\theta = 90°$, the field of such a dipole is given by

$$\underline{\overline{E}} = j\frac{\eta_0}{2\lambda}\underline{I}\frac{e^{-jkr}}{r}h\hat{\theta}, \tag{4.18}$$

where h is the total length of the dipole and η_0 is the free-space impedance ($\approx 377\ \Omega$). The expression for the field of any arbitrary antenna in the far field (equation (4.11)) shows that this can be written in a form that is an analog to equation (4.18) as

$$\underline{\overline{E}} = j\frac{\eta_0}{2\lambda}\underline{I}\frac{e^{-jkr}}{r}\underline{\overline{h}}_e(\theta, \phi), \tag{4.19}$$

where $\underline{I}$ is the input current. The complex vector $\underline{\overline{h}}_e$ defined by this equation is a characteristic property of the antenna and is called the *complex vector effective height*. For antennas radiating only one polarization, its magnitude is sometimes simply called

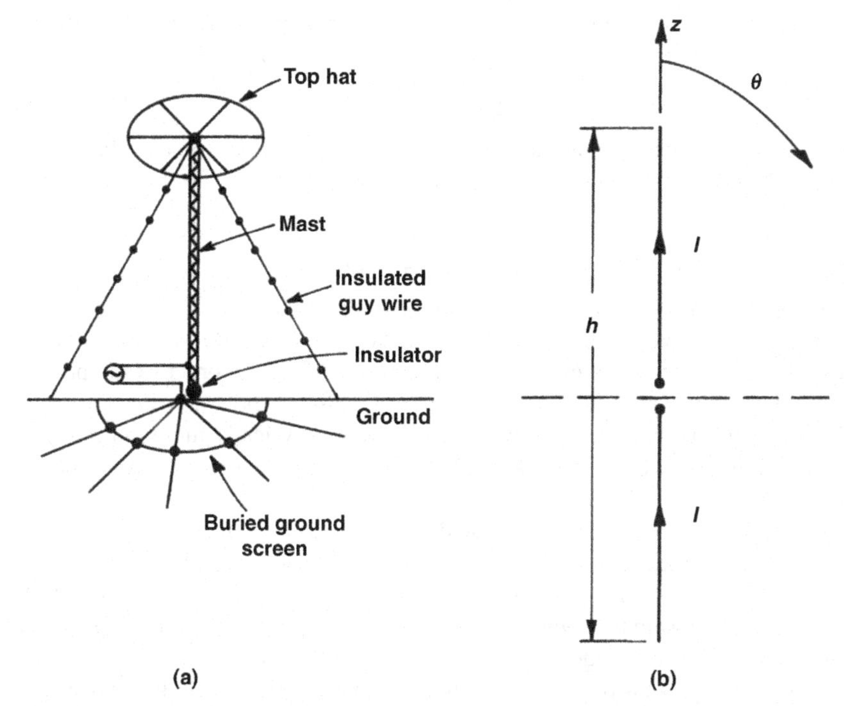

FIGURE 4.3 Origin of the "effective length" concept. (a) Typical low- frequency transmitting antenna. (b) Equivalent uniform current dipole.

the *effective height*. The term *effective length* and symbol $\underline{\bar{l}}_e$ are used interchangeably with effective height and $\underline{\bar{h}}_e$, respectively.

The performance of a receiving antenna is fully characterized by its transmitting properties provided it does not include nonreciprocal devices or materials, which are unusual. Assuming that an incoming field $\underline{\bar{E}}(\theta, \phi)$ is arriving from the direction (θ, ϕ), the power delivered by an antenna to a load Z_R connected to its terminals is given by the equivalent circuit as shown in Figure 4.4, where the Thevenin voltage $\underline{V}_{TH}$ is given by

$$\underline{V}_{TH} = \underline{\bar{E}}(\theta, \phi) \cdot \underline{\bar{h}}_e(\theta, \phi), \tag{4.20}$$

where Z_A represents the antenna impedance (i.e., the same impedance that the antenna would have as a transmitting antenna) and Z_R is the receiver input impedance. Thus,

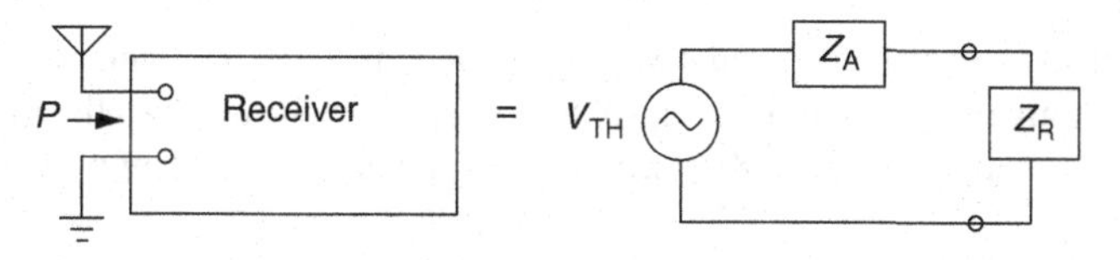

FIGURE 4.4 Equivalent circuit model for receiving antenna.

for an antenna that is responsive to only the same single polarization, for example, a dipole, the Thevenin generator voltage in Figure 4.4 is simply given by

$$|\underline{V}_{\mathrm{TH}}| = |\overline{\underline{E}}| l_{\mathrm{e}}. \tag{4.21}$$

Since we use peak value phasors, the time-averaged power delivered to the receiver can be calculated from the equivalent circuit in Figure 4.4 as

$$P_{\mathrm{R}} = \frac{|\overline{\underline{E}} \cdot \overline{\underline{h}}_{\mathrm{e}}|^2 R_{\mathrm{R}}}{2|Z_{\mathrm{A}} + Z_{\mathrm{R}}|^2}, \tag{4.22}$$

where the θ and ϕ arguments have been omitted for brevity.

At frequencies below about 30 MHz, long straight wires are commonly used as antennas, and h_{e}, the magnitude of $\overline{\underline{h}}_{\mathrm{e}}$, turns out to be on the order of the physical length of the wire, but always somewhat less. This interpretation in terms of an effective length is quite general, but it is of course physically more appealing for straight wires. As a result, it is used mostly in the frequency ranges where straight wires happen to be useful antennas.

At microwaves, straight wires are seldom used, and it is much more common to employ, for example, horns or reflector antennas. Such antennas are characterized by an *aperture* (a characteristic area) rather than a length. For this reason, it is useful to obtain a relationship between the effective height and antenna parameters, such as gain, in the microwave range.

To begin the process of deriving such a relationship, it is useful to express $\overline{\underline{h}}_{\mathrm{e}}$ as the product of a magnitude (real and positive) factor and a complex unit vector

$$\overline{\underline{h}}_{\mathrm{e}} = h_{\mathrm{e}} \hat{\underline{a}}_h, \tag{4.23}$$

where the magnitude factor is defined as

$$h_{\mathrm{e}} = \left(\overline{\underline{h}}_{\mathrm{e}} \cdot \overline{\underline{h}}_{\mathrm{e}}^* \right)^{1/2} \tag{4.24}$$

and the unit vector by

$$\hat{\underline{a}}_h = \overline{\underline{h}}_{\mathrm{e}} / h_{\mathrm{e}}. \tag{4.25}$$

Similarly, the electric field vector can be written as

$$\overline{\underline{E}} = E \hat{\underline{a}}_{\mathrm{e}}. \tag{4.26}$$

From equation (4.19), we can obtain the far-field power density expression from the peak value phasor field for any antenna as

$$S(\theta, \phi) = \frac{E^2}{2\eta_0} = \eta_0 |\underline{I}|^2 h_{\mathrm{e}}^2(\theta, \phi) / (8\lambda^2 r^2). \tag{4.27}$$

This flux density can also be calculated from input considerations. The power input to the antenna is given by equation (4.3), so that for an isotropic antenna the flux density at distance r would be $|\underline{I}|^2 R_{\mathrm{A}} / (8\pi r^2)$. Hence, for an arbitrary antenna we have from (4.14)

$$S(\theta, \phi) = |\underline{I}|^2 R_{\mathrm{A}} \upsilon D(\theta, \phi) / (8\pi r^2). \tag{4.28}$$

Equating equations (4.28) and (4.27) yields

$$h_e(\theta, \phi) = \lambda \left[R_A \upsilon/(\pi\eta_0)\right]^{1/2} [D(\theta, \phi)]^{1/2}, \tag{4.29}$$

which is the relationship we are looking for. Its use, together with equations (4.24) and (4.26), in equation (4.22) results in

$$P_R = \left(\frac{E^2(\theta, \phi)}{2\eta_0}\right) \left(\frac{\lambda^2 D(\theta, \phi)}{4\pi}\right) \upsilon\, |\hat{\underline{a}}_E \cdot \hat{\underline{a}}_h|^2 \left(\frac{4R_A R_R}{|Z_A + Z_R|^2}\right). \tag{4.30}$$

The first factor in the above expression is the flux density S of the wave arriving from direction (θ, ϕ). The second factor has the dimension of area. The remaining three factors are dimensionless. In particular, the factor involving the unit vectors depends only on the antenna polarization and the polarization of the arriving wave and represents the *polarization mismatch factor*, typically denoted as p. The polarization mismatch factor has a maximum value of unity when the antenna polarization is ideal for the impinging wave. It is not difficult to show (with proper attention to the relationship of the coordinate systems of the transmitting and receiving antennas) that the optimum receiving antenna polarization is the same as the transmitting antenna polarization. Thus a wave launched with a right circular polarized antenna should be received with a right circular polarized antenna. The last factor is the *impedance mismatch factor*, denoted as q. The impedance mismatch factor has a maximum value of unity when $Z_A = Z_R^*$. Equation (4.30) can therefore be written in the form

$$P_R = S(\theta, \phi)\, A_{em}(\theta, \phi) \upsilon\, p\, q, \tag{4.31}$$

where the maximum effective aperture or area A_{em} is given by

$$A_{em}(\theta, \phi) = \frac{\lambda^2 D(\theta, \phi)}{4\pi}, \tag{4.32}$$

with polarization mismatch factor

$$p = |\hat{\underline{a}}_E \cdot \hat{\underline{a}}_h|^2, \tag{4.33}$$

and impedance mismatch factor

$$q = \frac{4R_A R_R}{|Z_A + Z_R|^2}. \tag{4.34}$$

The product of the last four factors in equations (4.30) and (4.31) is sometimes called the *receiving aperture* or area. Since the maximum of the last three factors is unity, the receiving aperture cannot exceed A_{em} and hence the nomenclature *maximum effective aperture* for that quantity. The *effective aperture* A_e is the product

$$A_e = \upsilon A_{em} = \frac{\upsilon D \lambda^2}{4\pi} = \frac{G\lambda^2}{4\pi}. \tag{4.35}$$

For aperture-type antennas, the maximum effective aperture is usually close to, but somewhat less than, the physical area projected on a plane perpendicular to the propagation direction. This is generally not true for other antenna types. For example, a half-wavelength (0.5λ) dipole has a directivity of 1.64, which corresponds to a maximum effective aperture of $0.13\lambda^2$. Thus, it effectively absorbs power as though it had a width of $(0.13\lambda^2)/(0.5\lambda)$ or 0.26λ, even though the diameter of such a dipole is usually very much smaller.

The concepts of effective length, gain, and effective aperture will be especially useful in connecting propagation with the transmitter at one end and the receiver at the other, thus allowing propagation calculations to be used as one step in the overall process of estimating the performance of communication or radar systems in which wireless propagation is involved. Only an introductory treatment of antennas and their properties has been provided here; the reader is referred to many books on antennas, for example, references [1, 2], for additional information.

4.4 NOISE CONSIDERATIONS

The performance of systems that deal with information, such as communication links and radars, depends greatly on the signal-to-noise-ratio (SNR) at the receiver. Part of the noise originates from within the receiver system itself; this type of noise is called *internal noise*. All other noise (e.g., undesired interfering signals and static) is called *external noise*.

4.4.1 Internal Noise

All electronic devices produce small randomly varying electrical signals due to the thermal motions associated with their constituent atoms. These fluctuations, although they are small, are important when the signal is small. The receiver also receives noise generated by the antenna resistance R_A or by the internal resistance of a signal generator connected to its input. The standard equation for the thermal noise power emitted from a resistor at absolute temperature T in bandwidth B (in Hz) is given by

$$N_i = k_B T B, \tag{4.36}$$

where N_i is the emitted noise power and $k_B = 1.38 \times 10^{-23}$ J/K is Boltzmann's constant. Although equation (4.36) holds for a resistor at absolute temperature T, it is often used to define an equivalent "noise temperature" for any source of noise power N_i by $T = N_i/(k_B B)$. Thus the noise input from the antenna or signal source can be specified either as a noise power or as an equivalent noise temperature.

A receiver contains many electronic parts, and the use of (4.36) for each component is not feasible. Instead, a more easily measured quantity denoted as the *noise factor* is used generally to quantify amplifier performance with regard to noise. The noise factor (also known as the noise figure) of an amplifier is defined by

$$F = \frac{S_i/N_i}{S_O/N_O}, \tag{4.37}$$

where S_i and N_i are the available signal and noise input powers applied at the amplifier input and S_O and N_O are the available signal and noise output powers that result. The "available" implies that both the source and the load of the receiver or amplifier are terminated in a matched load, and this will be assumed throughout the remainder of this section. *Standard noise factor* measurements are made at a temperature $T_0 = 290\text{K}$, corresponding to $N_i = k_B T_0 B$ (W). This temperature is chosen as an approximation to the usual operating conditions; the standard noise factor would not be appropriate for a cryogenic receiver. Note that S_i/N_i is not the actual input SNR if there is a mismatch at the receiver input. Also, note that the noise figure is ≥ 1, with unity representing an idealized receiver that adds no noise, because the output SNR cannot exceed the input SNR.

If an "available power gain" is defined as

$$G_A = S_0/S_i, \tag{4.38}$$

then the noise figure can be written as

$$F = \frac{N_0}{G_A N_i}. \tag{4.39}$$

The output noise consists of the amplified input noise plus some additional noise N_N added by the receiver components, therefore

$$F = \frac{G_A N_i + N_N}{G_A N_i}. \tag{4.40}$$

It is sometimes useful to refer the additional noise to the input, resulting in a quantity known as the *excess noise* $N_e \equiv N_N/G_A$. Equation (4.40) then becomes

$$F = \frac{G_A N_i + G_A N_e}{G_A N_i} = \frac{N_i + N_e}{N_i}. \tag{4.41}$$

Recalling $N_i = k_B T_0 B$ under standard conditions and rearranging (4.41) gives

$$N_e = (F - 1) k_B T_0 B, \tag{4.42}$$

where F is the standard noise factor and N_e is the excess noise contributed by the receiver when referred to the input. In receivers containing several amplifier stages, most of the internal noise comes from the first stage, since noise generated by this stage will be amplified by the full gain of the receiver, while noise generated by later stages is subject to less amplification. Thus, a low-noise amplifier is most often found as the first stage in a receiving system.

In microwave systems both the input and the output of amplifiers are usually matched, so that $G = G_A$ (the true power gain) and S_0/N_0 is the true SNR at the output. Then

$$S_0/N_0 = \frac{S_i}{F N_i} = \frac{S_i}{F k_B T_0 B}, \tag{4.43}$$

which is applicable when the input noise N_i is the standard $k_B T_0 B$.

An equivalent circuit that models noise generated by the receiver or amplifier as being applied at the input, according to equation (4.43), is shown in Figure 4.5. In this figure N_e represents the excess noise, generated within the receiver or amplifier and referred to the input, while N_i represents the noise contribution from the signal source.

One method of measuring internal receiver noise directly is to orient a receiving antenna so that very low input signal and noise power are received (e.g., by directing

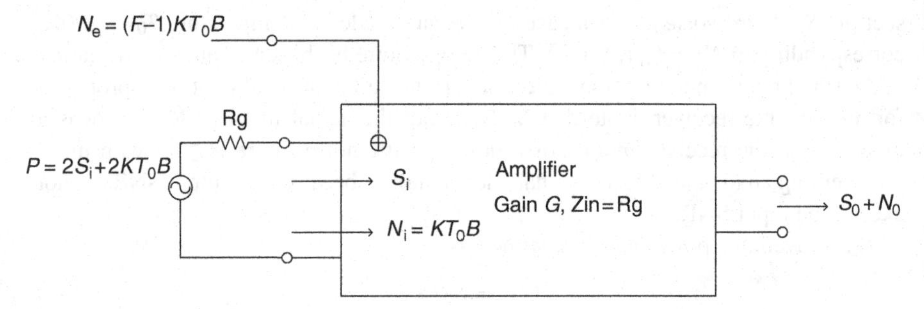

FIGURE 4.5 Equivalent circuit for noise contributions.

the antenna toward a quiet part of the sky if this is possible and the antenna has sufficient directivity). The resulting noise power at the output comes from the receiver and the antenna. The antenna loss noise can be neglected if the antenna is highly efficient; then the output noise can be divided by the gain to calculate the excess noise referred to the input. From Figure 4.5, we have

$$S_0/N_0 = \frac{S_i}{N_i + (F-1)k_B T_0 B}, \tag{4.44}$$

which allows the calculation of signal-to-noise ratio at the output of a receiver given the input signal (S_i) and noise (N_i) powers, as well as the standard receiver noise figure F.

4.4.2 External Noise

A somewhat loose distinction needs to be made here between external noise and interference. In a sense, they are alike in being unwanted signals that have deleterious effects on system performance. If the unwanted signal is man-made and has characteristics not too unlike that of the desired signal, it is likely to be termed interference; otherwise, it is termed noise. Here, our focus is on noise received by the antenna from external sources.

The fundamental measure of external noise is the noise brightness of the source, $B_s(f, \Omega)$, measured in watts per square meter per steradian per hertz. The direction at which the source is seen in the receiving coordinate system will be denoted by Ω, and $d\Omega$ is the increment of solid angle, $d\Omega = \sin\theta \, d\theta \, d\phi$. The noise brightness is a property of the source and does not depend on the distance from which it is viewed.

The analogy to vision may be useful here.[3] Concentrate on a single pixel, that is, the smallest solid picture element that can be resolved. The radiant flux density responsible for that pixel seems to come from a single direction, but actually it corresponds to the integration of brightness over the acceptance angle of the eye. A change

[3]Note that vision is an *imaging* system, which is not exactly what is being considered here. Nevertheless, the analogy is useful.

in the distance between source and receiver does not change the observed brightness, but it does change the received flux density because the angle subtended by the source changes with distance.

Similarly, the received external noise power at the output of a receiver is given by an integration over the view angle and frequency:

$$P_{\mathrm{N,out}} = \int_0^{\infty} |H(f)|^2 q(f) \upsilon(f) \left\{ \int_{\Omega=0}^{\Omega=4\pi} B_{\mathrm{s}}(f, \Omega) A_{\mathrm{em}}(f, \Omega) p(f, \Omega) d\Omega + k_{\mathrm{B}} \left[1 - \upsilon(f) \right] T_{\mathrm{p}} \right\} df. \tag{4.45}$$

In this equation, H denotes the (unitless) receiver system transfer function and T_{p} the physical temperature of the antenna in degrees Kelvin. As before, A_{em} is the maximum effective aperture of the antenna, q is the antenna impedance mismatch factor, p is the polarization mismatch factor, υ is the radiation efficiency of the antenna, and k_{B} is Boltzmann's constant. The term involving T_{p} gives the noise power due to antenna losses. It should be omitted if the antenna noise is included in the internal noise N_{i}. Its contribution is often small for well-designed antennas.

When the relationships between the antenna parameters are used,

$$\upsilon A_{\mathrm{em}} = A_{\mathrm{e}} = \frac{\lambda^2}{4\pi} G, \tag{4.46}$$

the result is

$$P_{\mathrm{N,out}} = \int_0^{\infty} |H(f)|^2 q(f) \left\{ \frac{\lambda^2}{4\pi} \int_{\Omega=0}^{\Omega=4\pi} B_{\mathrm{s}}(f, \Omega) G(f, \Omega) p(f, \Omega) d\Omega + k_{\mathrm{B}} \left[1 - \upsilon(f) \right] T_{\mathrm{p}} \right\} df. \tag{4.47}$$

Often, all the quantities in equation (4.47), with the exception of $H(f)$, do not vary appreciably over the receiver passband. The integration with respect to frequency can then be replaced with multiplication by a noise bandwidth B, defined as

$$\int_0^{\infty} |H(f)|^2 df = B G_{\mathrm{R}}, \tag{4.48}$$

where G_{R} is the nominal receiver gain, typically taken as the average value of $|H(f)|^2$ over the passband. Also, it is usual to specify the maximum available noise as referred to the input,

$$P_{\mathrm{N}} = \frac{P_{\mathrm{N,out}}}{q G_{\mathrm{R}}}. \tag{4.49}$$

The result is

$$P_{\mathrm{N}} = B \left\{ \frac{\lambda^2}{4\pi} \int_{\Omega=0}^{\Omega=4\pi} B_{\mathrm{s}}(\Omega) G(\Omega) p(\Omega) d\Omega + k_{\mathrm{B}} \left[1 - \upsilon \right] T_{\mathrm{p}} \right\}. \tag{4.50}$$

This simplification will be used hereafter, with the understanding that it is not applicable when the noise brightness of the source has a substantial frequency dependence within the receiver passband. The reader should be able to derive the corresponding expressions including frequency integration without difficulty.

4.4.2.1 ***Source Brightness Temperatures*** Often the source brightness is specified, not directly, but in terms of an equivalent brightness temperature, the temperature that a blackbody would need to have in order to radiate with the same brightness as the actual source in the antenna polarization. A "blackbody" is an idealized source that perfectly absorbs all incident radiation and reemits this radiation as thermal noise. For blackbody radiation, the brightness at frequency f is related to the physical temperature T by Planck's law

$$B_s = \frac{2hf}{\lambda^2} \frac{1}{e^{hf/k_B T} - 1}, \tag{4.51}$$

where h is Planck's constant. For reasonably warm sources, the exponent in this equation is small and the first two terms of a Maclaurin series expansion of the exponential suffice. Also, for blackbody radiation the polarization is random so that only half of the radiation is emitted with the antenna polarization, and therefore $p = 1/2$. When these considerations are used, the result is

$$k_B T_B(\Omega) = \lambda^2 B_s(\Omega) p(\Omega), \tag{4.52}$$

which gives

$$P_N = k_B B \left\{ \frac{1}{4\pi} \int_{\Omega=0}^{\Omega=4\pi} T_B(\Omega) G(\Omega) d\Omega + [1 - \upsilon]\, T_p \right\}. \tag{4.53}$$

For blackbody radiation, the "brightness temperature" is the physical temperature of the source. The concept is extended to other radiation sources by defining a brightness temperature by equation (4.52). The maximum noise power available at the antenna terminals, P_N, may also be represented as an equivalent antenna temperature T_A

$$P_N = k_B T_A B, \tag{4.54}$$

or as an equivalent standard receiver noise factor F_{ext}

$$P_N = F_{ext} k_B T_0 B, \tag{4.55}$$

where $T_0 = 290$K is the standard reference temperature. Thus, equation (4.53) can be written as

$$T_A = \frac{\upsilon}{4\pi} \int_{\Omega=0}^{\Omega=4\pi} T_B(\Omega) D(\Omega) d\Omega + [1 - \upsilon]\, T_p. \tag{4.56}$$

When the source brightness is constant over the entire antenna beam, T_B in equation (4.56) can be taken out of the integral, and the with use of

$$\int_{\Omega=0}^{\Omega=4\pi} D(\Omega) d\Omega = 4\pi, \tag{4.57}$$

it follows that

$$T_A = \upsilon T_B + (1 - \upsilon)T_p. \tag{4.58}$$

In this case, the antenna noise temperature is the weighted average of the source brightness temperature and the physical antenna temperature. It should be remembered that the three temperatures in this equation come from very different concepts. T_p is the physical temperature of the antenna. T_B is the brightness temperature of the source; for blackbody radiation, it relates directly to Planck's law, and for other radiation indirectly so. T_A is the temperature at which a resistor matching the receiver input would give the same noise output as that actually obtained. Three distinct concepts of temperature are involved here!

If the source subtends such a small angle at the receiver that it cannot be resolved, $G(\Omega)$ and $p(\Omega)$ can be held constant in the integration with respect to Ω in equation (4.50). The integration of brightness over view angle gives the flux density arriving from that range of view angles; thus, for an unresolved source in the direction Ω_0

$$B \int_{\Omega} B_s(\Omega) d\Omega = S_N(\Omega_0), \tag{4.59}$$

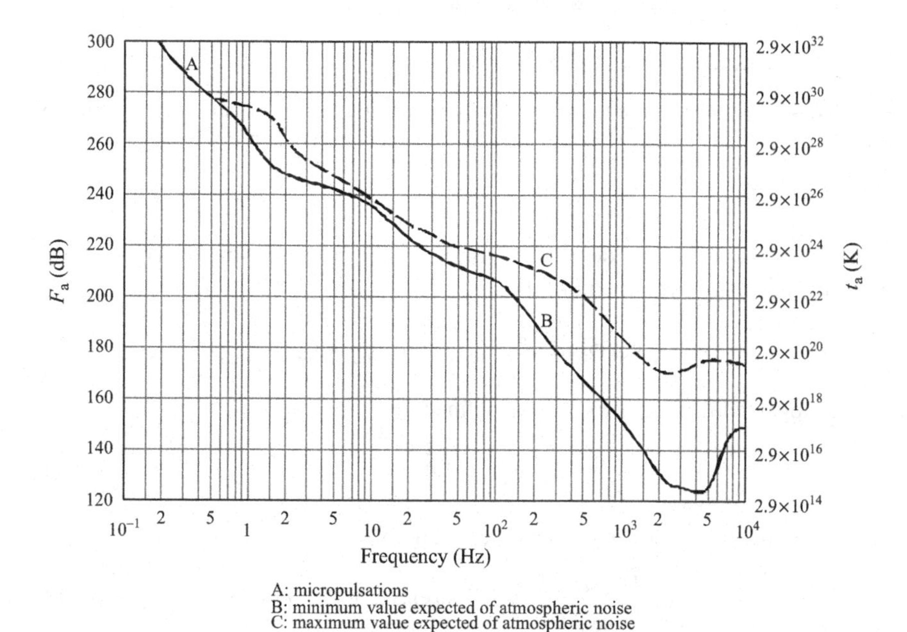

FIGURE 4.6 Typical values for equivalent external noise figure (F_a) and antenna temperature, t_a, from 0.1 Hz to 10 kHz. (*Source*: ITU-R Recommendation P.372-9 [3], used with permission.)

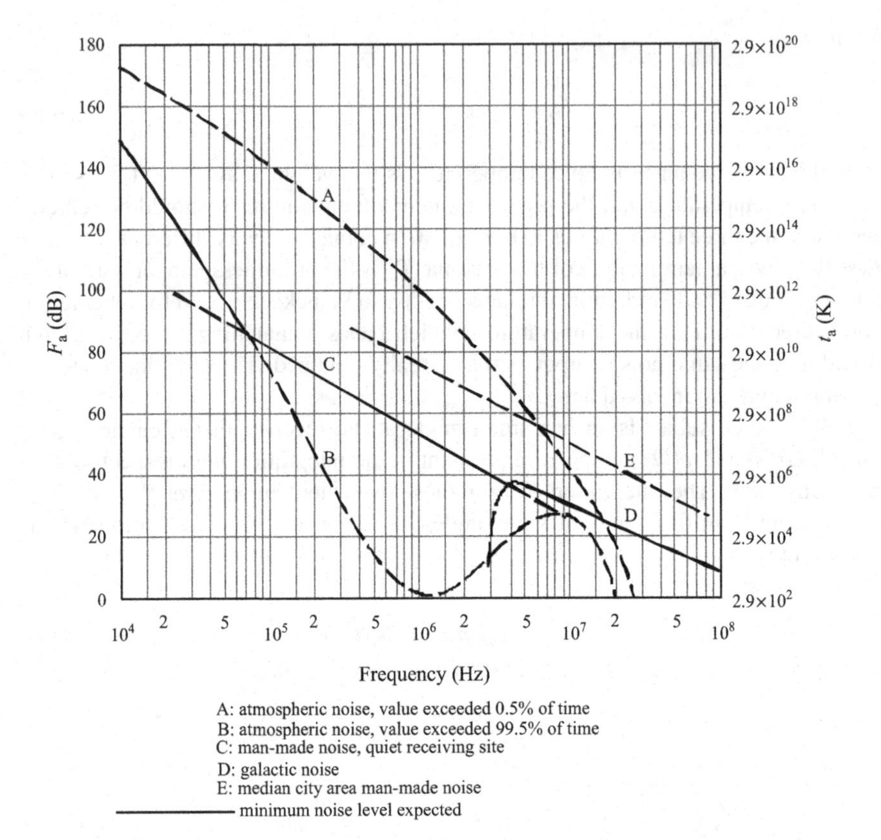

FIGURE 4.7 Typical values for equivalent external noise figure (F_a) and antenna temperature, t_a, from 10 kHz to 100 MHz. (*Source*: ITU-R Recommendation P.372-9 [3], used with permission.)

where S_N is the noise flux density (W/m^2) arriving from the source. The result of using this equality in equation (4.50) is

$$P_N = S_N(\Omega_0)\frac{\lambda^2 G(\Omega_0)}{4\pi}p(\Omega_0) + k_B B T_p(1-\upsilon). \tag{4.60}$$

An rms noise field strength E_N can be defined by

$$S_N = \frac{E_N^2}{\eta_0}, \tag{4.61}$$

where η_0 is the free-space impedance, resulting in

$$P_N = \frac{E_N^2(\Omega_0)}{\eta_0}\frac{\lambda^2 G(\Omega_0)}{4\pi}p(\Omega_0) + k_B B T_p(1-\upsilon) \tag{4.62}$$

with P_N the maximum available external noise power at the input when the antenna and receiver are impedance matched ($q = 1$). Comparing this result with equation (4.30)

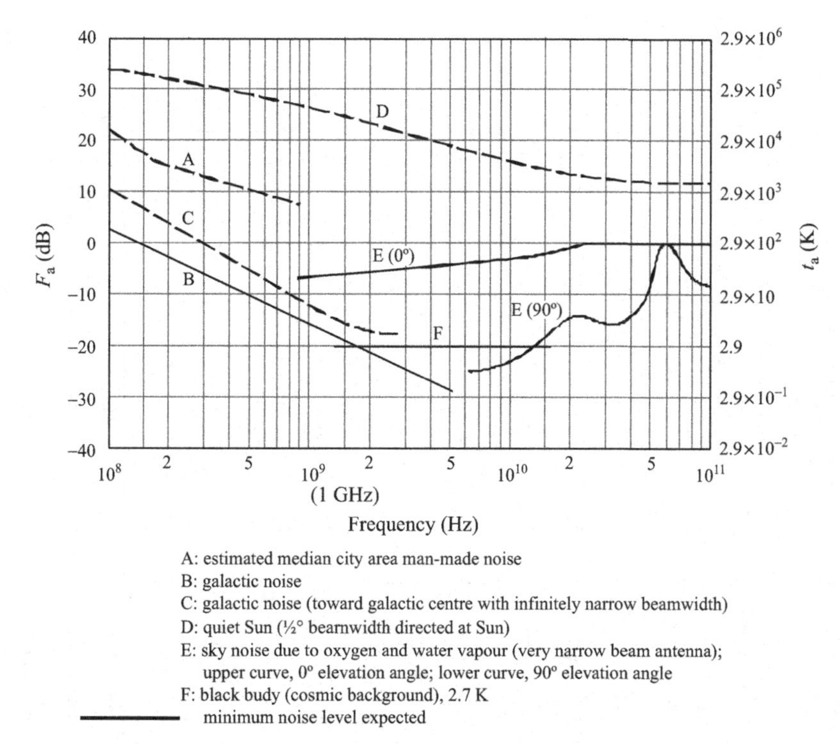

FIGURE 4.8 Typical values for equivalent external noise figure (F_a) and antenna temperature, t_a, from 100 MHz to 100 GHz. (*Source*: ITU-R Recommendation P.372-9 [3], used with permission.)

shows that the system treats noise that arrives as a plane wave from a single direction just as a signal smeared over the receiver bandwidth. This is what might well be expected!

Figures 4.6–4.8 plot typical external noise levels from 0.1 Hz to 10 kHz, 10 kHz to 100 MHz, and 100 MHz to 100 GHz, respectively, versus frequency. These plots are in terms of an "equivalent external noise figure" $F_a = 10\log_{10}(T_A/T_0)$ (left-hand side axis labels) and also in terms of T_A (right-hand side axis labels). For lower frequencies, the *atmospheric noise* in Figure 4.6, sometimes also called "atmosferics" or simply "sferics," is the noise radiated by lightning, which may be propagated ionospherically over long distances. At a single location, especially in the temperate climate zone, lightning may seem to be a moderately rare phenomenon; on a worldwide basis, however, lightning occurs almost continuously. The static heard between stations on the AM broadcast band in the United States (540–1700 kHz) is due to sferics. Sferics are usually the dominant noise source below about 10 MHz. Figures 4.6–4.8 illustrate the relative importance of various noise sources as a function of frequency and show a transition from lightning, galactic, and man-made noise at lower frequencies to "sky" and "solar" contributions at higher frequencies.

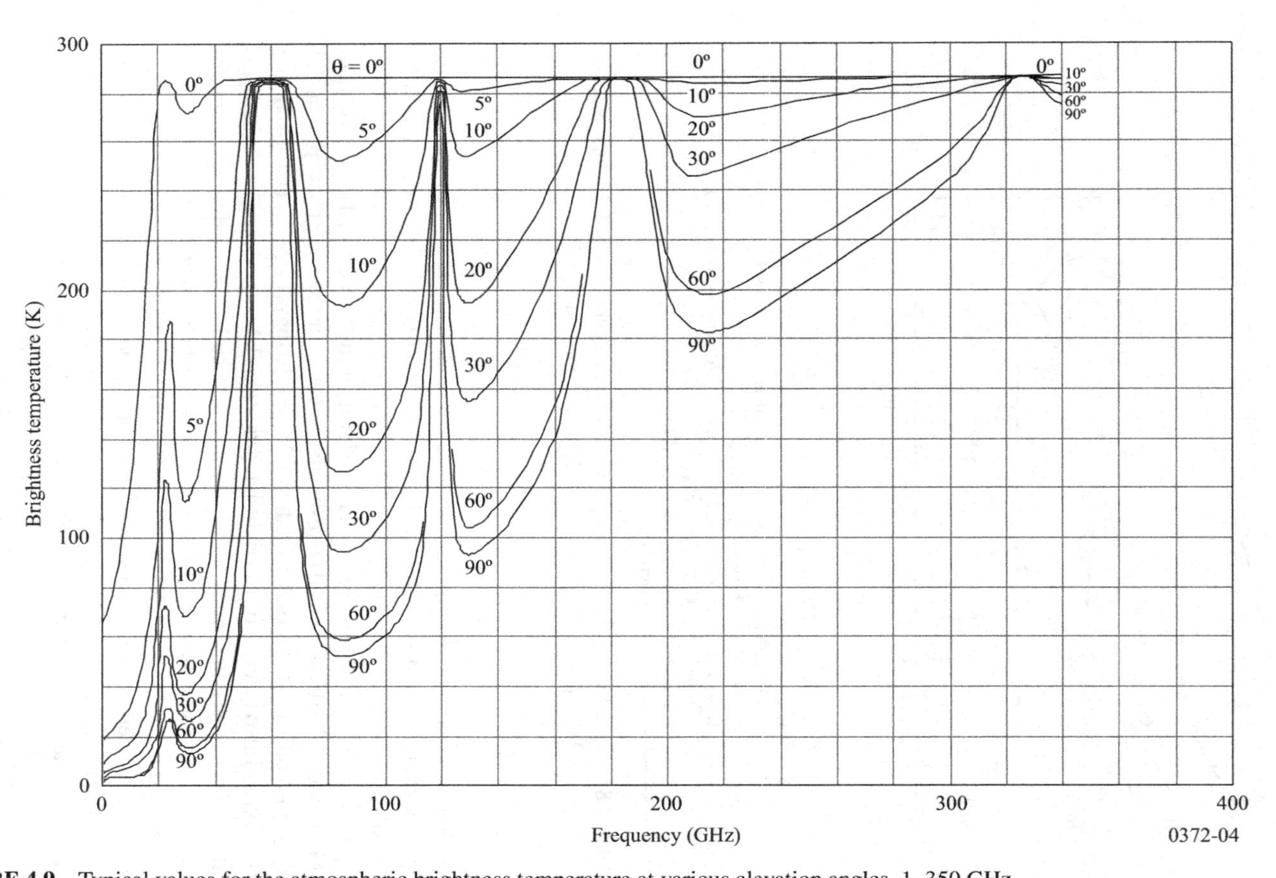

FIGURE 4.9 Typical values for the atmospheric brightness temperature at various elevation angles, 1–350 GHz. (*Source*: ITU-R Recommendation P.372-9 [3], used with permission.)

Figure 4.9 shows the sky brightness temperature due to emission from the gases that absorb in the microwave region, namely, water vapor and oxygen. We will see in Chapter 5 that the frequency bands of high naturally emitted thermal noise correspond to frequency bands of high absorption by atmospheric gases. The multiple curves in the figure are for various elevation angles at which the sky is viewed. These curves depend on meteorological factors, including vertical profiles of atmospheric temperature and water vapor, as will be discussed for Chapter 5 with regard to the total attenuation through the atmosphere. The fact that the radio noise emitted by the atmosphere depends on meteorology enables "remote sensing" of atmospheric properties by measuring emitted noise; sensors of this type are called "microwave radiometers."

Additional information on external noise values and their variations with geographic location and season is provided in ITU-R Recommendation P.372 [3].

REFERENCES

1. Kraus, J. D., and R. J. Marhefka, *Antennas for All Applications*, third edition, McGraw-Hill, 2001.
2. Stutzman W. L., and G. A. Thiele, *Antenna Theory and Design*, second edition, Wiley, 1997.
3. ITU-R Recommendation P.372-9, "Radio noise," International Telecommunication Union, 2007.

5

DIRECT TRANSMISSION

5.1 INTRODUCTION

In general, the propagation of a transmitted signal may be affected by interactions with the environment in which the signal propagates (such as reflection and refraction as will be discussed in Chapter 6). In many systems of practical importance, interactions with the environment do not cause significant changes in the signal propagation. The term "direct transmission" is meant to designate such situations; this propagation mechanism is dominant for many satellite communication and radar systems.

In this chapter, the Friis transmission formula for prediction of signal strength under ideal direct transmission conditions is derived. Calculation of attenuation due to atmospheric gases is then discussed, on both vertical and slant paths. A model for predicting rain attenuation and the expected system outage time for a given "rain margin" follows. Methods of improving system performance through diversity and the scintillation phenomena that occur on paths through the atmosphere are then discussed. Finally, a derivation of the look angle from a specified point on the Earth's surface to a geostationary satellite is given in the appendix of this chapter because of the importance of the direct transmission mechanism in satellite communications applications.

Radiowave Propagation: Physics and Applications. By Curt A. Levis, Joel T. Johnson, and Fernando L. Teixeira

5.2 FRIIS TRANSMISSION FORMULA

Consider an isotropic radiator, which is a conceptual antenna that radiates energy equally in all directions. Such antennas do not exist in practice; nevertheless, the concept is useful. Note that an omnidirectional antenna is one that radiates nearly equally well in all the directions of one plane, for example, in azimuth, but that has a directional, though generally broad, radiation pattern in the orthogonal plane. Omnidirectional antennas are used in practice. It is important not to confuse "isotropic" and "omnidirectional."

Consider an isotropic radiator in an ideally simple medium. Since neither the antenna nor the medium imparts any directionality to the radiated power, the radiated power flux density will be a function only of the distance R and, by power conservation, is given by

$$S(R) = \frac{P_T\, \upsilon_T}{4\pi R^2}, \tag{5.1}$$

where P_T is the power input to the transmitting antenna and υ_T is its radiation efficiency. The subscript T is used above because both a transmitting and a receiving antenna will be needed in an application.

The directional gain $D(\theta, \phi)$ of an antenna is defined as the ratio of the power flux density radiated in the direction (θ, ϕ) to that which would be radiated by an isotropic radiator. Thus, the power density radiated by a directive antenna is obtained from equation (5.1) as

$$S(R, \theta_T, \phi_T) = \frac{P_T\, \upsilon_T}{4\pi R^2} D_T(\theta_T, \phi_T) = \frac{P_T\, G_T(\theta_T, \phi_T)}{4\pi R^2}, \tag{5.2}$$

where the last step follows from the relation between gain G and directional gain D.

For transmitting and receiving antennas separated sufficiently so that far-field conditions (4.1) are satisfied, according to equation (4.31) the power delivered to the terminals of the receiving antennas will be

$$P_R = S(R, \theta_T, \phi_T)\, A_R(\theta_R, \phi_R) = \frac{P_T\, G_T(\theta_T, \phi_T)}{4\pi R^2} A_R(\theta_R, \phi_R), \tag{5.3}$$

where A_R is the receiving aperture (or receiving area) of the receiving antenna. Note that the coordinates (θ_T, ϕ_T) are those at which the receiving antenna appears when viewed in the coordinate system of the transmitting antenna, while (θ_R, ϕ_R) are those of the transmitting antenna in the coordinate system of the receiving antenna.

In microwave systems, it is customary to match the receiving antenna to its receiver and also to match the receiving antenna polarization to the received field (the polarizations may be mismatched in interference calculations.) If both the impedance and the polarization are matched, the receiving aperture A_R becomes equal to the effective area A_e of the receiving antenna,

$$A_R = A_e = \frac{G\, \lambda^2}{4\pi}, \tag{5.4}$$

where equation (4.35) has been used for the second equality. Equation (5.3) now becomes

$$P_R = \frac{P_T\, G_T(\theta_T, \phi_T)\, G_R(\theta_R, \phi_R)\, \lambda^2}{(4\pi)^2 R^2}, \tag{5.5}$$

where the wavelength λ and the distance R must be in the same units. This equation is known as the Friis transmission equation. It is the basis for many system calculations. Note that the arguments (θ, ϕ) of G_T and G_R are often omitted.

Since the ratio of P_T to P_R is generally large, decibel notation is often used for system calculations. In decibel notation, with the use of

$$\lambda = c/f, \tag{5.6}$$

the result is

$$P_{R,\mathrm{dbW}} = P_{T,\mathrm{dbW}} + G_{T\mathrm{db}} + G_{R\mathrm{db}} - 20\log_{10} R_{\mathrm{km}} - 20\log_{10} f_{\mathrm{MHz}} - 32.44. \tag{5.7}$$

In this expression, the notation $P_{T,\mathrm{dbW}}$, $P_{R,\mathrm{dbW}}$ indicates that both the transmitted and received powers are specified in decibels relative to 1 W. Often decibels relative to 1 mW are used instead; in this case, the notation $P_{T,\mathrm{dbm}}$, $P_{R,\mathrm{dbm}}$ is used. Note that the units for distance and frequency are km and MHz, respectively.

5.2.1 Including Losses in the Friis Formula

The Friis transmission equation takes no account of any losses in the transmission medium that may attenuate the signal. In the atmosphere, loss can generally be neglected in the frequency range from 100 MHz to 5 GHz and, except for heavy rain, even up to 10 GHz. In the frequency range above 10 GHz, two loss effects become important. The first is due to gases, specifically water vapor and oxygen, whose molecules exhibit resonances at these frequencies and can therefore convert energy from the transmitted wave into heat, thus causing the wave to be attenuated. The second is due to hydrometeors, that is, liquid or solid water, such as rain, snow, and fog, with rain generally the most important because liquid water absorbs more strongly than ice in the microwave frequency range. Finally, the Friis transmission equation has neglected impedance and polarization mismatches. Taking all these considerations into account transforms equation (5.5) into

$$P_R = \frac{P_T\, G_T(\theta_T, \phi_T)\, G_R(\theta_R, \phi_R)\, \lambda^2}{(4\pi)^2 R^2}\, T_{\mathrm{gas}}\, T_{\mathrm{rain}}\, T_{\mathrm{imp}}\, T_{\mathrm{pol}}\, T_{\mathrm{coup}}. \tag{5.8}$$

The T_{imp} and T_{pol} factors represent impedance and polarization mismatches, as in equation (4.30). The transmission factor due to gases T_{gas} can be calculated from a "specific extinction coefficient" for gas absorption α_{gas} as

$$T_{\mathrm{gas}} = \exp\left(-\int_{\mathrm{path}} \alpha_{\mathrm{gas}}\, dl\right), \tag{5.9}$$

where α_{gas} is measured in nepers per unit length and T_{gas} is the fractional transmission (e.g., 0.5 for 50% transmission).

Similarly the rain transmission factor T_{rain} can be obtained from a specific extinction coefficient due to rain as

$$T_{\text{rain}} = \exp\left(-\int_{\text{path}} \alpha_{\text{rain}}\, dl\right). \tag{5.10}$$

It is easy to see that transmission factors are multiplied to obtain the total transmission of several layers. If the first layer transmits 90% of the power and the second layer transmits 80% of what it receives, then the total transmission is $0.9 \times 0.8 = 0.72$, or 72%. The logarithmic nature of α_{gas} converts the multiplication to addition, and the integral represents the addition of infinitely thin layers. From (5.9) it is seen that power density decreases exponentially in a homogeneous medium.

The antenna-to-medium coupling factor T_{coup} comes into play when the transmission medium modifies the usual antenna pattern, for example, when $G_{\text{T}}(\theta_{\text{T}}, \phi_{\text{T}})$ represents a free-space antenna pattern, but the antenna is used for ionospheric transmission. For direct transmission $T_{\text{coup}} = 1$.

It is more common to express the relations of equation (5.8) in decibel notation, similar to equation (5.7),

$$P_{\text{R,dbW}} = P_{\text{T,dbW}} - L_{\text{sys,db}}, \tag{5.11}$$

where the system loss $L_{\text{sys,db}}$ is given by

$$\begin{aligned} L_{\text{sys,db}} &= 20\log_{10} R_{\text{km}} + 20\log_{10} f_{\text{MHz}} + 32.44 + L_{\text{gas,db}} + L_{\text{rain,db}} + L_{\text{pol,db}} \\ &\quad + L_{\text{imp,db}} + L_{\text{coup,db}} - G_{\text{Tdb}} - G_{\text{Rdb}}. \end{aligned} \tag{5.12}$$

The losses in equation (5.12) can be obtained from the corresponding transmission factors in equation (5.8) by

$$L_{\text{db}} = -10\log_{10} T. \tag{5.13}$$

The attenuation due to gases in decibels (defined to be a positive number if loss occurs) can also be found from a specific attenuation in dB as

$$L_{\text{gas,db}} = \int_{\text{path}} \alpha_{\text{gas,db}}\, dl \tag{5.14}$$

and that for rain from

$$L_{\text{rain,db}} = \int_{\text{path}} \alpha_{\text{rain,db}}\, dl. \tag{5.15}$$

It is easy to show that the relationship between the specific extinction coefficient, α, usually measured in nepers/km, and the specific attenuation, α_{db}, usually measured in dB/km, is

$$\begin{aligned} \alpha_{\text{gas,db}} &= 4.34\,\alpha_{\text{gas}} \\ \alpha_{\text{rain,db}} &= 4.34\,\alpha_{\text{rain}}. \end{aligned} \tag{5.16}$$

Because of the simpler form of equations (5.14)–(5.15) compared to equations (5.9)–(5.10), it is generally preferred to use the specific attenuations $\alpha_{\text{gas,db}}$ and $\alpha_{\text{rain,db}}$, defined as the attenuation in decibels per unit distance, usually specified in dB/km.

These quantities are discussed in more detail in the following sections. Polarization and impedance mismatch losses are obtained from Chapter 4 as

$$L_{\text{pol,db}} = -20\log_{10}|\hat{a}_{\text{e}} \cdot \hat{a}_{h}|, \tag{5.17}$$

$$L_{\text{imp,db}} = -10\log_{10}\frac{4R_{\text{A}}R_{\text{R}}}{|Z_{\text{A}} + Z_{\text{R}}|^2}. \tag{5.18}$$

In using the logarithmic form, it is useful to have a feel for the range of values to be expected. When a calculation falls out of the expected range, one should suspect an error in the calculation! A brief example will demonstrate practical system loss limits. A large transmitter might transmit a megawatt. A sensitive receiver with 100 Ω input impedance might detect 1 μV (rms). For these values, a system loss of up to $10\log_{10} 10^6 / \left((10^{-6})^2/100\right) = 200$ dB would be tolerable. Thus, the practicality of $P_{\text{T}}/P_{\text{R}}$ ratios much greater than 200 dB should be questioned.

In this section, the distinction between natural and decibel quantities has been denoted carefully by subscripts, for example, T_{gas} and $L_{\text{gas,db}}$, for describing the attenuation due to atmospheric gases. Such a convention will also be followed in the rest of this chapter. In the literature, such a careful distinction is seldom found; the reader is expected to understand which is meant from the units or the context.

5.3 ATMOSPHERIC GAS ATTENUATION EFFECTS

The specific attenuation $\alpha_{\text{gas,db}}$ in dB/km for a standard atmosphere at sea level (surface water vapor content of 7.5 g/m^3 and surface temperature 20° C) is plotted in Figure 5.1, using the methods described in ITU-R Recommendation P.676-7 [1]. The

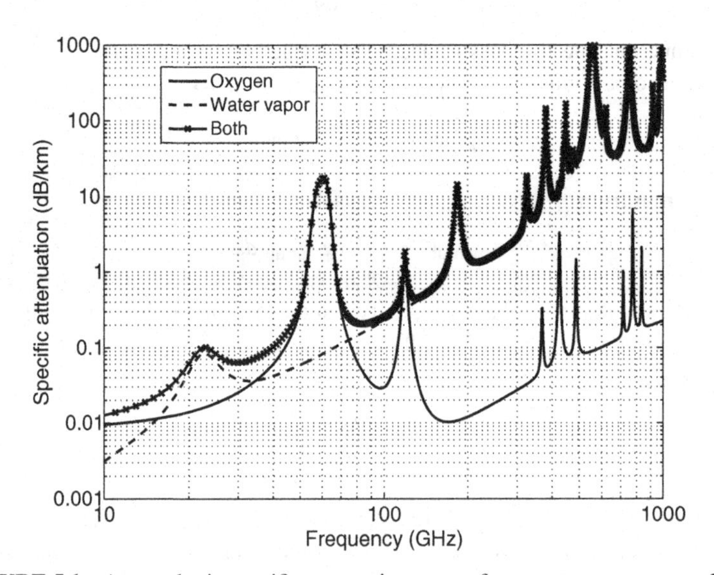

FIGURE 5.1 Atmospheric specific attenuation versus frequency, at or near sea level.

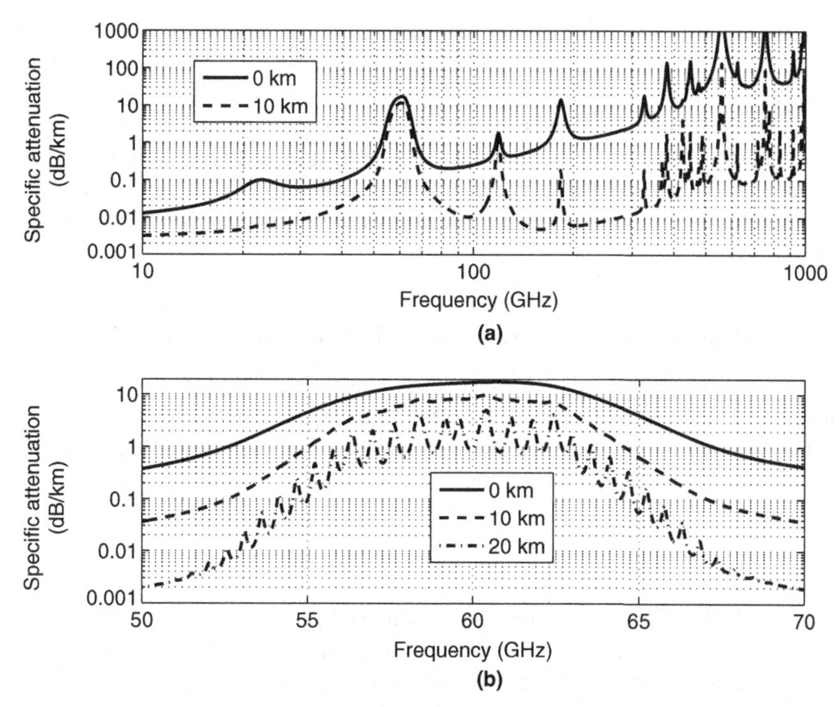

FIGURE 5.2 (a) Specific attenuation at varying altitudes. (b) Specific attenuation by atmospheric oxygen at varying altitudes.

multiple curves in the plot describe the contributions of atmospheric water vapor and oxygen separately, as well as the combined attenuation of these two effects. Note that water vapor dominates at and near sea level for frequencies greater than 100 GHz, so that the "both" and "water vapor" curves coincide. The peaks at 22.2, 183.3, 325.4 GHz, and higher frequencies are due to water vapor and would be greatly reduced for a very dry climate; those at 60 and 118.7 GHz are due to oxygen. The attenuation near the peaks can be so high as to make the atmosphere virtually opaque; the frequency regions where the attenuation is not excessive are often referred to as "windows" in the atmosphere, for example, the windows near 40 and 90 GHz. Attenuation due to water vapor shows an increasing trend with frequency, so the impact of water vapor (when present) is significant at all frequencies greater than 100 GHz.

Figure 5.2a compares the total atmospheric specific attenuation at sea level with that at altitude 10 km. Two features are important to observe here: first, the decreased attenuation observed in the 10 km case is largely due to the decreased concentration of water vapor at higher altitudes. Second, the absorption peaks (or "resonances"), for example, near 118.7 or 183.3 GHz, are sharper at 10 km than at sea level due to a mechanism called "pressure broadening." Collisions with other atmospheric molecules can broaden the resonant behavior of atmospheric gas absorption; these effects are more significant at lower than higher altitudes due to the increased atmospheric pressure at lower altitudes. Both these effects make clear that it is important to take the altitude dependence into account (as described in ITU-R P.676-7) when computing the specific attenuation due to atmospheric gases.

A more vivid example of pressure broadening effects is provided in Figure 5.2b, which illustrates the specific attenuation due to atmospheric oxygen near the 60 GHz resonance. The curves illustrated for multiple altitudes show that the broad peak at sea level actually is due to many fine lines that merge due to pressure broadening at the lower altitudes.

5.3.1 Total Attenuation on Horizontal or Vertical Atmospheric Paths

When a long vertical or slant path through the atmosphere is under consideration (for example, to a satellite), the total attenuation must be ascertained by integrating the specific attenuation over the path. The total attenuation due to atmospheric gases for horizontal paths is obtained by integrating the specific attenuation over the path

$$L_{\text{gas,db}} = \int \alpha_{\text{gas,db}} \, dl = \alpha_{\text{gas,db}} L, \tag{5.19}$$

where L is the path length, and the last equality stems from the fact that the specific attenuation seldom varies greatly in the horizontal direction. Such a simple relationship for vertical or slant paths is not possible, however, due to atmospheric variations with altitude.

Precisely vertical paths are uncommon in practice, but they provide a useful starting point for considering the common case of slant paths through the atmosphere, for example, in satellite communications. Results of a calculation for a vertical path beginning at various altitudes are shown in Figure 5.3. An interesting feature of these curves is that the resonance absorption peaks narrow for increasing altitude due to the collision broadening mechanism discussed earlier.

To calculate the curves of Figure 5.3, distributions of oxygen and water vapor densities as a function of height must be known or assumed. One approximation is an exponential distribution; for example, for water vapor density the function

$$\rho = \rho_0 \exp\left(-h/h_{s,w}\right) \tag{5.20}$$

is often used, where ρ_0 is the water vapor density at sea level, h is the altitude of interest, and $h_{\text{s,w}}$ is called the scale height for water vapor, with a typical value of approximately 2 km. Use of an exponential function of altitude for many atmospheric properties is justified by the "barometric law" discussed in Chapter 10. For water vapor, such approximations can be good in a statistical sense for altitudes up to about 8 km and can also usually be used on paths from a low altitude through the atmosphere because nearly all the attenuation occurs at low altitudes. On the other hand, use of equation (5.20) would certainly be a poor assumption for calculating the attenuation from a stratospheric balloon to a satellite! Another approach is the use of a "model atmosphere," that is, an empirical model of atmospheric properties versus altitude based on theory and measurements.

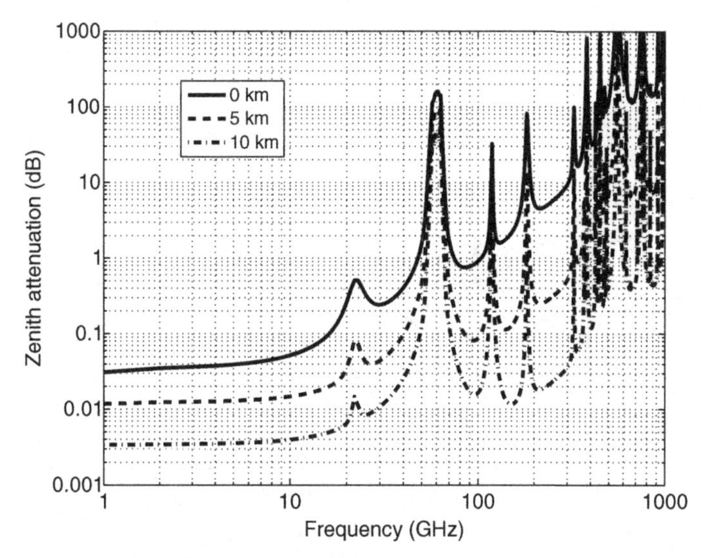

FIGURE 5.3 Total attenuation from specified height to top of atmosphere, ITU-R Standard Atmosphere (7.5 g/m^3 water vapor concentration at sea level).

The actual calculation of the attenuation due to gases on a zenith path through the atmosphere can then be performed by integrating the specific attenuation for the specified atmospheric composition over height. Results are commonly presented as an equivalent height by which the specific attenuation due to the gas at the lowest altitude must be multiplied to yield the actual attenuation. More information on such calculations can be found in ITU-R Recommendation P.676-7 [1]. The complexity of such calculations is seldom required. Useful estimates of total attenuation on a vertical path through the atmosphere can be obtained from Figure 5.3.

5.3.2 Total Attenuation on Slant Atmospheric Paths

On slant paths with elevation angles greater than about 10° above the horizon, the curvature of the Earth may be neglected. In that case, the relationship between vertical and slant attenuation in dB can be derived readily from the geometry of Figure 5.4. For the vertical path, the attenuation is given by

$$L_{\text{gas,vert}}(h_1, h_2) = \int_{h_1}^{h_2} \alpha_{\text{gas,db}}(h)\, dh \tag{5.21}$$

in dB, while for the slant path, the corresponding expression is

$$L_{\text{gas,slant}}(h_1, h_2) = \int_{s_1}^{s_2} \alpha_{\text{gas,db}}(s)\, ds = \int_{h_1}^{h_2} \alpha_{\text{gas,db}}(h) \csc\theta\, dh \tag{5.22}$$

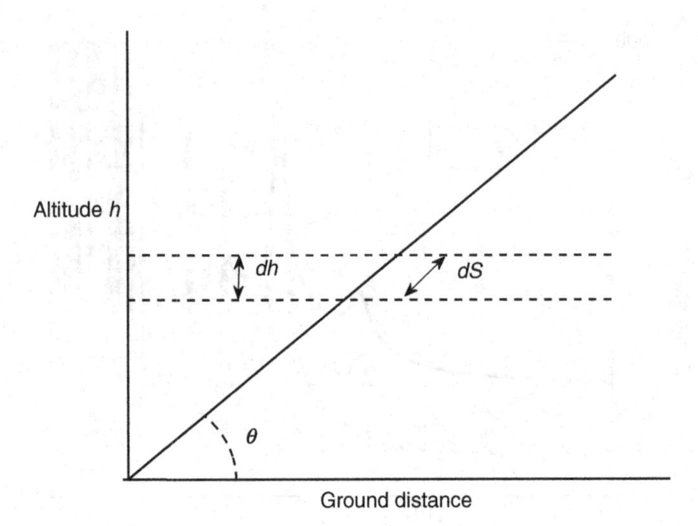

FIGURE 5.4 Slant path geometry.

in dB. Since θ is a constant over the path, the cosecant may be taken outside the integral, and the result is

$$L_{\text{gas,slant}} = L_{\text{gas,vert}} \csc\theta \tag{5.23}$$

in dB. As a corollary, the attenuations for two different elevation angles θ_1 and θ_2 are related by

$$\frac{L_g(\theta_1)}{L_g(\theta_2)} = \frac{\csc\theta_1}{\csc\theta_2} = \frac{\sin\theta_2}{\sin\theta_1}. \tag{5.24}$$

The cosecant law and its corollary are not valid for elevation angles less than about 10°. For lower elevations, Earth curvature must be taken into account, and so must refraction since a small angular deviation may cause a considerable change in the path length. Expressions for treating this case may be found in ITU-R Recommendation P.676-7 [1].

5.3.3 Attenuation at Higher Frequencies and Further Information Sources

The preceding discussion has been centered primarily on the microwave portion of the spectrum, where the atmosphere is largely transparent except for the oxygen and water vapor line frequencies. However, the molecular components of the atmosphere have a much larger number of resonances at frequencies greater than 300 GHz, and greater consideration must be given to atmospheric gas attenuation. Even Figure 5.3, which considers only oxygen and water vapor attenuation, shows that the low atmospheric attenuation obtained for frequencies below around 300 GHz (with the exception of the oxygen and water line frequencies) increases rapidly beyond this limit. Communication and radar systems intended to exploit these frequencies must therefore be designed to overcome substantial atmospheric attenuation.

Calculations based on the information given above yield only rough estimates of atmospheric attenuation. ITU-R P.676-7 gives detailed instructions for more precise calculations in the frequency range 1–1000 GHz, with a simplified procedure for 1–350 GHz. A computer program is also available from the ITU-R. Propagation at higher frequencies including the infrared, visible, and ultraviolet spectrum is beyond the scope of this book. Computer codes for these spectral regions include the FASCODE and MODTRAN programs for moderate resolution and the proprietary Hitran-PC code for high resolution, as is needed for laser applications. They are based on the HITRAN database for molecular resonance line parameters. Access to these codes has changed over the past few years; the reader is encouraged to use the Internet for more details, where references to other specialized codes in these spectral regions may also be found.

5.4 RAIN ATTENUATION

The water vapor discussed in the preceding section is water in its gaseous state. Water can also exist in its liquid or solid state in the atmosphere, in the form of rain, snow, fog, clouds, ice, and so on. "Hydrometeors" is the general term for nongaseous water in the atmosphere. Of these, rain is the most significant factor in propagation analyses for two reasons. The first is that liquid water absorbs more electromagnetic energy at microwave frequencies than ice or snow. The second is the fact that the sizes of liquid water particles for rain are much larger than those in clouds or fog (which usually have radii smaller than 100 μm), leading to increased attenuation effects. We therefore focus on rain attenuation in the following discussion. Information on attenuation by other hydrometeors is provided in ITU-R Recommendation P.840-3 [2], and would generally be of concern only when small attenuation values are of interest, and at frequencies greater than approximately 30 GHz.

Rain attenuation is highly variable in time, in its vertical and horizontal extent, and from location to location. Its impact can usually be neglected at frequencies below 5 GHz, but becomes increasingly important at higher frequencies. Because it remains impossible to predict the occurrence of rain with certainty, rain attenuation is usually described by statistical means: that is, the rain attenuation for a given location is modeled as a random variable. A more formal description of random variables and their properties is provided in Chapter 8. For the purposes of this chapter, it is sufficient to understand that our goal is to obtain an estimate of the number of hours per year (or equivalently the annual percentage of time) that a given amount of rain attenuation will be exceeded at a given location. Such information can be used in system design to plan for enough transmit power so that rain attenuation causes insufficient received power for only a very low percentage of the time. Calculations of this sort are known as "rain margin" or "system reliability" analyses.

Figure 5.5 provides an example of a rain attenuation distribution, in this case computed for an Earth–space path where the Earth station is at sea level and at latitude 40°, and for specified location rain statistics (discussed in what follows). The horizontal axis represents the number of hours per year (on average) that the

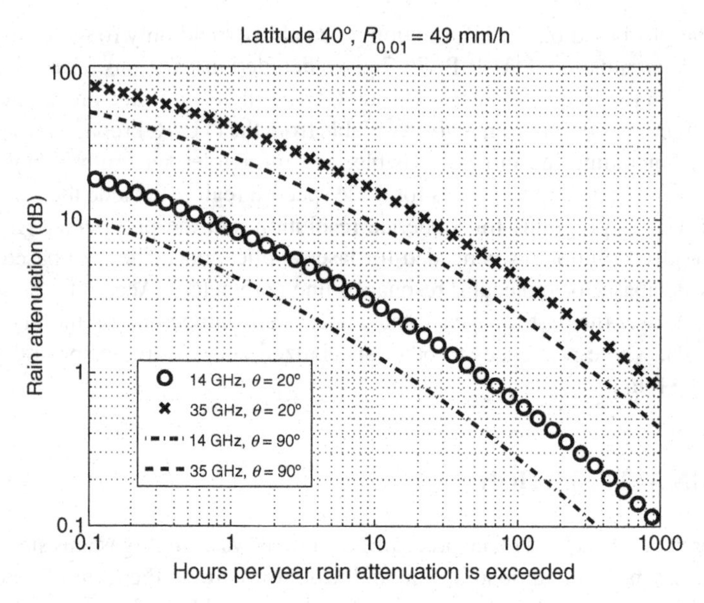

FIGURE 5.5 Sample rain attenuation distributions for station latitude 40°, station height 0 km, and $R_{r,0.01} = 49$ mm/h.

rain attenuation indicated on the vertical axis is exceeded. For example, for path elevation angle θ of 20°, and at 14 GHz, we can see that a rain attenuation of 3 dB is expected to be exceeded around 10 h per year. Such information is very useful in assessing the impact of rain on system performance. It is not surprising that the curves in Figure 5.5 decrease from left to right, as higher rain attenuations (very heavy rain) are encountered less frequently than smaller rain attenuations (light rain). The dependencies on frequency and path elevation angle shown in Figure 5.5 will become clearer further on. The curves of Figure 5.5 are applicable for a specific location and must be recalculated for different locations of the Earth station to take into account local climate effects on rain properties.

In order to reach the goal of producing rain attenuation distributions, we begin with a physical description of the rain itself, consider the mechanisms of rain specific attenuation, and present a simplified model for this quantity. We then consider the requirements for computing the total rain attenuation over a specified path, but again will find it necessary to resort to an empirical model for this process, due to the difficulty of obtaining the statistics of the physical processes involved. For example, it is difficult to measure raindrop shape and size distributions, and such measurements have been performed only at relatively few locations and over limited time spans. In contrast, measurement of the attenuation of a satellite signal is relatively simple and, within limits, the attenuation can also be inferred from radiometric measurements of the noise emitted from the rain. Thus, a relatively large database of rain attenuation statistics is available for fitting an empirical model. Methods for describing statistics of rain attenuation at a given location are also discussed.

5.4.1 Describing Rain

The simplest description of rain is an atmosphere that contains a given number of water particles, all of the same shape, size, and orientation, uniformly spread through a region of space. Unfortunately, this description is not sufficient for rain, as raindrops vary in all three of these properties, and variations in raindrop size, in particular, have an important impact on attenuation. Neglecting shape and orientation effects for the moment by modeling the raindrops as spherical, it is possible to describe rain in terms of a "drop size distribution": $n(r)dr$ is the number of raindrops contained in a unit volume of the atmosphere that are within an interval dr of the drop radius r. According to this definition, the total number of raindrops per unit volume is then

$$N = \int_0^\infty n(r)\, dr, \tag{5.25}$$

and $n(r)$ is usually reported in units of "per cubic meter per millimeter" with r in millimeters, resulting in N being the number of drops per cubic meter. It is also possible to compute the fraction of the atmospheric volume that is occupied by liquid water as

$$m_{\rm w} = \int_0^\infty \frac{4\pi}{3} \left(\frac{r}{1000}\right)^3 n(r)\, dr, \tag{5.26}$$

where the factor of 1000 is included assuming the units of $n(r)$ are per cubic meter per millimeter and r is in millimeters.

Numerous measurements of the drop size distribution have been performed. One model that is widely used (but not necessarily the ideal model in all situations) is that of Marshall and Palmer: [3, 4]

$$n(r) = N_0 e^{-\Lambda r}, \tag{5.27}$$

where $N_0 = 16000$ and $\Lambda = 8.2 R_{\rm r}^{-0.21}$ when $n(r)$ has units of per cubic meter per millimeter and r is in millimeters.

Here, $R_{\rm r}$ (the only parameter needed by this distribution to specify rain properties) is the "rain rate" measured in units of millimeters per hour. This quantity specifies the number of millimeters of rain that would be captured in an open, vertically oriented, container of uniform cross section over a period of 1 h. Rain rates of 5 mm/h or less represent a light rain, while rain rates up to 100 mm/h or more can occur for brief periods during strong thunderstorms.

One issue in using the $R_{\rm r}$ parameter involves the duration of the measurement. For example, it would be highly unlikely to observe an $R_{\rm r}$ of 100 mm/h or more if the rain measurement is performed over a duration of 1 h because such heavy rains occur only over shorter periods of time. Nevertheless, such short periods of time can be important in examining system reliability! For this reason, it is common in propagation analyses to use rain rates $R_{\rm r}$ in mm/h that represent measurements performed over a period of 1 min.

Figure 5.6 provides a plot of the Marshall–Palmer drop size distribution for two rain rates. Both curves show a decrease in the number of particles per unit volume for

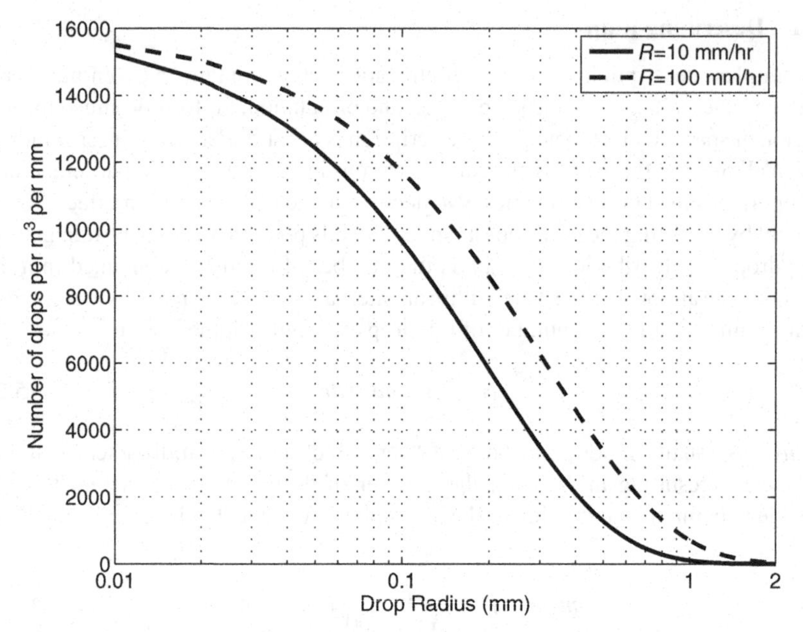

FIGURE 5.6 Marshall–Palmer drop size distribution for two rain rates.

larger drop sizes, but the high rain rate case shows a shift toward larger drop sizes in general. Using the Marshall–Palmer distribution, we can find

$$N = \frac{N_0}{\Lambda} = (1951.2)\, R_r^{0.21}, \tag{5.28}$$

$$m_w = 10^{-9}\frac{8\pi N_0}{\Lambda^4} = \left(8.9 \times 10^{-8}\right) R_r^{0.84}, \tag{5.29}$$

showing that, even at high rain rates, the fractional volume occupied by liquid water in the atmosphere is very small.

If the assumptions to this point are accepted, then the description of rain properties reduces to a description of the rain rate R_r as a function of both horizontal and vertical space. Rain rate at the Earth's surface is a commonly measured meteorological quantity, and extensive data sets are available for this parameter, for example, from the National Weather Service or the National Oceanographic and Atmospheric Administration (NOAA) in the United States. Such information can be used to represent rain rate at a given location as a random variable, and the property of most interest in propagation analysis is the percentage of time that a given rain rate is expected to be exceeded at a specific location. Rain rate analyses using globally compiled data sets have shown that rain statistics can often be adequately modeled as a combination of two processes: normal rain and thunderstorm rain, with the latter contributing mostly to the higher rain rates. Current models for rain attenuation effects usually consider only the rain rate that is exceeded 0.01% of the time (i.e., 52.6 min/year)

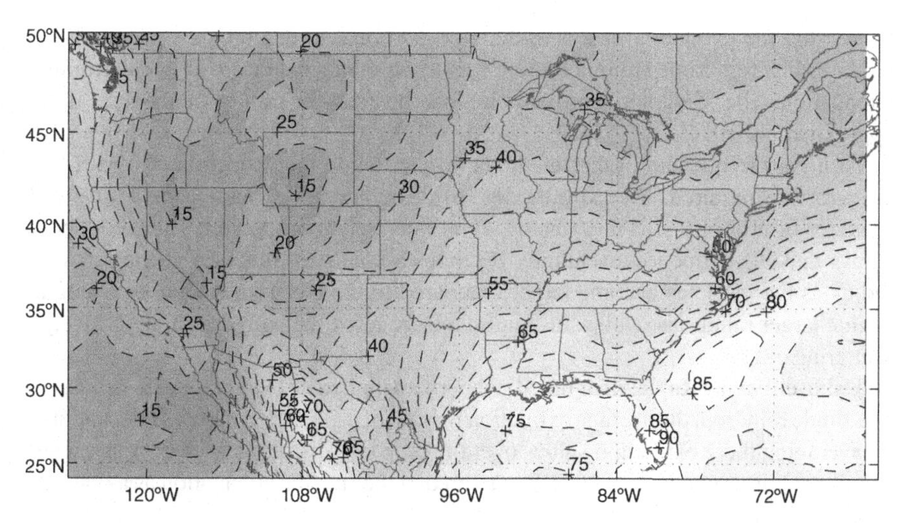

FIGURE 5.7 Shaded contour plot of $R_{r,0.01}$ (mm/h) for the continental United States, using data from ITU-R P. 837-5.

at a given location, called $R_{r,0.01}$. ITU-R Recommendation P. 837-5 [5] provides a global database for estimating $R_{r,0.01}$; local measurements of such information are preferable if they are available. Figure 5.7 provides a contour map of $R_{r,0.01}$ for the continental United States.

5.4.2 Computing Rain Specific Attenuation

The extremely low fractional volume of raindrops in the atmosphere enables computation of rain specific attenuation using an "independent scatterer" assumption. This method assumes that the power lost by a radiowave propagating through rain is simply the sum of all the powers lost in the individual raindrops; that is, there is no shadowing of one raindrop by another or multiple scattering. Determining the power loss per unit volume then reduces to an integration of the power lost for each drop over the drop size distribution, since drops of different sizes absorb different amounts of power.

When a radiowave encounters a single raindrop, two effects cause loss of power. The first is absorption of power inside the raindrop. Since water is a lossy medium at microwave frequencies, this absorption is an important cause of attenuation in rain. The second effect is scattering of the radiowave by the raindrop that causes some of the radiowave energy to be redirected into propagation directions other than the original direction. The sum of these two effects is called the "extinction" caused by the raindrop.

Computing the scattering and absorption of incident radiowave power for a given scatterer is a classical electromagnetic problem. For scatterers that are very small compared to the wavelength of the radiowave, the Rayleigh approximation [6] shows that the extinction of a particle is proportional to its volume squared. Thus, larger radius raindrops are much more important than smaller drops in determining attenuation. The Rayleigh approximation can be used for rain at frequencies up to approximately 10 GHz.

At higher frequencies, other methods for predicting the extinction of a specified scatterer are required. The Mie theory [6] describes the extinction of a spherical scatterer of arbitrary size in terms of an infinite series summation. The Mie theory shows that the extinction eventually becomes a quadratic function of the scatterer radius as the particles become larger compared to the electromagnetic wavelength, so that larger raindrops, while still important, are less dominant than in the Rayleigh scattering case.

Using either the Rayleigh or the Mie theory as applicable to compute the extinction of a single raindrop, the average extinction for rain of a given rain rate can be computed by averaging these extinction values over the drop size distribution. Such approaches usually require a numerical integration to predict the specific attenuation as a function of the rain rate.

To this point, we have assumed for simplicity that the raindrops all have a spherical shape. In fact, raindrops are nonspherical and have a typical shape that depends on the drop size. Once drops are modeled as nonspherical, their orientation also must be specified because drops often fall at an angle (called "canting") with respect to the vertical, especially in higher wind conditions. Two primary effects are caused by nonspherical raindrops. First, the specific attenuation becomes sensitive to the polarization of the radiowave. This effect can be captured using the process described in this section as long as an appropriate "Mie-like" theory is used for the nonspherical raindrop shape. Second, canting effects can cause a change in the polarization of a radiowave as it propagates, due to the fact that the rain medium is an anisotropic medium when nonspherical raindrops are present (similar effects will be treated in greater detail in Chapter 11). Rain impact on the polarization of a radiowave is not discussed further here, but is important to consider when it is desired to communicate distinct information using horizontal and vertical polarizations. A parameter called the "cross-polarization discrimination" (often abbreviated XPD) is used to describe the maximum isolation between polarizations that can be achieved in rain; see ITU-R Recommendation P. 618-9 [7] for an empirical XPD model.

5.4.3 A Simplified Form for Rain Specific Attenuation

Detailed analyses of computations using drop size distributions and Mie extinction theories, as well as extensive empirical measurements, have shown that rain specific attenuation in dB/km can be described reasonably well by the simpler relationship

$$\alpha_{\mathrm{rain,db}} = k_{\mathrm{r}} R_{\mathrm{r}}^{a_{\mathrm{r}}}, \tag{5.30}$$

where R_{r} is in mm/h, and k_{r} and a_{r} are empirical coefficients that are tabulated as a function of frequency for horizontal and vertical polarizations. Figures 5.8 and 5.9 plot k_{r} and a_{r} values versus frequency for horizontal and vertical polarizations, as obtained from ITU-R P. 838-3 [8]. For other polarizations, the required k_{r} and a_{r}

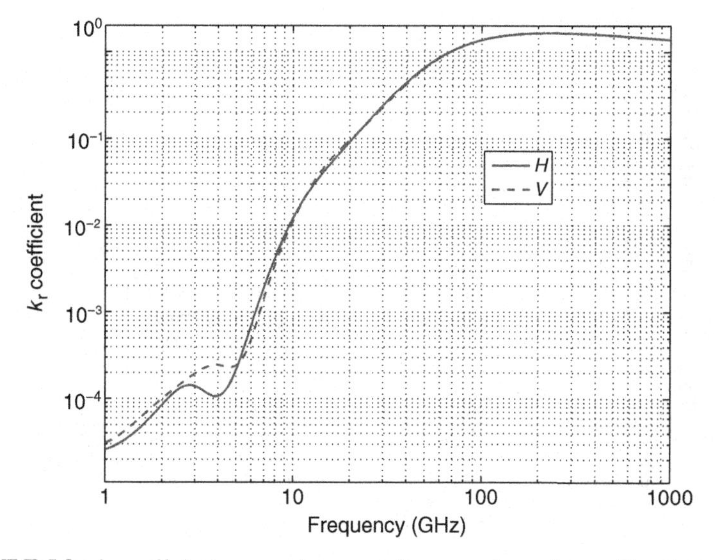

FIGURE 5.8 k_r coefficients versus frequency for computing rain specific attenuation.

values can be computed from

$$k_r = \frac{1}{2}\left[k_H + k_V + (k_H - k_V)\cos^2\theta\cos 2\tau\right], \tag{5.31}$$

$$a_r = \frac{1}{2k_r}\left[k_H a_H + k_V a_V + (k_H a_H - k_V a_V)\cos^2\theta\cos 2\tau\right], \tag{5.32}$$

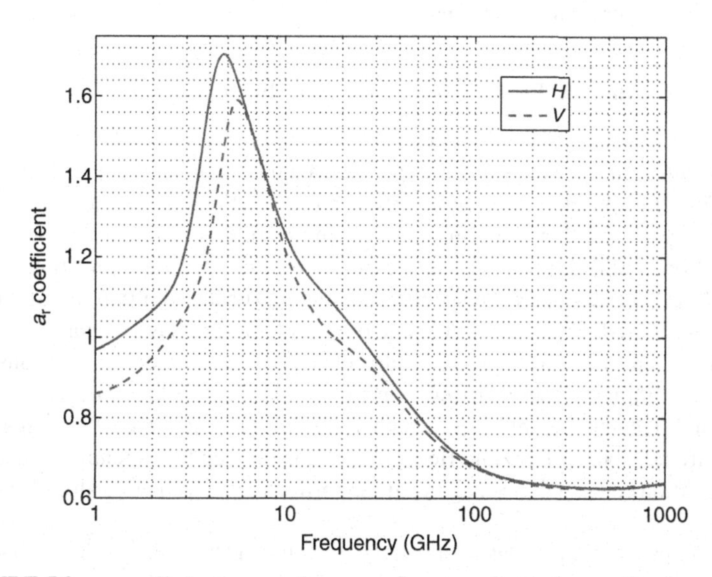

FIGURE 5.9 a_r coefficients versus frequency for computing rain specific attenuation.

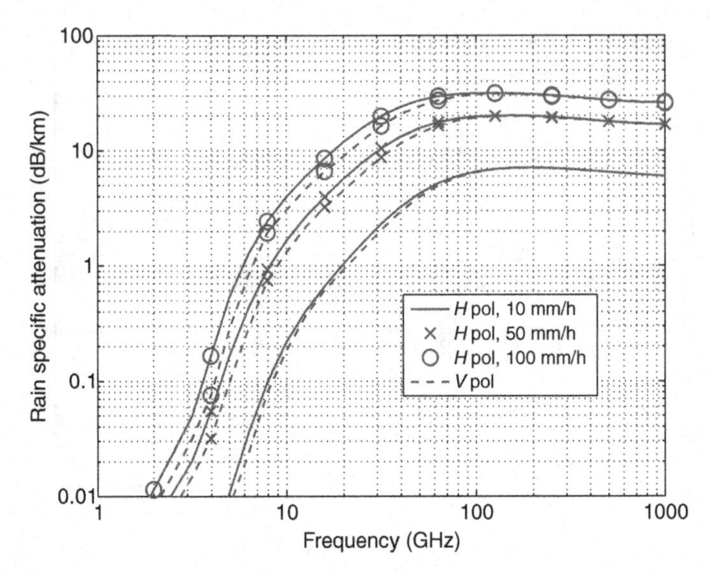

FIGURE 5.10 Rain specific attenuation versus frequency for selected rain rates.

given k_H, k_V, a_H, and a_V from the figures for horizontal and vertical polarizations. Here, θ represents the path elevation angle (with respect to the horizontal) of interest and τ is the polarization tilt angle with respect to horizontal (45° for circular). Note that the phase relationship between the vertical and horizontal field components is not relevant. Figure 5.10 plots rain specific attenuations versus frequency using these equations for a few rain rates and shows the importance of rain attenuation at frequencies greater than approximately 5 GHz.

5.4.4 Computing the Total Path Attenuation Through Rain

As with atmospheric gases, computing the total attenuation (in dB) of rain along a specified path requires integrating the specific attenuation (in dB per km) over the path length. For atmospheric gases, only vertical path variations were important because it could be assumed that the atmospheric composition is uniform horizontally, so that equation (5.23) applied. However, this is not true for rain, and both vertical and horizontal variations of the specific attenuation (or the rain rate) along the path must be considered. In general, it is quite difficult to describe horizontal and vertical variations of the rain rate that are applicable to a wide range of conditions, even at a fixed location. A useful quantity in this process is the "zero degree isotherm height" or "freezing height", h_R, which is the altitude at which the atmospheric temperature falls below freezing. Because ice causes much less attenuation than rain, rain attenuation can usually be neglected for portions of paths having altitudes greater than h_R.

The freezing height is a meteorological parameter and therefore varies with season, location, and so on. ITU-R Recommendation P.839-3 [9] provides global maps of expected h_R values. An approximation for h_R that is useful over many global land areas is

$$h_R = 5 \qquad |\phi| < 23, \tag{5.33}$$

$$h_R = 5 - 0.075\,(|\phi| - 23) \qquad |\phi| \geq 23, \tag{5.34}$$

where h_R is in km and $|\phi|$ is the absolute value of the location latitude in degrees. This approximation shows that the freezing height ranges from approximately 5 km at the equator to near 0 km at the poles.

Even given a prediction for h_R, a description of the vertical and horizontal variations of the rain rate is required to compute the total attenuation over a specific path. While studies that propose descriptions of rain rate spatial properties are not uncommon, their predictions are limited in applicability due to the difficulty in obtaining sufficient data on rain properties.

Consequently, empirical methods are the most practical approach for determining the total rain attenuation. Numerous empirical studies have been performed by placing a transmission of known transmit power (called a "beacon") on either satellite or ground station platforms, and recording over a long time period (usually at least several years) the power received from this beacon in conjunction with local rain rates. Investigators across the globe have performed measurements of this sort since the 1960s, and an extensive database of measurements is available for developing empirical models.

Two methods are commonly used to represent empirical path attenuation information. In the first, the total attenuation along a path is described as the specific attenuation at ground level multiplied by an effective path length:

$$L_{\text{rain,db}} = \alpha_{\text{rain,db}}\, L_E = \left(k_r R_r^{a_r}\right) L_E, \tag{5.35}$$

where R_r is the rain rate at ground level and L_E is the effective path length that is determined from empirical $L_{\text{rain,db}}$ measurements. An alternate approach uses

$$L_{\text{rain,db}} = \tilde{\alpha}_{\text{rain,db}}\, L_{\text{true}} = \left(k_r \tilde{R}_r^{a_r}\right) L_{\text{true}}, \tag{5.36}$$

where $\tilde{\alpha}_{\text{rain,db}}$ or $\tilde{R}_r$ are effective specific attenuations or rain rates, respectively, that when multiplied by the true path length L_{true} yield the measured $L_{\text{rain,db}}$ data. The true path length L_{true} is computed as the path distance between the station and the freezing height.

A significant number of descriptions of both L_E and of $\tilde{\alpha}_{\text{rain,db}}$ have been proposed. The ITU-R continues to examine and cross-compare such models, but at present recommends in ITU-R P.618-9 [7] an approach (given below) that should provide rain attenuation predictions (in decibels) accurate to within approximately 20% on Earth to satellite paths. The ITU-R also specifies 20% as the expected variability in the rain attenuation from year to year.

The following ITU-R model is recommended for use on a global basis at frequencies up to 55 GHz; however, given the wide variations in meteorology that occur globally, it is unlikely that a globally averaged model will yield high accuracy for all locations. For this reason, it is almost always preferable to use local attenuation data if they are available when forecasting propagation behaviors, even if at a different frequency or polarization: methods for "scaling" such data sets to other frequencies and polarizations exist as will be described later. In the absence of such information, the following approach can be used.

5.4.4.1 ITU Model for Earth to Satellite Rain Attenuation Given the surface rain rate that is exceeded 0.01% of the time ($R_{r,0.01}$) at a ground station of height h_S km above sea level, and a path elevation angle θ, the path length below the rain height is

$$L_S = \csc\theta\,(h_R - h_S) \qquad \theta \geq 5^\circ, \tag{5.37}$$

$$L_S = 2\frac{h_R - h_S}{\left(\sin^2\theta + \frac{2(h_R - h_S)}{R_e}\right)^{1/2} + \sin\theta} \qquad \theta < 5^\circ, \tag{5.38}$$

where R_e is the effective Earth radius 8500 km (as will be discussed further in Chapter 6); the lower equation accounts for refraction effects in the Earth's atmosphere at small elevation angles, which will also be discussed further in Chapter 6. The projection of L_S onto the Earth's surface is also used:

$$L_G = \cos\theta\, L_S. \tag{5.39}$$

Note if $h_S > h_R$, the predicted rain attenuation is zero.

The effective path length L_E is specified as a product of two factors

$$L_E = V_{0.01}\, L_R\,. \tag{5.40}$$

The computation of L_R, the modified true path length, involves the following steps:

- First compute the "horizontal reduction factor" $r_{0.01}$:

$$r_{0.01} = \left[1 + 0.78\sqrt{\frac{\alpha_{\text{rain,db}}\, L_G}{f_{\text{GHz}}}} - 0.38\left(1 - e^{-2L_G}\right)\right]^{-1}, \tag{5.41}$$

  where $\alpha_{\text{rain,db}} = k_r R_{r,0.01}^{a_r}$ is the specific attenuation exceeded 0.01% of the time at ground level. The factor $r_{0.01}$ is defined so that $r_{0.01}\, L_G$ can be interpreted as an estimate of the horizontal extent of the rain cell.

- Next an angle ζ is computed as

$$\zeta = \tan^{-1}\left(\frac{h_R - h_S}{r_{0.01}\, L_G}\right). \tag{5.42}$$

- L_R is then specified depending on ζ:

$$\begin{aligned} L_R &= \sec\theta\,(r_{0.01}\, L_G) & \zeta \geq \theta, \\ &= \csc\theta(h_R - h_S) & \zeta < \theta. \end{aligned} \tag{5.43}$$

  The above equations ensure that the modified true path length is always less than or equal to the original L_S.

The computation of $V_{0.01}$, which scales the modified true path length into the effective path length L_{E}, involves the following steps:

- First define the angle χ as

$$\chi = 36 - |\phi| \qquad |\phi| < 36^\circ, \\ = 0 \qquad |\phi| \geq 36^\circ, \tag{5.44}$$

where ϕ is the ground station latitude in degrees.

- Then

$$V_{0.01} = \left\{1 + \sqrt{\sin\theta}\left(31\left[1 - e^{(-\theta/(1+\chi))}\right]\frac{\sqrt{L_{\mathrm{R}}\alpha_{\mathrm{rain,db}}}}{f_{\mathrm{GHz}}^2} - 0.45\right)\right\}^{-1}. \tag{5.45}$$

Figure 5.11 illustrates L_{E} under this model for varying frequency, path elevation angle, rain rate exceeded 0.01% of the time, and station latitude parameters, assuming a station height of 0 km and circular polarization. Figure 5.11a (for a vertical path and for station latitude 40°) shows that the effective path length increases as a function of frequency: this is due to the use of the surface rain rate in the computation, so that

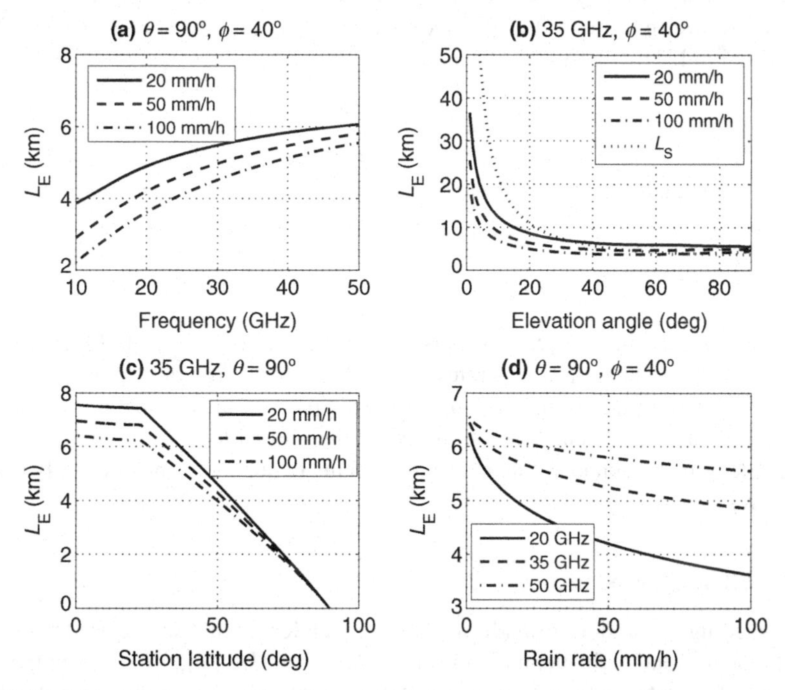

FIGURE 5.11 Earth to satellite effective path length for computing rain attenuation for various path elevation angle (θ), frequency, rain rate exceeded 0.01% of the time, and station latitude (ϕ) parameters.

L_E must vary with frequency to account for vertical variations in the local specific attenuation. In Figure 5.11a, the lower rain rates exceeded 0.01% of the time pertain to drier climates. Thunderstorms are rare in such climates; the rain is usually from stratiform clouds that extend over relatively large distances; hence the L_E values are greater for such climates.

Figure 5.11b (at frequency 35 GHz and station latitude 40°) illustrates the dependence of the effective path length on the elevation angle. The true path length L_S is also included for comparison. The results show that the effective path length varies substantially from L_S at low elevations, with greater L_E values again obtained for drier climates.

Figure 5.11c considers the impact of the station latitude for three rain rates that are exceeded 0.01% of the time for vertical paths at 35 GHz. L_E decreases significantly at higher latitudes primarily due to the decreased freezing height at such latitudes.

Figure 5.11d considers variations in L_E with the rain rate that is exceeded 0.01% of the time for three frequencies; all curves show a decreasing trend of L_E with that rain rate. The decrease is more rapid at lower frequencies.

Combined, these plots show an effective path length ranging from 0 to 8 km on vertical paths; these path lengths increase dramatically at lower elevation angles as seen in Figure 5.11b.

5.4.4.2 ITU-R Model for Rain Attenuation Between Earth Stations For links between two Earth stations, the effective path length L_E to be used in equation (5.35) is specified in ITU-R P. 530-12 [10] as

$$L_E = \frac{d}{1 + d/d_0}, \tag{5.46}$$

where d is the path distance in km and

$$\begin{aligned} d_0 &= 35e^{-0.015R_{r,0.01}} \qquad R_{r,0.01} \leq 100, \\ d_0 &= 7.81 \qquad R_{r,0.01} > 100 \end{aligned} \tag{5.47}$$

with $R_{r,0.01}$ in mm/h. The quantity d_0 is an average horizontal extent of the rain and decreases from 35 km at low rain rates to 7.81 km at very high rain rates.

L_E in equation (5.46) ranges from d when the path distance is much less than d_0 (so that the entire path length is multiplied by the ground level $\alpha_{\text{rain,db}}$) to d_0 when the path length is much larger than d_0, so that rain attenuation is included only in the raining region.

5.4.5 Attenuation Statistics

The preceding discussions provide an effective path length that can be used to obtain an estimate of total path attenuation using $L_{\text{rain},0.01} = \left(k_r R_{r,0.01}^{a_r}\right) L_E$. Given the use of $R_{r,0.01}$ (the surface rain rate exceeded 0.01% of the time for the ground station location) in the computations, the predicted attenuation $L_{\text{rain},0.01}$ is the rain attenuation (in dB) that is expected to be exceeded 0.01% of the time. Since the ITU-R model for L_E can be used only for $L_{\text{rain},0.01}$, the expected attenuation exceeded in other percentiles must be determined through an additional procedure. The ITU-R recommends that the rain attenuation exceeded at other probabilities be determined directly from $L_{\text{rain},0.01}$ by a "probability scaling" method. The recommended procedures are described in ITU-R P.618-9 [7] for Earth–space paths and in ITU-R P.530-12 [10] for ground-to-ground paths. Given $L_{\text{rain},0.01}$, the rain attenuation exceeded p percent of the time, $L_{\text{rain},p}$, on Earth–space paths is obtained from

$$L_{\text{rain},p} = L_{\text{rain},0.01}\left(\frac{p}{0.01}\right)^{-[0.655+0.033\ln(p)-0.045\ln(L_{\text{rain},0.01})-\beta(1-p)\sin\theta]}, \tag{5.48}$$

where

$$\beta = 0 \qquad p \geq 1 \text{ or } |\phi| \geq 36^\circ,$$

$$\beta = -0.005\left(|\phi| - 36^\circ\right) \qquad p < 1 \text{ and } |\phi| < 36^\circ \text{ and } \theta \geq 25^\circ, \tag{5.49}$$

$$\beta = -0.005\left(|\phi| - 36^\circ\right) + 1.8 - 4.25\sin\theta \qquad \text{otherwise.} \tag{5.50}$$

This approach is specified as accurate for p ranging from 0.001% to 5%.

For ground-to-ground paths and for p ranging from 0.001% to 1%, the probability scaling equations are

$$L_{\text{rain},p} = 0.12\, L_{\text{rain},0.01}\, p^{-(0.546+0.01867\ln(p))} \qquad |\phi| \geq 30^\circ,$$

$$L_{\text{rain},p} = 0.07\, L_{\text{rain},0.01}\, p^{-(0.855+0.06037\ln(p))} \qquad |\phi| < 30^\circ. \tag{5.51}$$

Using these equations the rain attenuations exceeded at a specified percentage of the time can be determined; this makes it possible to produce rain rate attenuation distribution plots as in Figure 5.5. This approach assumes that the probability of exceeding a given surface rain rate versus the rain rate has a similar functional form in all global regions; again local information on attenuation statistics is preferred, if available, to this globally averaged approach.

5.4.6 Frequency Scaling

In some cases, local information on rain attenuation may be available at one frequency, but information is sought on an alternate nearby frequency. ITU-R P. 618-9 [7] states that for otherwise identical situations, for frequencies from 7 to 55 GHz, and at the same probability,

$$\frac{L_{\text{rain},p}(f_2)}{L_{\text{rain},p}(f_1)} = \left(\frac{g(f_2)}{g(f_1)}\right)^{1-H}, \tag{5.52}$$

where $L_{\text{rain},p}(f_1)$ is the known rain attenuation exceeded p percent of the time at frequency f_1, and $L_{\text{rain},p}(f_2)$ is the desired rain attenuation exceeded p percent of the time at frequency f_2. In this equation,

$$g(f) = \frac{f_{\text{GHz}}^2}{1 + 10^{-4} f_{\text{GHz}}^2} \tag{5.53}$$

and

$$H = 1.12 \times 10^{-3} \left(\frac{g(f_2)}{g(f_1)} \right)^{0.5} \left[g(f_1)\, L_{\text{rain},p}(f_1) \right]^{0.55}. \tag{5.54}$$

5.4.7 Rain Margin Calculations: An Example

Consider a 12 GHz Earth to satellite link with the ground station location near Columbus, OH (latitude $\approx 40^\circ$, station height 0 km). The satellite is in a geostationary orbit (see appendix to this chapter) and is observed at range 38,000 km from the ground station at elevation angle 40°. Suppose the satellite transmitter transmits 120 W of power in a 24 MHz bandwidth using circular polarization and has an antenna gain of 46 dBi, while the receiver has a noise figure of 6 dB and the receive antenna gain is 30 dBi. The signal-to-noise ratio required for the system to function is 10 dB. Our goal is to assess the number of hours per year (on average) that the system would be expected to fail due to insufficient signal-to-noise ratio. It will be assumed throughout that the systems are impedance and polarization matched and that the antennas are directed to achieve maximum received power.

We begin by using the Friis formula to compute the signal-to-noise ratio neglecting atmospheric gas and rain attenuation:

$$\begin{aligned} P_{\text{R,dbW}} &= P_{\text{T,dbW}} + G_{\text{Tdb}} + G_{\text{Rdb}} - 20 \log_{10} R_{\text{km}} - 20 \log_{10} f_{\text{MHz}} - 32.44 \\ &= 20.8 + 46 + 30 - 91.6 - 81.6 - 32.44 = -108.8 \text{ dbW}, \end{aligned}$$

while the thermal noise power level at the receiver (assuming a 290K external noise temperature) is

$$\begin{aligned} P_{\text{N,Watts}} &= F k_{\text{B}} T_0 B \\ &= 4 \left(1.38 \times 10^{-23} \right) (290) \left(24 \times 10^6 \right) = 3.84 \times 10^{-13} \text{ W} \end{aligned}$$

or -124.1 dBW. The signal-to-noise ratio achieved by neglecting rain and gas attenuation is then 15.3 dB, so 5.3 dB margin is achieved over the 10 dB signal-to-noise threshold required for operation.

Examination of Figure 5.3 shows the zenith gas attenuation through the entire atmosphere to be approximately 0.06 dB, so the slant path attenuation is estimated as $0.06 \csc 40^\circ = 0.09$ dB. Gas attenuation at 12 GHz is generally negligible.

To investigate rain attenuation effects, begin by using Figure 5.7 to determine that the rain rate exceeded 0.01% of the time near Columbus, OH, is approximately 45 mm/h. Also, use Figures 5.8 and 5.9 to determine the appropriate k_{r} values (0.024, 0.024) and a_{r} values (1.12, 1.18) for vertical and horizontal polarizations, respectively, and equation (5.32) to determine k_{r} and a_{r} values of 0.024 and 1.15, respectively, for circular polarization ($\theta = 40^\circ$, $\tau = 45^\circ$).

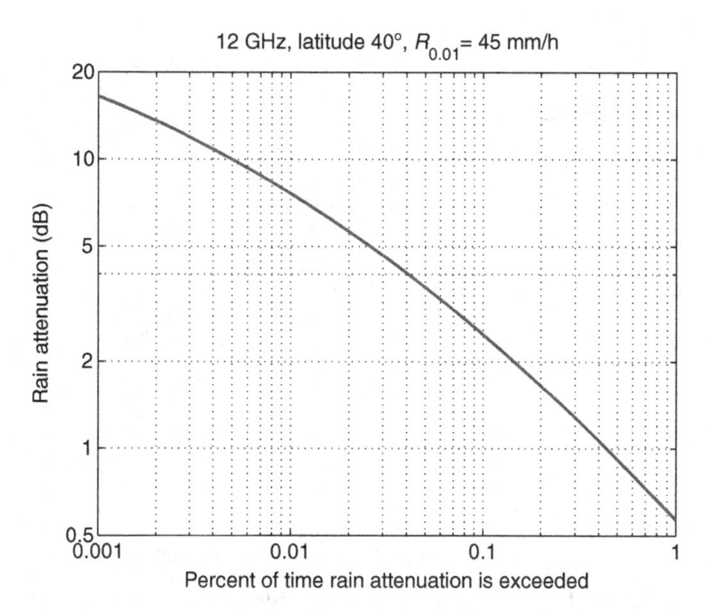

FIGURE 5.12 Rain attenuation distribution function obtained in example problem.

The procedure of Section 5.4.4.1 is next applied to determine $L_{\mathrm{E}} = 3.93$ km, so the attenuation in decibels exceeded 0.01% of time is

$$L_{\mathrm{rain},0.01} = 3.93 \times \left(0.024 \times 45^{1.15}\right) = 7.5 \text{ dB}. \tag{5.55}$$

Attenuations exceeded at other percentages of the time can be determined using the probability scaling approach of equation (5.48).

Unfortunately, equation (5.48) is not easy to invert to determine the percentage of time given a specific attenuation level, as we are seeking here. Instead, a graphical procedure can be used by plotting the predictions of equation (5.48) as shown in Figure 5.12. From the plot, we find that greater than 5.3 dB of rain attenuation occurs approximately 0.025% of the time, or approximately 2.2 h per year, on the average. If this level of reliability is not acceptable, additional transmit power could be added or other system properties modified in order to improve performance.

5.4.8 Site Diversity Improvements

Given the large effect that rain attenuation can have on system reliability, especially at higher frequencies, methods for overcoming these effects are sometimes necessary if reliable links are to be obtained. "Site diversity" refers to the use of multiple antennas that are separated by appreciable distances for signal reception, with some method of choosing or combining the multiple signals received to improve the performance over that of a single antenna. Site diversity should be expected to yield system improvements because the large rain rates that lead to the largest attenuations often have

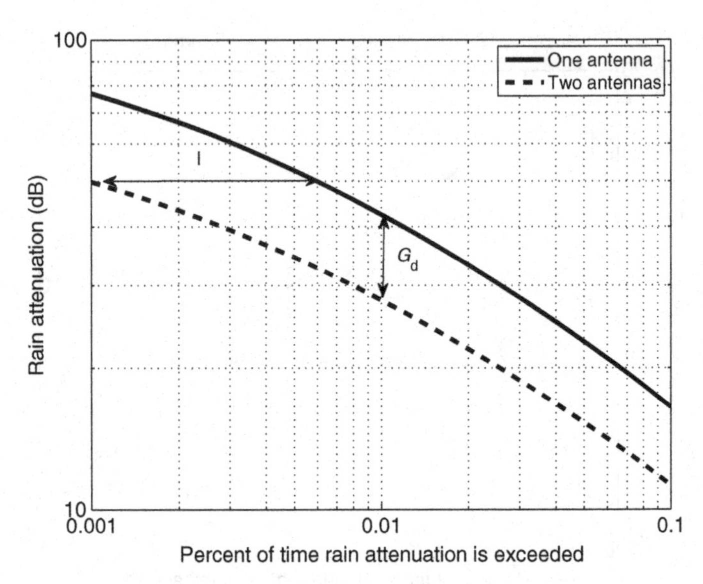

FIGURE 5.13 Hypothetical rain attenuation distributions and diversity effects.

rather small horizontal extents, so another antenna located outside of the rain cell would have a greatly reduced attenuation.

Figure 5.13 illustrates two hypothetical rain attenuation distributions obtained for a single antenna and a diversity system. The single antenna curve shows attenuations greater than 40 dB for more than 0.01% of the time. However, the diversity or "two-antenna" curve shows better performance, with less than 30 dB attenuation exceeded 0.01% of the time. The "diversity gain" obtained is defined as G_d in the figure and is the difference between attenuations (in decibels) in the single antenna and diversity systems for a specified probability. The "diversity improvement" is labeled I in the figure, and is defined as the ratio of the probabilities for a specified attenuation.

An important issue in diversity systems is the method that is chosen for combining the signals from different antennas. An approach that is sometimes applied is to switch in a binary fashion to the signal receiving the most power, in which case the diversity system output is that of the antenna with higher power. However, this approach can require rapid switching between systems and becomes impractical at higher data rates. A second approach is to always combine signals from different antennas in a ratio proportional to their signal-to-noise ratios.

Of course, the expected improvement of a diversity system depends strongly on the distance separating the antennas—a zero separation would clearly not yield any improvement! Figure 5.14 plots the two-antenna diversity gain versus separation distance as described in ITU-R Recommendation P.618-9 [7]. The different curves on this plot were obtained for varying rain attenuations (or "fade depths") in the worst antenna (indicated in the figure legend) and illustrate that diversity gain is also clearly associated with the fade strength observed. Note that the most improvement is

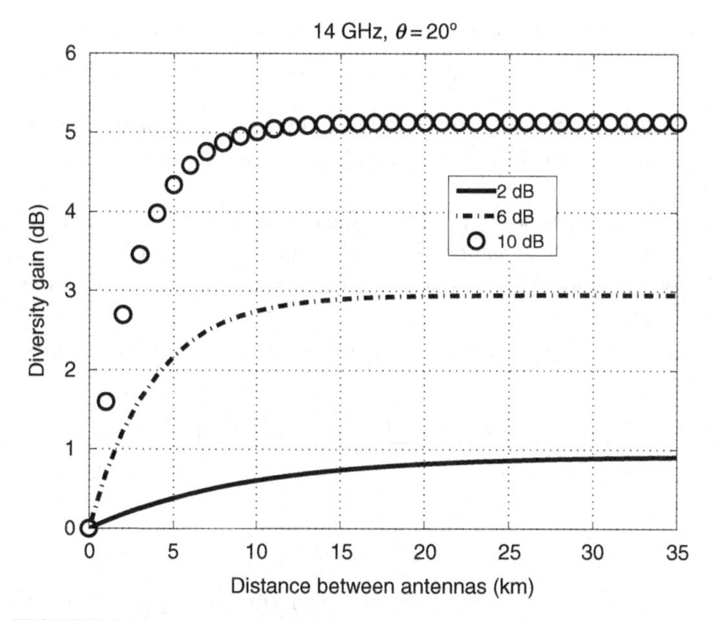

FIGURE 5.14 Sample plots of diversity gain versus separation distance.

obtained in the highest fade situations, which should be expected since it is here that rain attenuation is playing the major role. The saturation in these curves as distance increases gives some idea of the typical horizontal extent of a rain event, with a separation of 8 km clearly larger than the extent of the more significant rain events observed.

ITU-R Recommendation P.618-9 provides an empirical model of the diversity gain G_d in dB as

$$G_\mathrm{d} = G_\mathrm{s}\, G_\mathrm{f}\, G_\theta\, G_\delta. \tag{5.56}$$

The site separation factor, G_s is

$$G_\mathrm{s} = a\left(1 - e^{-b\,d}\right), \tag{5.57}$$

where d is the antenna separation in km and

$$a = 0.78\,A - 1.94\left(1 - e^{0.11\,A}\right), \tag{5.58}$$

$$b = 0.59\left(1 - e^{-0.1\,A}\right), \tag{5.59}$$

and A is the single-site rain attenuation in dB. The frequency factor G_f is

$$G_\mathrm{f} = e^{-0.025\,f_\mathrm{GHz}}. \tag{5.60}$$

The elevation factor G_θ is

$$G_\theta = 1 + 0.006\,\theta, \tag{5.61}$$

where θ is the elevation angle in degrees, and the baseline orientation factor G_δ is

$$G_\delta = 1 + 0.002\,\delta, \tag{5.62}$$

where δ is the acute angle between the great circle connecting the two stations and the projection onto the Earth surface of the path to the satellite from one of the stations, in degrees.

A more accurate method for predicting the outage probability due to rain attenuation in a system with two-antenna site diversity is also given in ITU-R P.618-9; this method however is more complex and provides less physical insight than the simpler approach presented above.

A model for the diversity improvement is also provided in earlier versions of ITU-R P.618 as

$$I = \frac{p_1}{p_2} = \frac{1}{1+\beta^2}\left(1 + \frac{100\beta^2}{p_1}\right), \tag{5.63}$$

where

$$\beta^2 = 10^{-4}\, d^{1.33}, \tag{5.64}$$

and where d is again the antenna separation in km. Here, p_1 and p_2 refer to the percentages of time that a given rain attenuation is observed for the one- and two-antenna cases, respectively. It is interesting to note that, unlike the diversity gain, the diversity improvement is modeled as independent of the rain attenuation level.

5.5 SCINTILLATIONS

Scintillation on a radiowave path is the name given to rapid fluctuations of the amplitude, phase, or angle of arrival of the wave passing through a medium with small-scale dielectric constant irregularities that cause changes in the transmission path with time. Scintillation effects, sometimes also referred to as atmospheric multipath fading, can be produced in both the troposphere and the ionosphere, with the major ionospheric effects occurring at frequencies below 2 GHz (ionospheric scintillations will be discussed in Chapter 11). To a first approximation, the dielectric constant structure can be considered as horizontally stratified, and variations appear as thin layers that change with altitude. Slant paths at low elevation angles tend to be affected most by scintillation conditions. Tropospheric scintillation is typically produced in the first few kilometers of altitude by high humidity gradients and temperature inversion layers. The effects are seasonally dependent and vary from day to day, as well as with the local climate.

Extensive observations of amplitude scintillations have been performed, and ITU-R Recommendation 618-9 [7] provides an empirical attenuation distribution for tropospheric attenuation caused by scintillation effects. Generally, frequencies from 2 to 20 GHz show scintillations less than 0.5–1 dB at high elevation angles (from 20° to 30°) in a temperate climate. Information on the expected rates of fading is also included in Recommendation 618 and shows that both "fast" (i.e., <1 s) and slow (periods of 10 s or more) fading components can occur, with the latter more important at low elevation angles. At lower frequencies, scintillations in the ionosphere must also be considered.

The measurements show a very definite increase in the magnitude of scintillation effects as the elevation angle is reduced below 10°. Deep fades of 20 dB or more with a few seconds duration have been observed; such fades are indicative of multipath contributions to scintillation effects. Multipath fading mechanisms will be discussed in more detail in Chapter 8. In some cases, scintillation effects can be attributed to variations in the angle of arrival of the signal transmitted from a satellite; to a high-gain receiving antenna, such variations in the angle of arrival cause variations in the received power, and therefore appear to be amplitude variations.

It is important to be aware of potential atmospheric scintillation effects, especially at lower elevation angles, so that they will not be confused with system errors or system noise!

APPENDIX 5.A LOOK ANGLES TO GEOSTATIONARY SATELLITES

Since our methods for predicting propagation under direct transmission conditions are frequently applied to satellite systems, it is useful to have a method for computing the elevation and azimuth angles at which a satellite appears to the Earth station. For example, if the Earth station receives the satellite signal, then these angles are the θ_R, ϕ_R of equations (5.3) and (5.5).

When a satellite is injected into a circular orbit 35,900 km (22,300 miles) above the surface of the Earth, its period equals that of the Earth's rotation. If the orbit is in the Earth's equatorial plane, the satellite will rotate in synchronism with the Earth and appear to be stationary. Such "geostationary" satellites are very useful for communications because Earth station antennas do not need elaborate steering mechanisms to track the satellites. In practice, complete stationarity is not realized because the Earth is not a homogeneous sphere, and the satellite orbit is affected by the gravitational anomalies that exist. As a result, satellites in geostationary orbits appear to drift in a small diurnal pattern, generally subtending a small fraction of a degree of arc from an Earth station, so that low-gain antennas often need not be steered at all, while the tracking requirements of high-gain antennas are minimized. Of course, geostationary orbits are not the only possible satellite orbits! We will limit ourselves to this situation only for simplicity.

Earth stations communicating with satellites in geostationary orbits need to be able to orient antennas in the proper direction, specified by the elevation and azimuth angles to the satellite. A derivation of these angles based on the latitude and longitude of the Earth station and subsatellite points on the Earth's surface (the point on the Earth's surface on a line between the center of the Earth and the satellite) is as follows: For an Earth station located at point B (latitude θ', longitude ϕ') and subsatellite point A (latitude 0°, longitude ϕ_0), as shown in Figure 5.15, spherical trigonometry shows

$$\cos\delta = \cos\theta' \cos\left(\phi' - \phi_0\right), \tag{5.65}$$

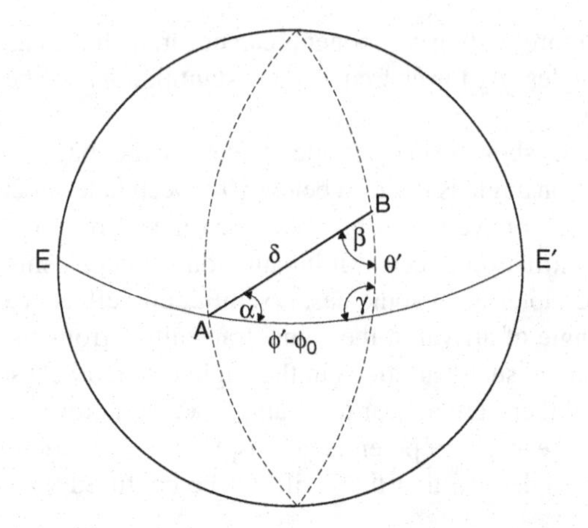

FIGURE 5.15 Geometry for defining look angles.

$$\cos\beta = \tan\theta' \cot\delta. \tag{5.66}$$

The azimuth bearing of the satellite as seen from a station in the Northern hemisphere is therefore $\pi + \beta$ for a satellite west of the station and $\pi - \beta$ for a satellite east of the station.

For a station in the Southern hemisphere, the corresponding relations are

$$\cos\delta = \cos\theta' \cos\left(\phi' - \phi_0\right), \tag{5.67}$$

$$\cos\beta = \tan|\theta'| \cot\delta, \tag{5.68}$$

and the azimuth bearing is $2\pi - \beta$ for a satellite west of the station and β for a satellite to the east of the station.

In Figure 5.16, the plane of the page is determined by the subsatellite point A, the station location B, and the center of the Earth O. Since A is the intersection of OS

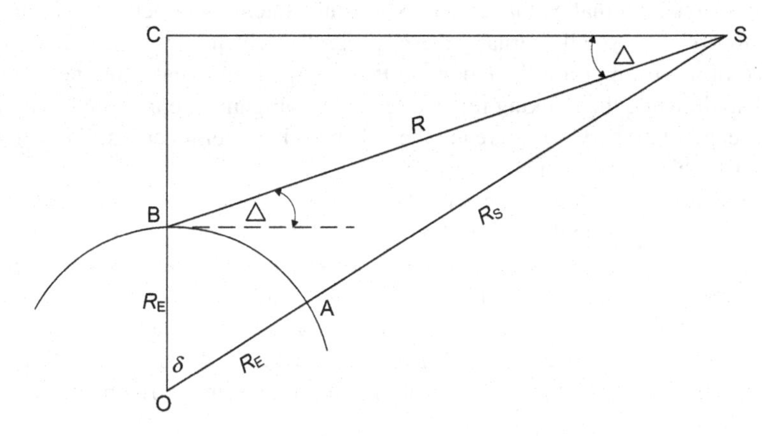

FIGURE 5.16 Geometry for finding elevation angle.

with the Earth surface, S is also in the plane of the page. The central angle δ measures the great-circle arc AB, as implied by Figure 5.15. Applying the law of cosines to the plane triangle OBS gives

$$R^2 = R_\epsilon^2 + (R_\epsilon + R_S)^2 - 2R_\epsilon(R_\epsilon + R_S)\cos\delta. \tag{5.69}$$

The distance CB is given by

$$\mathrm{CB} = R\sin\Delta \tag{5.70}$$

and also by

$$\mathrm{CB} = \mathrm{CO} - \mathrm{OB} = (R_\epsilon + R_S)\cos\delta - R_\epsilon. \tag{5.71}$$

Equating the previous two equations allows solution for the elevation angle Δ as

$$\sin\Delta = \frac{(R_\epsilon + R_S)\cos\delta - R_\epsilon}{R}, \tag{5.72}$$

which is valid for both the Northern and Southern hemispheres.

REFERENCES

1. ITU-R Recommendation P.676-7, "Attenuation by atmospheric gases," International Telecommunication Union, 2007.
2. ITU-R Recommendation P.840-3, "Attenuation due to clouds and fog," International Telecommunication Union, 1999.
3. Marshall, J. S. and W. M. K. Palmer, "The distribution of raindrops with size," *J. Meteor.*, vol. 5, pp. 165–166, 1948.
4. Williams, C. R. and K. S. Gage, "Raindrop size distribution variability estimated using ensemble statistics," *Ann. Geophys.*, vol. 27, pp. 555–567, 2009.
5. ITU-R Recommendation P.837-5, "Characteristics of precipitation for propagation modeling," International Telecommunication Union, 2007.
6. Van de Hulst, H. C., *Light Scattering by Small Particles*, Dover Publications, 1981.
7. ITU-R Recommendation P.618-9, "Propagation data and prediction methods required for the design of Earth–space telecommunication systems," International Telecommunication Union, 2007.
8. ITU-R Recommendation P.838-3, "Specific attenuation model for rain for use in prediction methods," International Telecommunication Union, 2005.
9. ITU-R Recommendation P.839-3, "Rain height model for prediction methods," International Telecommunication Union, 2001.
10. ITU-R Recommendation P.530-12, "Propagation data and prediction methods required for the design of terrestrial line-of-sight systems," International Telecommunication Union, 2007.

6

REFLECTION AND REFRACTION

6.1 INTRODUCTION

In many situations, Earth terrain and/or the atmosphere may be modeled as consisting of layers of dielectric media separated by planar or spherical interfaces. Analytical models for such configurations are well established [1–3]: the basic formalisms of reflection and refraction for a planar boundary are reviewed in this chapter to prepare for their application to path loss predictions in subsequent chapters. The basic theory of refraction derived in this chapter is also applied to study the propagation of electromagnetic waves in an inhomogeneous atmosphere above a spherical Earth, and the ducting phenomena that can result are investigated.

6.2 REFLECTION FROM A PLANAR INTERFACE: NORMAL INCIDENCE

Consider an interface between two linear, homogeneous, isotropic, and time-invariant media located in the xy plane as shown in Figure 6.1. Regions 1 and 2 are located above and below the interface, respectively, and are characterized by ϵ_i, μ_i, and σ_i where $i = 1, 2$. Our experience tells us that a wave that encounters a boundary with a different medium will have some portion of its energy reflected from the boundary while the remainder is transmitted; for example, in a piece of glass one can see both reflected light (forming an image of the observer) and light transmitted from the

Radiowave Propagation: Physics and Applications. By Curt A. Levis, Joel T. Johnson, and Fernando L. Teixeira

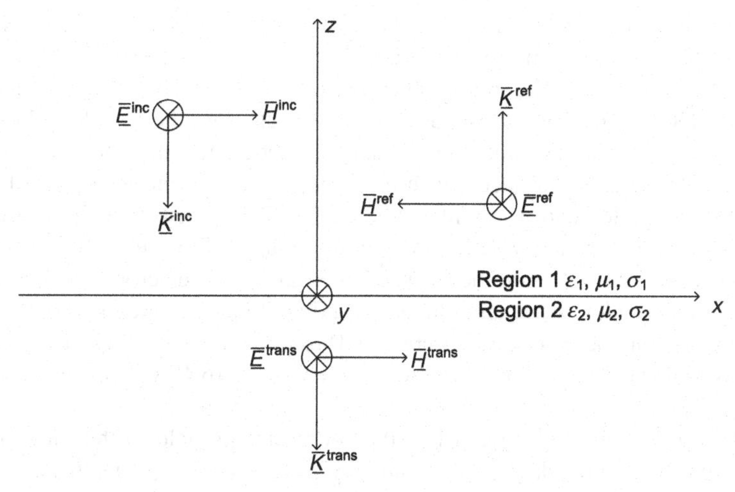

FIGURE 6.1 Geometry for normal incidence reflection.

other side. A determination of how much energy is reflected given the properties of the two media, however, requires use of electromagnetic theory and is considered in this section for a plane wave propagating in a direction normal to the interface. In this case, an azimuthal symmetry exists about the z axis since the interface is a plane. Thus, the final solution to the problem should have no dependence on the x or y coordinates, and the resulting reflected and transmitted fields should have the same amplitude regardless of their polarization.

To represent a general linear polarization, let us choose the incident electric field to be polarized in the $\hat{y}$ direction. Propagation in the $-\hat{z}$ direction then implies that the incident magnetic field is in the $\hat{x}$ direction. As stated previously, a reflected plane wave should result in region 1 while a transmitted plane wave results in region 2. The total electric and magnetic fields in regions 1 and 2 can then be written as

$$\begin{aligned}\overline{\underline{E}}_1^{tot} &= \overline{\underline{E}}^{inc} + \overline{\underline{E}}^{ref} \\ &= \hat{y}E_0 e^{j\underline{k}_1 z} + \hat{y}\underline{\Gamma}E_0 e^{-j\underline{k}_1 z}, \end{aligned} \tag{6.1}$$

$$\begin{aligned}\overline{\underline{H}}_1^{tot} &= \overline{\underline{H}}^{inc} + \overline{\underline{H}}^{ref} \\ &= \hat{x}\frac{E_0}{\underline{\eta}_1} e^{j\underline{k}_1 z} - \hat{x}\frac{\underline{\Gamma}E_0}{\underline{\eta}_1} e^{-j\underline{k}_1 z}, \end{aligned} \tag{6.2}$$

$$\begin{aligned}\overline{\underline{E}}_2^{tot} &= \overline{\underline{E}}^{trans} \\ &= \hat{y}E_0 \underline{T} e^{j\underline{k}_2 z}, \end{aligned} \tag{6.3}$$

$$\begin{aligned}\overline{\underline{H}}_2^{tot} &= \overline{\underline{H}}^{trans} \\ &= \hat{x}\frac{E_0 \underline{T}}{\underline{\eta}_2} e^{j\underline{k}_2 z}, \end{aligned} \tag{6.4}$$

where E_0 is an arbitrary amplitude, and the propagation constant of plane waves in different dielectric media has been incorporated through $\underline{k}_i = \omega\sqrt{\mu_i(\epsilon_i - j\sigma_i/\omega)} = \omega\sqrt{\mu_i\underline{\epsilon}_i^e}$, $\underline{\eta}_i = \sqrt{\mu_i/\underline{\epsilon}_i^e}$. The appropriate propagation and field polarization directions are indicated. The field polarization directions result from the fact that Maxwell's equations dictate that tangential electric and magnetic fields must be continuous at a sourceless dielectric interface. Since the incident electric field is polarized in the $\hat{y}$ direction, reflected and transmitted electric fields must also be polarized in this same direction if tangential $\overline{E}$ fields are to be continuous. The coefficients $\underline{\Gamma}$ and $\underline{T}$ in the above equations represent the reflected and transmitted electric field amplitudes relative to the incident wave amplitude E_0, and are known as the Fresnel reflection and transmission coefficients, respectively. These coefficients are as yet unknown, but applying the boundary conditions to equations (6.1)–(6.4) will enable them to be determined.

Matching tangential electric fields (the entire electric field in this case since $\hat{y}$ is tangential to the boundary) at the boundary $z = 0$ from equations (6.1) and (6.3) yields

$$1 + \underline{\Gamma} = \underline{T}, \tag{6.5}$$

while matching tangential magnetic fields from equations (6.2) and (6.4) yields

$$\frac{1}{\underline{\eta}_1} - \frac{\underline{\Gamma}}{\underline{\eta}_1} = \frac{\underline{T}}{\underline{\eta}_2}. \tag{6.6}$$

Solving these two equations yields

$$\underline{\Gamma} = \frac{\underline{\eta}_2 - \underline{\eta}_1}{\underline{\eta}_2 + \underline{\eta}_1}, \tag{6.7}$$

$$\underline{T} = 1 + \underline{\Gamma} = \frac{2\underline{\eta}_2}{\underline{\eta}_2 + \underline{\eta}_1}. \tag{6.8}$$

An examination of equation (6.7) shows that no reflected wave will be produced (i.e., $\underline{\Gamma} = 0$) if $\underline{\eta}_2 = \underline{\eta}_1$, that is, $\mu_2/\underline{\epsilon}_2^e = \mu_1/\underline{\epsilon}_1^e$. While this equation is true if we have the same medium on both sides of the boundary, that is, $\mu_2 = \mu_1$, $\underline{\epsilon}_2^e = \underline{\epsilon}_1^e$ as one would expect, no reflected wave can be obtained if there are different media on both sides of the boundary but the intrinsic impedances are equal (requires magnetic materials).

Also note that the $\underline{\Gamma}$ derived in equation (6.7) is an *electric field* reflection coefficient, so the reflected *power* density relative to the incident power density is $|\underline{\Gamma}|^2$.

6.3 REFLECTION FROM A PLANAR INTERFACE: OBLIQUE INCIDENCE

The reflection problem becomes more involved if the incident field is not propagating in a direction normal to the boundary because the azimuthal symmetry is lost in this case, so that the problem is no longer invariant with respect to incident field polarization. In general, any electromagnetic wave can be written as the sum of two orthogonal field components in a plane normal to the direction of propagation. Since Maxwell's equations are linear, the reflection problem can be solved separately for each of these two field components, and then the solution for a wave having any polarization can be written as the sum of the two component solutions multiplied by the corresponding component amplitudes of the specified incident wave.

6.3.1 Plane of Incidence

The standard choice for the two field polarization components for which we will solve the reflection problem involves the "plane of incidence," defined to be a plane containing both the incident field propagation direction and the normal to the surface. The plane of incidence is the plane that is usually drawn when considering oblique incidence reflection problems, as shown in Figure 6.2; that is, the plane of the page is the plane of incidence for this figure. Clearly, any arbitrary field polarization can be written as the sum of one component that is parallel to the plane of incidence and perpendicular to the propagation direction and a second component that is perpendicular to the plane of incidence (i.e., pointing in the $\hat{y}$ direction into the page.) Standard notation defines a field as having parallel polarization if its electric field vector lies in the plane of incidence, and perpendicular polarization if its electric field vector is perpendicular to the plane of incidence. These two polarizations are also called

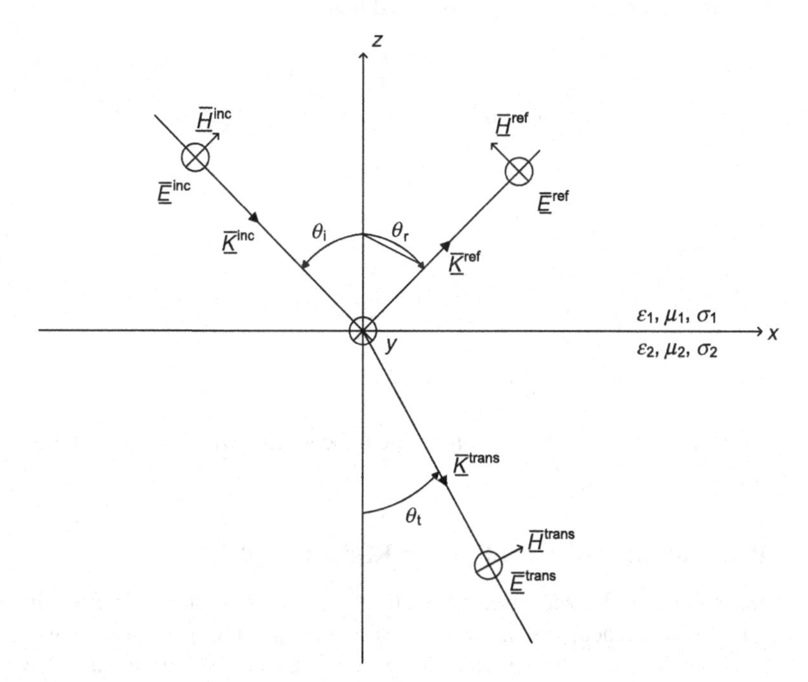

FIGURE 6.2 Geometry for oblique incidence reflection: perpendicular polarization.

vertical and horizontal, respectively, in problems where the boundary represents a horizontal plane such as the ground, or TM and TE, for transverse magnetic or transverse electric, respectively, depending on which field component is perpendicular to the plane of incidence.

As an example of representing an arbitrarily polarized wave in terms of its parallel and perpendicular polarization components, consider the following incident field propagating in the $\hat{k} = (\hat{x} - \hat{z})/\sqrt{2}$ direction from free space onto a "horizontal" boundary in the xy plane:

$$\overline{\underline{E}}^{\text{inc}} = 377\,(-\hat{x} + \hat{y} - \hat{z})\,e^{-jk_0(x-z)/\sqrt{2}}, \tag{6.9}$$

$$\overline{\underline{H}}^{\text{inc}} = \hat{k} \times \overline{\underline{E}}^{\text{inc}}/\eta_0 = \frac{\hat{x} + 2\hat{y} + \hat{z}}{\sqrt{2}} e^{-jk_0(x-z)/\sqrt{2}}. \tag{6.10}$$

The first two components of $\overline{\underline{E}}^{\text{inc}}$ are "horizontal," so one may be tempted to take them and the associated magnetic field as the perpendicular (or horizontal) polarized electric field component. Similarly, the first two components of $\overline{\underline{H}}^{\text{inc}}$ are "horizontal," so one might be tempted to take them and the associated electric field as being parallel (or vertical) polarization. Such a decomposition is not useful because the decomposed fields are not transverse to the direction of propagation and therefore do not correspond to solutions of Maxwell's equations in this problem.

The correct way to decompose the field is to use the plane of incidence defined by $\hat{k} \times \hat{z} = -\hat{y}$, that is, a plane normal to the $-\hat{y}$ direction or the xz plane. Then the perpendicular (therefore horizontal) polarized field is

$$\overline{\underline{E}}_{\perp}^{\text{inc}} = 377\hat{y}e^{-jk_0(x-z)/\sqrt{2}}, \tag{6.11}$$

and the associated magnetic field is

$$\overline{\underline{H}}_{\perp}^{\text{inc}} = \frac{\hat{x} + \hat{z}}{\sqrt{2}} e^{-jk_0(x-z)/\sqrt{2}}, \tag{6.12}$$

while the parallel polarized fields are

$$\overline{\underline{E}}_{\parallel}^{\text{inc}} = 377\sqrt{2}\left(\frac{-\hat{x} - \hat{z}}{\sqrt{2}}\right) e^{-jk_0(x-z)/\sqrt{2}}, \tag{6.13}$$

$$\overline{\underline{H}}_{\parallel}^{\text{inc}} = \sqrt{2}\hat{y}e^{-jk_0(x-z)/\sqrt{2}}. \tag{6.14}$$

We now proceed to solve the reflection problem, beginning with perpendicular polarization.

6.3.2 Perpendicular Polarized Fields in Regions 1 and 2

In the perpendicular polarized case, the electric field vector is polarized perpendicular to the plane of incidence, that is, in the $\hat{y}$ direction in Figure 6.2. Defining $\theta_{\rm i}$, $\theta_{\rm r}$ and $\theta_{\rm t}$ as the angles between the z axis and the propagation directions (as shown in Figure 6.2) allows the incident, reflected, and transmitted $\overline{\underline{k}}$ vectors to be written, respectively, as

$$\overline{\underline{k}}^{\rm inc} = \hat{x}\underline{k}_x^{\rm inc} - \hat{z}\underline{k}_z^{\rm inc} = \omega\sqrt{\mu_1\underline{\epsilon}_1^{\rm e}}\,(\hat{x}\sin\theta_{\rm i} - \hat{z}\cos\theta_{\rm i})\,, \tag{6.15}$$

$$\overline{\underline{k}}^{\rm ref} = \hat{x}\underline{k}_x^{\rm ref} + \hat{z}\underline{k}_z^{\rm ref} = \omega\sqrt{\mu_1\underline{\epsilon}_1^{\rm e}}\,(\hat{x}\sin\theta_{\rm r} + \hat{z}\cos\theta_{\rm r})\,, \tag{6.16}$$

$$\overline{\underline{k}}^{\rm trans} = \hat{x}\underline{k}_x^{\rm trans} - \hat{z}\underline{k}_z^{\rm trans} = \omega\sqrt{\mu_2\underline{\epsilon}_2^{\rm e}}\,(\hat{x}\sin\theta_{\rm t} - \hat{z}\cos\theta_{\rm t})\,, \tag{6.17}$$

where the fact that

$$\overline{\underline{k}}^{\rm inc}\cdot\overline{\underline{k}}^{\rm inc} = \overline{\underline{k}}^{\rm ref}\cdot\overline{\underline{k}}^{\rm ref} = \omega^2\mu_1\underline{\epsilon}_1^{\rm e}, \tag{6.18}$$

$$\overline{\underline{k}}^{\rm trans}\cdot\overline{\underline{k}}^{\rm trans} = \omega^2\mu_2\underline{\epsilon}_2^{\rm e} \tag{6.19}$$

has been used. Note that $\theta_{\rm r}$ and $\theta_{\rm t}$ are as yet unknown. A consideration of the boundary conditions results in a "phase matching" condition that will enable these angles to be determined.

Total electric fields in regions 1 and 2 can then be written as

$$\begin{aligned}\overline{\underline{E}}_1^{\rm tot} &= \overline{\underline{E}}_\perp^{\rm inc} + \overline{\underline{E}}_\perp^{\rm ref}\\ &= \hat{y}E_{0,\perp}e^{-j\overline{\underline{k}}^{\rm inc}\cdot\overline{r}} + \hat{y}\underline{\Gamma}_\perp E_{0,\perp}e^{-j\overline{\underline{k}}^{\rm ref}\cdot\overline{r}},\end{aligned} \tag{6.20}$$

$$\begin{aligned}\overline{\underline{E}}_2^{\rm tot} &= \overline{\underline{E}}_\perp^{\rm trans}\\ &= \hat{y}E_{0,\perp}\underline{T}_\perp e^{-j\overline{\underline{k}}^{\rm trans}\cdot\overline{r}},\end{aligned} \tag{6.21}$$

and their associated magnetic fields derived from equation (3.26). Again, the $\hat{y}$ polarization assumed for the reflected and transmitted waves reflects the fact that the boundary conditions require tangential electric fields to be continuous, so that the sole $\hat{y}$ component of the incident fields dictates that the reflected and transmitted fields also have only a $\hat{y}$ component.

6.3.3 Phase Matching and Snell's Law

The boundary conditions for reflection from a boundary between two dielectric media require that tangential electric and magnetic field components are continuous at the boundary; that is, that the same value of the field is approached as one moves toward the boundary from either side. Examining the incident electric field for this oblique incidence case shows that there is a phase variation along the boundary (i.e., the plane $z = 0$), given by

$$\overline{\underline{E}}_\perp^{\rm inc}|_{z=0} = \hat{y}E_{0,\perp}e^{-j\underline{k}_x^{\rm inc}x}. \tag{6.22}$$

If total tangential fields at the boundary are to be continuous, it is clear that reflected and transmitted fields must retain this same phase variation. Thus, the x components of the $\overline{\underline{k}}$ vectors must be the same for the incident, reflected, and transmitted waves, or more generally, the components of the $\overline{\underline{k}}$ vectors tangential to the boundary must be the same. This requirement is known as the "phase matching" condition.

Phase matching has some immediate consequences in terms of the reflected and transmitted wave propagation directions. For the reflected wave, phase matching requires $\underline{k}_x^{\text{ref}} = \underline{k}_x^{\text{inc}}$, which from equations (6.15) and (6.16) implies $\theta_{\text{r}} = \theta_{\text{i}}$. This is known as the "law of reflection."

Phase matching also allows the direction of the transmitted wave to be determined. Beginning with the transmitted $\overline{\underline{k}}$ vector

$$\overline{\underline{k}}^{\text{trans}} = \hat{x}\underline{k}_x^{\text{trans}} - \hat{z}\underline{k}_z^{\text{trans}} \tag{6.23}$$

phase matching requires

$$\underline{k}_x^{\text{trans}} = \underline{k}_x^{\text{inc}}, \tag{6.24}$$

while equation (6.19) states that

$$\left(\underline{k}_x^{\text{trans}}\right)^2 + \left(\underline{k}_z^{\text{trans}}\right)^2 = \omega^2 \mu_2 \underline{\epsilon}_2^{\text{e}}. \tag{6.25}$$

Equations (6.24) and (6.25) can then be used to obtain

$$\left(\underline{k}_z^{\text{trans}}\right)^2 = \omega^2 \mu_2 \underline{\epsilon}_2^{e} - \left(\underline{k}_x^{\text{inc}}\right)^2, \tag{6.26}$$

allowing both components of the transmitted $\overline{\underline{k}}$ vector to be determined once $\underline{k}_x^{\text{inc}}$ and the medium properties are known. When both media are lossless (i.e., $\sigma_i = 0$) and clear propagation angles exist, this relationship is known as "Snell's law." In this case, $k_x^{\text{trans}} = k_x^{\text{inc}}$ and $k_x^{\text{trans}} = \omega\sqrt{\mu_2\epsilon_2}\sin\theta_{\text{t}}$, yielding

$$\sqrt{\mu_2\epsilon_2}\sin\theta_{\text{t}} = \sqrt{\mu_1\epsilon_1}\sin\theta_{\text{i}}, \tag{6.27}$$

$$\sin\theta_{\text{t}} = \sqrt{\frac{\mu_1\epsilon_1}{\mu_2\epsilon_2}}\sin\theta_{\text{i}}, \tag{6.28}$$

which is a more typical form for Snell's law. Equation (6.28) shows that when propagating from a less dense into a more dense medium, that is, $\mu_1\epsilon_1 < \mu_2\epsilon_2$, the transmitted angle is smaller than the incident angle, that is, $\sin\theta_{\text{t}} < \sin\theta_{\text{i}}$, indicating that the wave propagates along a direction closer to the normal. Conversely, when a wave propagates from a more dense to a less dense medium, $\sin\theta_{\text{t}} > \sin\theta_{\text{i}}$ and the transmitted wave is bent away from the normal. This bending of the rays is known as *refraction*. With nonmagnetic media on both sides of the boundary, Snell's law reduces to

$$\sqrt{\epsilon_2}\sin\theta_{\text{t}} = \sqrt{\epsilon_1}\sin\theta_{\text{i}}, \tag{6.29}$$

or simply

$$n_2\sin\theta_{\text{t}} = n_1\sin\theta_{\text{i}}, \tag{6.30}$$

where n_{i} is the *index of refraction* or *refractive index*, given by $n_{\text{i}} = \sqrt{\frac{\mu_i\underline{\epsilon}_i^{\text{e}}}{\mu_0\epsilon_0}}$ $\left(= \sqrt{\frac{\underline{\epsilon}_i^{\text{e}}}{\epsilon_0}}\right.$ in nonmagnetic media). Refraction has important consequences for propagation through the Earth's atmosphere and will be discussed in more detail later in this chapter.

6.3.4 Perpendicular Reflection Coefficient

Now that the components of the $\underline{\overline{k}}$ vectors have been determined, total magnetic fields in regions 1 and 2 can be written and the boundary conditions applied to determined $\underline{\Gamma}_\perp$ and $\underline{T}_\perp$. Using the fact that $\underline{\overline{H}} = \frac{1}{\omega\mu}\underline{\overline{k}} \times \underline{\overline{E}}$, we obtain

$$\begin{aligned}\underline{\overline{H}}_1^{\text{tot}} &= \underline{\overline{H}}_\perp^{\text{inc}} + \underline{\overline{H}}_\perp^{\text{ref}} \\ &= \frac{\hat{x}\underline{k}_z^{\text{inc}} + \hat{z}\underline{k}_x^{\text{inc}}}{\omega\mu_1} E_{0,\perp} e^{-j\underline{\overline{k}}^{\text{inc}}\cdot\overline{r}} + \frac{-\hat{x}\underline{k}_z^{\text{inc}} + \hat{z}\underline{k}_x^{\text{inc}}}{\omega\mu_1} \underline{\Gamma}_\perp E_{0,\perp} e^{-j\underline{\overline{k}}^{\text{ref}}\cdot\overline{r}},\end{aligned} \tag{6.31}$$

$$\begin{aligned}\underline{\overline{H}}_2^{\text{tot}} &= \underline{\overline{H}}_\perp^{\text{trans}} \\ &= \frac{\hat{x}\underline{k}_z^{\text{trans}} + \hat{z}\underline{k}_x^{\text{inc}}}{\omega\mu_2} E_{0,\perp} \underline{T}_\perp e^{-j\underline{\overline{k}}^{\text{trans}}\cdot\overline{r}}.\end{aligned} \tag{6.32}$$

Setting tangential electric and magnetic fields equal at $z = 0$ yields

$$1 + \underline{\Gamma}_\perp = \underline{T}_\perp, \tag{6.33}$$

$$\frac{\underline{k}_z^{\text{inc}}}{\omega\mu_1}\left(1 - \underline{\Gamma}_\perp\right) = \underline{T}_\perp \frac{\underline{k}_z^{\text{trans}}}{\omega\mu_2}, \tag{6.34}$$

where only the x component of the magnetic field has been used since the z component is not tangential to the boundary. Solving these two equations yields

$$\underline{\Gamma}_\perp = \frac{\mu_2 \underline{k}_z^{\text{inc}} - \mu_1 \underline{k}_z^{\text{trans}}}{\mu_2 \underline{k}_z^{\text{inc}} + \mu_1 \underline{k}_z^{\text{trans}}}, \tag{6.35}$$

$$\underline{T}_\perp = \frac{2\mu_2 \underline{k}_z^{\text{inc}}}{\mu_2 \underline{k}_z^{\text{inc}} + \mu_1 \underline{k}_z^{\text{trans}}}. \tag{6.36}$$

Sometimes it is useful to consider oblique incidence on a perfectly conducting medium. As in the case of normal incidence, one finds $\underline{\Gamma}_\perp = -1$, and there is no transmission into the perfectly conducting medium as the magnetic field is terminated by a surface current.

This concludes the solution for the perpendicular polarized case. Note that again the fraction of power density reflected is $|\underline{\Gamma}_\perp|^2$.

6.3.5 Parallel Polarized Fields in Regions 1 and 2

In the parallel polarized case, the electric field vector is polarized in the plane of incidence, as shown in Figure 6.3, which requires the magnetic field vector to be polarized perpendicular to the plane of incidence. In this case, it is easier to begin with the magnetic fields since they have only a $\hat{y}$ component. Total magnetic fields in

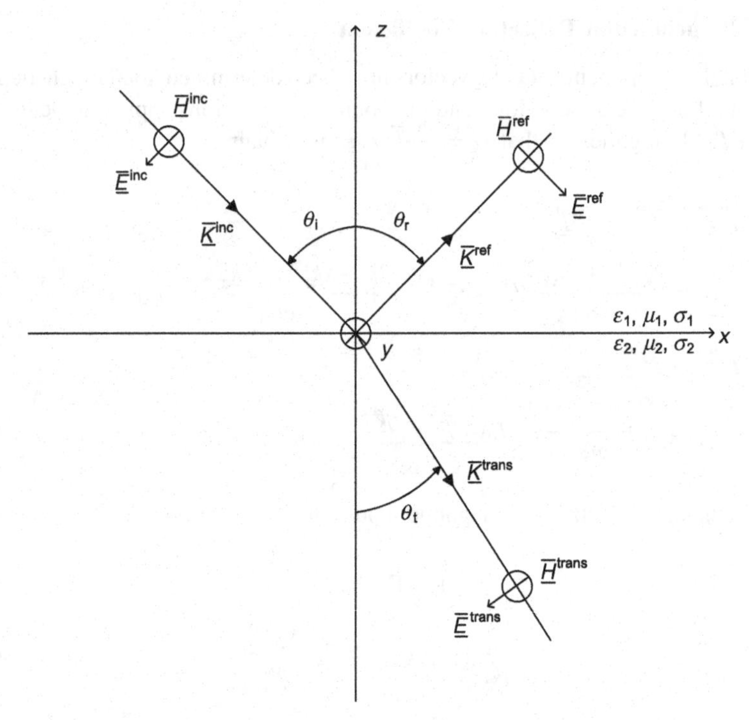

FIGURE 6.3 Geometry for oblique incidence reflection: parallel polarization.

regions 1 and 2 analogous to equations (6.20)–(6.21) can be written as

$$\begin{aligned}\underline{\overline{H}}_1^{\text{tot}} &= \underline{\overline{H}}_{\|}^{\text{inc}} + \underline{\overline{H}}_{\|}^{\text{ref}} \\ &= \hat{y}\frac{E_{0,\|}}{\underline{\eta}_1}e^{-j\underline{\overline{k}}^{\text{inc}}\cdot\overline{r}} + \hat{y}\frac{E_{0,\|}\underline{\Gamma}_{\|}}{\underline{\eta}_1}e^{-j\underline{\overline{k}}^{\text{ref}}\cdot\overline{r}},\end{aligned} \tag{6.37}$$

$$\begin{aligned}\underline{\overline{H}}_2^{\text{tot}} &= \underline{\overline{H}}_{\|}^{\text{trans}} \\ &= \hat{y}\frac{E_{0,\|}\underline{T}_{\|}}{\underline{\eta}_2}e^{-j\underline{\overline{k}}^{\text{trans}}\cdot\overline{r}}.\end{aligned} \tag{6.38}$$

Since the boundary conditions are the same as in the perpendicular polarization case, the phase matching condition still holds and the incident, reflected, and transmitted $\underline{\overline{k}}$ vectors are the same as those of the previous section, so that Snell's law (6.28) holds also for parallel polarization for lossless media. Electric fields for the parallel polarized case can be derived using $\underline{\overline{E}} = -\frac{1}{\omega\underline{\epsilon}}\underline{\overline{k}} \times \underline{\overline{H}}$ to be

$$\begin{aligned}\underline{\overline{E}}_1^{\text{tot}} &= \underline{\overline{E}}_{\|}^{\text{inc}} + \underline{\overline{E}}_{\|}^{\text{ref}} \\ &= \frac{-\hat{x}\underline{k}_z^{\text{inc}} - \hat{z}\underline{k}_x^{\text{inc}}}{\underline{k}_1}E_{0,\|}e^{-j\underline{\overline{k}}^{\text{inc}}\cdot\overline{r}} + \frac{\hat{x}\underline{k}_z^{\text{inc}} - \hat{z}\underline{k}_x^{\text{inc}}}{\underline{k}_1}\underline{\Gamma}_{\|}E_{0,\|}e^{-j\underline{\overline{k}}^{\text{ref}}\cdot\overline{r}},\end{aligned} \tag{6.39}$$

$$\overline{\underline{E}}_2^{\text{tot}} = \overline{\underline{E}}_{\|}^{\text{trans}}$$
$$= \frac{-\hat{x}\underline{k}_z^{\text{trans}} - \hat{z}\underline{k}_x^{\text{inc}}}{\underline{k}_2} E_{0,\|}\underline{T}_{\|} e^{-j\overline{\underline{k}}^{\text{trans}} \cdot \overline{r}}, \tag{6.40}$$

where $\underline{k}_{\text{i}} = \omega\sqrt{\mu_i \underline{\epsilon}_{\text{i}}^{\text{e}}}$.

6.3.6 Parallel Reflection Coefficient

Setting tangential electric and magnetic fields equal at $z = 0$ yields

$$1 + \underline{\Gamma}_{\|} = \underline{T}_{\|}\frac{\underline{\eta}_1}{\underline{\eta}_2}, \tag{6.41}$$

$$\frac{\underline{k}_z^{\text{inc}}}{\underline{k}_1}\left(1 - \underline{\Gamma}_{\|}\right) = \underline{T}_{\|}\frac{\underline{k}_z^{\text{trans}}}{\underline{k}_2}, \tag{6.42}$$

where only the x component of the electric field has been used since the z component is not tangential to the boundary. Solving these two equations yields

$$\underline{\Gamma}_{\|} = \frac{\underline{\epsilon}_2^{\text{e}}\underline{k}_z^{\text{inc}} - \underline{\epsilon}_1^{\text{e}}\underline{k}_z^{\text{trans}}}{\underline{\epsilon}_2^{\text{e}}\underline{k}_z^{\text{inc}} + \underline{\epsilon}_1^{\text{e}}\underline{k}_z^{\text{trans}}}, \tag{6.43}$$

$$\underline{T}_{\|} = \frac{2\underline{\epsilon}_2^{\text{e}}\underline{k}_z^{\text{inc}}\underline{\eta}_2/\underline{\eta}_1}{\underline{\epsilon}_2^{\text{e}}\underline{k}_z^{\text{inc}} + \underline{\epsilon}_1^{\text{e}}\underline{k}_z^{\text{trans}}}. \tag{6.44}$$

This concludes the solution for the parallel polarized case. Note that again the fraction of the power density reflected is $|\underline{\Gamma}_{\|}|^2$. While the use of a reflection coefficient for the magnetic field for parallel polarization is convenient in the derivation, a reflection coefficient for the electric field is often used in the literature and will be used in Chapters 7 and 9. From (6.39), the reflected electric field tangential and normal components are $-\underline{\Gamma}_{\|}$ and $\underline{\Gamma}_{\|}$, respectively, times the corresponding incident field components. The case of normal incidence can be treated as either parallel or perpendicular polarization, with the same result for the ratio of reflected to incident electric fields, even though $\underline{\Gamma}_{\|}$ and $\underline{\Gamma}_{\perp}$ are defined differently and differ by a minus sign in this limit.

6.3.7 Summary of Reflection Problem

The solutions we have obtained so far for the parallel and perpendicular polarization cases also enable the solution for any arbitrarily polarized incident field to be constructed. Returning to the example given in equations (6.9)–(6.14), we have

$$\overline{\underline{E}}^{\text{inc}} = 377(-\hat{x} + \hat{y} - \hat{z})\, e^{-jk_0(x-z)/\sqrt{2}} = \overline{\underline{E}}_{\perp}^{\text{inc}}\left(E_{0,\perp} = 377\right)$$
$$+ \overline{\underline{E}}_{\|}^{\text{inc}}\left(E_{0,\|} = 377\sqrt{2}\right), \tag{6.45}$$

illustrating the decomposition of an arbitrarily polarized wave into its parallel and perpendicular components along with the determination of the appropriate field component amplitudes (the E_0 values given in parenthesis) associated with a particular set of definitions in the problem solution. The reflected wave for this incident wave is then given by

$$\overline{\underline{E}}^{\mathrm{ref}} = \overline{\underline{E}}_{\perp}^{\mathrm{ref}}\left(E_{0,\perp} = 377\right) + \overline{\underline{E}}_{\parallel}^{\mathrm{ref}}\left(E_{0,\parallel} = 377\sqrt{2}\right), \tag{6.46}$$

where $\overline{\underline{E}}_{\perp}^{\mathrm{ref}}$ and $\overline{\underline{E}}_{\parallel}^{\mathrm{ref}}$ are as specified in the rightmost terms of equations (6.20) and (6.39), respectively. Note that the reflection coefficients for perpendicular and parallel polarizations are usually not equal. As a result, the original polarization of the incident wave can be modified upon reflection.

The reflection coefficients derived can also be simplified for application to cases of particular interest. Noting that magnetic media are uncommon in typical propagation problems, we set $\mu_2 = \mu_1$ to obtain

$$\underline{\Gamma}_{\perp} = \frac{\underline{k}_z^{\mathrm{inc}} - \underline{k}_z^{\mathrm{trans}}}{\underline{k}_z^{\mathrm{inc}} + \underline{k}_z^{\mathrm{trans}}}, \tag{6.47}$$

$$\underline{\Gamma}_{\parallel} = \frac{\underline{\epsilon}_2^{\mathrm{e}}\underline{k}_z^{\mathrm{inc}} - \underline{\epsilon}_1^{\mathrm{e}}\underline{k}_z^{\mathrm{trans}}}{\underline{\epsilon}_2^{\mathrm{e}}\underline{k}_z^{\mathrm{inc}} + \underline{\epsilon}_1^{\mathrm{e}}\underline{k}_z^{\mathrm{trans}}}. \tag{6.48}$$

Furthermore, using $\underline{k}_z^{\mathrm{inc}} = \omega\sqrt{\mu_1\underline{\epsilon}_1^{\mathrm{e}}}\cos\theta_{\mathrm{i}}$ and $\underline{k}_z^{\mathrm{trans}} = \sqrt{\omega^2\left(\mu_2\underline{\epsilon}_2^{\mathrm{e}} - \mu_1\underline{\epsilon}_1^{\mathrm{e}}\sin^2\theta_{\mathrm{i}}\right)}$ from equation (6.26), we obtain

$$\underline{\Gamma}_{\perp} = \frac{\cos\theta_{\mathrm{i}} - \sqrt{\frac{\underline{\epsilon}_2^{\mathrm{e}}}{\underline{\epsilon}_1^{\mathrm{e}}} - \sin^2\theta_{\mathrm{i}}}}{\cos\theta_{\mathrm{i}} + \sqrt{\frac{\underline{\epsilon}_2^{\mathrm{e}}}{\underline{\epsilon}_1^{\mathrm{e}}} - \sin^2\theta_{\mathrm{i}}}}, \tag{6.49}$$

$$\underline{\Gamma}_{\parallel} = \frac{\frac{\underline{\epsilon}_2^{\mathrm{e}}}{\underline{\epsilon}_1^{\mathrm{e}}}\cos\theta_{\mathrm{i}} - \sqrt{\frac{\underline{\epsilon}_2^{\mathrm{e}}}{\underline{\epsilon}_1^{\mathrm{e}}} - \sin^2\theta_{\mathrm{i}}}}{\frac{\underline{\epsilon}_2^{\mathrm{e}}}{\underline{\epsilon}_1^{\mathrm{e}}}\cos\theta_{\mathrm{i}} + \sqrt{\frac{\underline{\epsilon}_2^{\mathrm{e}}}{\underline{\epsilon}_1^{\mathrm{e}}} - \sin^2\theta_{\mathrm{i}}}}. \tag{6.50}$$

Magnitudes of $\underline{\Gamma}_{\perp}$ and $\underline{\Gamma}_{\parallel}$ are plotted versus θ_{i} for a boundary between free space and "wet ground" (Figure 2.5) in Figure 6.4 for three different frequencies. From these plots, it is apparent that $|\underline{\Gamma}_{\perp}| \geq |\underline{\Gamma}_{\parallel}|$ for all incidence angles when $|\underline{\epsilon}_2| > |\underline{\epsilon}_1|$, a fact that can be proven for all dielectric media from equations (6.49) and (6.50). It is also observed that the magnitude of both reflection coefficients approaches unity as θ_{i} approaches 90° (also known as "grazing incidence").

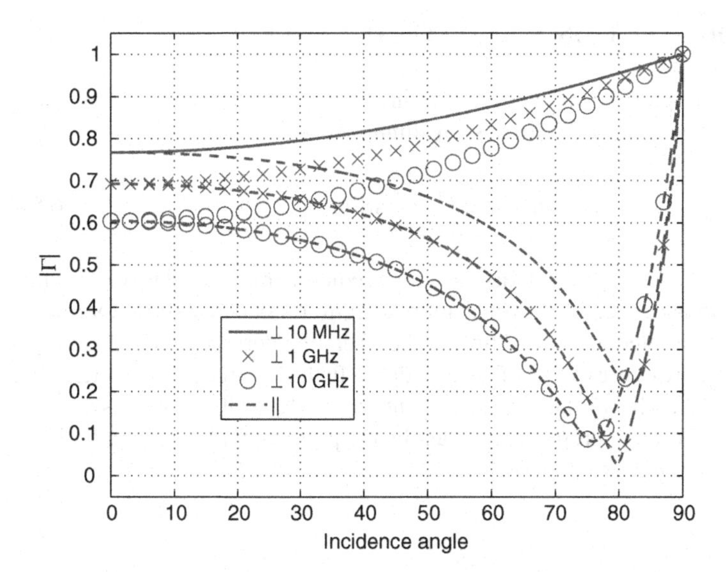

FIGURE 6.4 Typical reflectivities at a boundary between free space and "wet ground" versus incidence angle.

An examination of equation (6.49) shows that $\underline{\Gamma}_\perp = 0$ can be obtained only if $\underline{\epsilon}_2^e = \underline{\epsilon}_1^e$, in which case no reflection is expected. For $\underline{\Gamma}_\parallel = 0$, however, an equation

$$\frac{\underline{\epsilon}_2^e}{\underline{\epsilon}_1^e}\cos\theta_i = \sqrt{\frac{\underline{\epsilon}_2^e}{\underline{\epsilon}_1^e} - \sin^2\theta_i} \tag{6.51}$$

is obtained, which can be solved for the "Brewster angle," θ_B as

$$\theta_B = \tan^{-1}\sqrt{\frac{\underline{\epsilon}_2^e}{\underline{\epsilon}_1^e}}. \tag{6.52}$$

Thus, a parallel (or vertically) polarized wave will generate no reflected wave if it is incident at angle θ_B. Note that θ_B is a real angle only if $\underline{\epsilon}_2^e$ and $\underline{\epsilon}_1^e$ are purely real; that is, there is no loss in either region. However, a minimum reflection still exists in the lossy medium case near the "pseudo-Brewster angle" predicted by using $|\underline{\epsilon}_2^e/\underline{\epsilon}_1^e|$ in equation (6.52).

The phases of $\underline{\Gamma}_\perp$ and $\underline{\Gamma}_\parallel$ near grazing are also of importance. Equation (6.49) shows that $\underline{\Gamma}_\perp$ approaches -1 at grazing, indicating a zero total tangential electric field on the boundary. The analysis of the behavior of $\underline{\Gamma}_\parallel$ at grazing is not as immediately clear. The phase of $\underline{\Gamma}_\parallel$ near grazing depends on $\underline{\epsilon}_2^e/\underline{\epsilon}_1^e$ in (6.50), but it can be shown that $\underline{\Gamma}_\parallel$ also approaches -1 at near grazing angles. In fact, the phase of $\underline{\Gamma}_\parallel$ is positive for incidence angles less than the Brewster or pseudo-Brewster angle, but becomes negative once the incidence angle exceeds the Brewster angle.

6.4 TOTAL REFLECTION AND CRITICAL ANGLE

We now return to a closer examination of an interesting consequence of Snell's law, given by equation (6.28) for the lossless medium case as

$$\sin\theta_t = \sqrt{\frac{\mu_1\epsilon_1}{\mu_2\epsilon_2}}\sin\theta_i. \tag{6.53}$$

We have previously discerned that this equation predicts a transmitted wave traveling closer to the normal direction if propagation is from a less dense to a more dense medium, or a transmitted wave traveling further from the normal direction if propagation is from a more dense to a less dense medium. In the latter case some interesting phenomena can occur that reveal surprising solutions of Maxwell's equations that play an important role in wave propagation. Clearly, the problem with equation (6.53) in this case occurs when

$$\sqrt{\frac{\mu_1\epsilon_1}{\mu_2\epsilon_2}}\sin\theta_i > 1 \tag{6.54}$$

because then there is no real-valued solution for $\sin\theta_t$. The angle when this first occurs is defined as the "critical angle," θ_c, and is found by replacing the $>$ sign with equality in equation (6.54) to obtain

$$\theta_c = \sin^{-1}\sqrt{\frac{\mu_2\epsilon_2}{\mu_1\epsilon_1}}. \tag{6.55}$$

What happens in the reflection problem at or beyond the critical angle? It is easier to examine these effects by returning to the phase matching condition $k_x^{\text{trans}} = k_x^{\text{inc}} = \omega\sqrt{\mu_1\epsilon_1}\sin\theta_i$ and equation (6.26)

$$\begin{aligned}\left(\underline{k}_z^{\text{trans}}\right)^2 &= \omega^2\mu_2\epsilon_2 - \left(\underline{k}_x^{\text{inc}}\right)^2 \\ &= \omega^2\mu_2\epsilon_2\left(1 - \frac{\mu_1\epsilon_1}{\mu_2\epsilon_2}\sin^2\theta_i\right),\end{aligned} \tag{6.56}$$

which is seen from equation (6.54) to yield a negative value for $\left(\underline{k}_z^{\text{trans}}\right)^2$ when the critical angle is exceeded. This implies that $\underline{k}_z^{\text{trans}}$ becomes imaginary in the transmitted medium, so that instead of propagating away from the boundary, the field decays exponentially. Note that we are still considering lossless media only, so that the resulting decay is not a consequence of a medium conductivity. Rather, this is another type of solution to Maxwell's equations, a wave that decays in a lossless medium. This type of solution is called an "evanescent wave."

The reflection coefficients given by equations (6.35)

$$\underline{\Gamma}_\perp = \frac{\mu_2 k_z^{\text{inc}} - \mu_1\underline{k}_z^{\text{trans}}}{\mu_2 k_z^{\text{inc}} + \mu_1\underline{k}_z^{\text{trans}}} \tag{6.57}$$

and (6.43)

$$\underline{\Gamma}_{\parallel} = \frac{\epsilon_2 k_z^{\text{inc}} - \epsilon_1 \underline{k}_z^{\text{trans}}}{\epsilon_2 k_z^{\text{inc}} + \epsilon_1 \underline{k}_z^{\text{trans}}} \tag{6.58}$$

can also be simplified in this case because all elements in these equations are assumed to be real except for $\underline{k}_z^{\text{trans}}$, which is purely imaginary when the critical angle is exceeded. Both of these reflection coefficients can then be written in the form

$$\underline{\Gamma} = \frac{A - jB}{A + jB} = \frac{Ce^{j\phi}}{Ce^{-j\phi}}, \tag{6.59}$$

for which

$$|\underline{\Gamma}| = 1, \tag{6.60}$$

which means that the wave is totally reflected. This phenomenon is known as "total reflection," indicating that a wave incident from a more dense medium to a less dense medium beyond the critical angle will be completely reflected. A phase shift of $e^{j2\phi}$, known as the Goos–Hanchen phase shift, does result, however.

Let us now consider the form of the electric and magnetic fields in the transmitted medium. Defining $\underline{k}_z^{\text{trans}} = -j\alpha$, where α is a positive real number, to account for the imaginary component, transmitted fields for the perpendicular polarized case are

$$\overline{\underline{E}}_2^{\text{tot}} = \hat{y} E_{0,\perp} \underline{T}_{\perp} e^{-j\underline{k}_x^{\text{inc}} x + \alpha z}, \tag{6.61}$$

$$\overline{\underline{H}}_2^{\text{tot}} = \frac{-j\hat{x}\alpha + \hat{z} k_x^{\text{inc}}}{\omega \mu_2} E_{0,\perp} \underline{T}_{\perp} e^{-j\underline{k}_x^{\text{inc}} x + \alpha z}. \tag{6.62}$$

Note that decay of the fields away from the boundary is expressed in these equations since in region 2, z becomes more negative as one moves away from the boundary. This type of evanescent wave is also referred to as a "surface wave" because of such properties; that is, the field is strongest near the surface and decreases as one moves away from the surface. An examination of equation (6.56) shows that we could also have chosen $k_z^{\text{trans}} = j\alpha$, where α is a positive real, since only $\left(k_z^{\text{trans}}\right)^2$ is specified in equation (6.56). However, this choice would result in an increasing field amplitude as one moves away from the boundary that grows without bound since the transmitted medium is unbounded in the $-z$ direction. Because this would require infinite energy, this choice is discarded as being nonphysical.

The fact that the incident wave has undergone "total reflection" while a transmitted field still remains may cause some confusion. However, if the Poynting vector of the transmitted wave is calculated (peak value phasors), we obtain

$$\frac{1}{2}\text{Re}\left[\overline{\underline{E}}_2^{\text{tot}} \times \left(\overline{\underline{H}}_2^{\text{tot}}\right)^*\right] = \hat{x} \frac{k_x^{\text{inc}} |E_{0,\perp} \underline{T}_{\perp}|^2}{2\omega\mu_2} e^{2\alpha z}, \tag{6.63}$$

showing that the surface wave carries power *along* the surface but that there is no power flow perpendicular to the surface in the steady state. Thus, an observer moving far away from the boundary in the (lossless) transmitted medium would measure no power at all, while an observer in the reflected medium would observe total power reflection.

Of course, the creation of the evanescent field in the transmitted medium would originally require some power flow across the boundary in a transient state, but once the steady state is reached, the transmitted field is solely an energy storing field.

The surface wave in this reflection problem is of importance, but is not quite the same as the "surface wave" or "ground wave" typically discussed in propagation problems. However, the ground wave considered in chapter 9 has properties similar to these evanescent waves. Hence, from this point forward, we should be aware that, in addition to the standard propagating wave solutions of Maxwell's equations, evanescent wave solutions are possible as well, with the sole requirement that

$$\underline{\overline{k}} \cdot \underline{\overline{k}} = \omega^2 \mu \underline{\epsilon}^{\text{e}}, \tag{6.64}$$

and no restriction otherwise on the complex components of the $\underline{\overline{k}}$ vector.

6.5 REFRACTION IN A STRATIFIED MEDIUM

Now that we have derived Snell's law for a single planar interface, we can consider refraction effects in a medium consisting of many planar interfaces, each individual interface with only slightly different dielectric constants so that reflections may be considered negligible. The basic configuration is shown in Figure 6.5 and consists of a series of nonmagnetic media having refractive indices (defined below equation (6.30)) $n_0, n_1, \ldots$. A plane wave is incident on the boundary between layers 0 and 1 with incidence angle θ_0. Note that since we are neglecting reflections in this problem and considering only refraction, the polarization of the incident wave is not of importance. Snell's law as stated in equation (6.30) at the boundary between layers 0 and

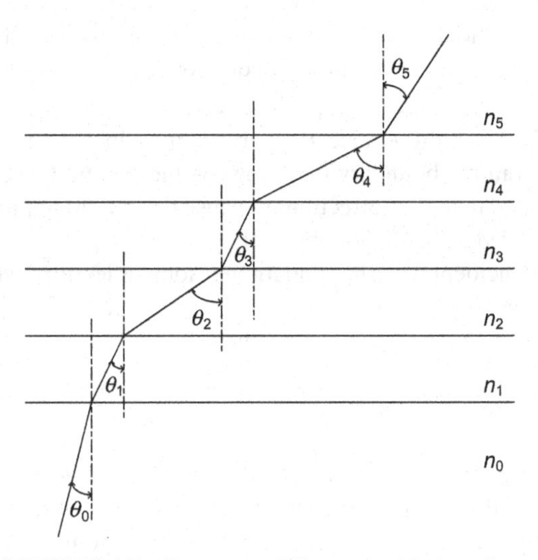

FIGURE 6.5 Geometry of planar stratified medium.

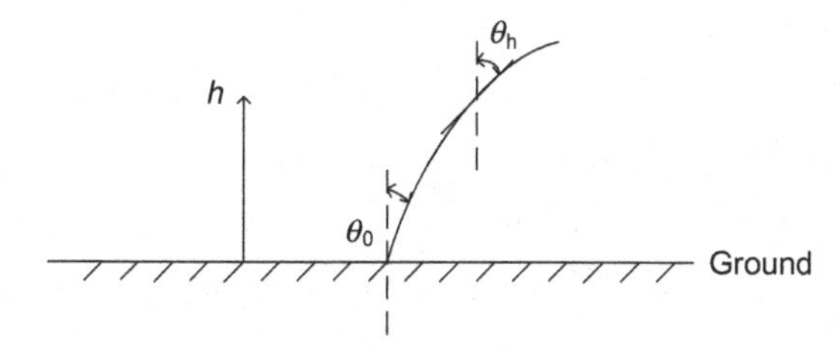

FIGURE 6.6 Variation in propagation angle with height.

1 requires

$$n_0 \sin \theta_0 = n_1 \sin \theta_1. \tag{6.65}$$

In a similar manner, we can obtain at the boundary between layers 1 and 2

$$n_1 \sin \theta_1 = n_2 \sin \theta_2. \tag{6.66}$$

Combining (6.65) and (6.66), we obtain

$$n_0 \sin \theta_0 = n_2 \sin \theta_2, \tag{6.67}$$

and by continuing the same procedure, we can obtain for region r

$$n_r \sin \theta_r = n_0 \sin \theta_0. \tag{6.68}$$

A continuously horizontally stratified medium can be thought of as consisting of the limit of a discretely stratified medium as the layers become thinner and increase indefinitely in number, as shown in Figure 6.6. For such a medium, n becomes a continuous function of height, and therefore at any height h, the tangent to the ray of the propagating wave will have a direction given by

$$n(h) \sin \theta(h) = n_0 \sin \theta_0. \tag{6.69}$$

Our neglect of reflections requires that $n(h)$ should vary slowly with respect to the electromagnetic wavelength in order to make reflections negligible.

6.6 REFRACTION OVER A SPHERICAL EARTH

The previous section considered refraction in a stratified, planar medium. However, propagation over the surface of the Earth requires consideration of the Earth's spherical shape: the Earth's atmosphere is stratified in spherical, rather than planar layers, as illustrated in Figure 6.7.

To derive the refraction properties of such a medium, we return again initially to a discretely stratified medium before considering the continuous case. For the discrete

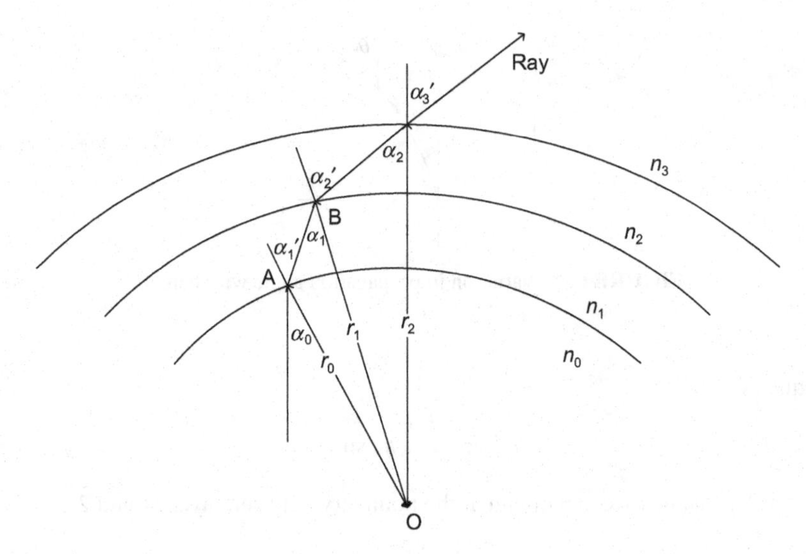

FIGURE 6.7 Geometry of spherically stratified medium.

medium,

$$n(r) = \begin{cases} n_0 \text{ for } & r < r_0, \\ n_1 \text{ for } & r_0 < r < r_1, \\ \ldots\ldots & \ldots \\ n_m \text{ for } & r_{m-1} < r < r_m. \end{cases} \tag{6.70}$$

From Figure 6.7, applying Snell's law at A results in

$$n_0 \sin\alpha_0 = n_1 \sin\alpha_1'. \tag{6.71}$$

Figure 6.8 shows the region in the vicinity of AB of Figure 6.7 in more detail; also, a rectangle BDAC has been constructed using the fact that α_1 is the angle with respect to the normal direction defined at point B. We find

$$\sin\alpha_1 = \frac{\text{AD}}{\text{AB}} = \frac{\text{AO}\sin\theta}{\text{AB}} = \frac{r_0 \sin\theta}{\text{AB}}, \tag{6.72}$$

where r_0 is the radius of the Earth. Similarly, a rectangle AEBF can be defined as in Figure 6.9 using the fact that α_1' is the angle with respect to the normal direction defined at point A. Using this triangle, we find

$$\sin\alpha_1' = \frac{\text{EB}}{\text{AB}} = \frac{\text{BO}\sin\theta}{\text{AB}} = \frac{r_1 \sin\theta}{\text{AB}}. \tag{6.73}$$

Therefore,

$$\sin\alpha_1' = \frac{r_1}{r_0}\sin\alpha_1, \tag{6.74}$$

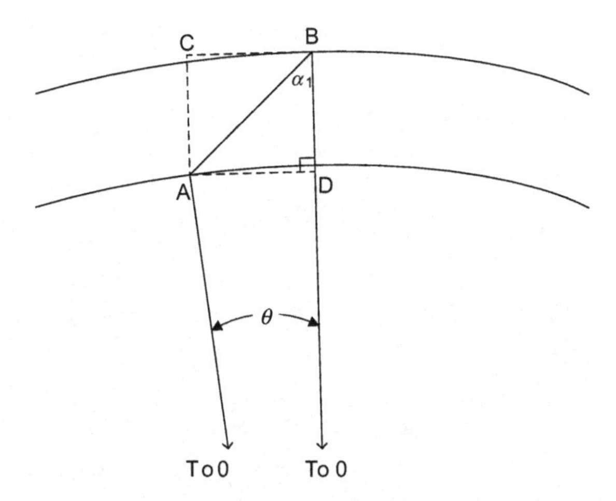

FIGURE 6.8 Detail of rectangle ABCD.

and substitution of this into (6.71) yields

$$n_0 r_0 \sin \alpha_0 = n_1 r_1 \sin \alpha_1, \tag{6.75}$$

which is Snell's law for a spherically stratified medium. In the same way, one can show

$$n_1 r_1 \sin \alpha_1 = n_2 r_2 \sin \alpha_2 = n_3 r_3 \sin \alpha_3 = \cdots \tag{6.76}$$

so that at all boundaries in the medium

$$n_0 r_0 \sin \alpha_0 = n_r r_r \sin \alpha_r \tag{6.77}$$

is satisfied.

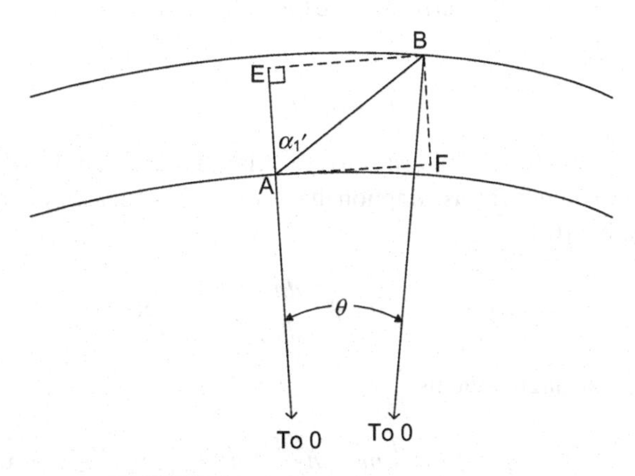

FIGURE 6.9 Detail of rectangle AEBF.

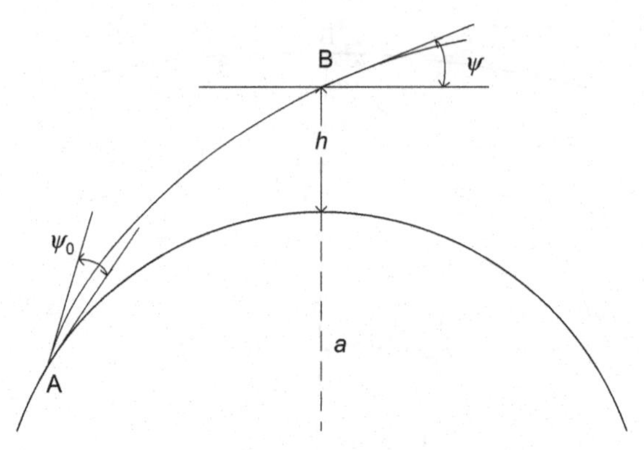

FIGURE 6.10 Definition of ψ.

Media that vary continuously are more characteristic of nature than discretely layered media; again these can be viewed as the limiting case as the number of layers increases and the radial distances between layers decrease to zero. In this case, the limiting form of equation (6.77) becomes

$$n_0 r_0 \sin \alpha_0 = n(r)\, r \sin \alpha. \tag{6.78}$$

With some exceptions, such as in communication links between aircraft and satellites or those involving ionosphere reflections, propagation paths generally occur relatively close to the Earth surface, and the angles α and α_0 approach $\pi/2$ rad. It is then more convenient to use the complementary angles ψ and ψ_0 (i.e., the "elevation angles") shown in Figure 6.10 because they are small so that small-angle approximations can be used. Equation (6.78) then transforms to

$$n_0 r_0 \cos \psi_0 = n(r)\, r \cos \psi. \tag{6.79}$$

Let the subscript 0 refer to the surface of the Earth. Then $r_0 = a = 6370$ km, the Earth radius, and r is given by

$$r = a + h, \tag{6.80}$$

where h is the height above the Earth surface. Also, assume that the refractive index varies linearly with height, an assumption that is often (but not always) approximately true. Then equation (6.79) becomes

$$n_0 a \cos \psi_0 = \left(n_0 + h \frac{dn}{dh} \right) (a + h) \cos \psi. \tag{6.81}$$

Expanding and rearranging yields

$$n_0 a \left(\cos \psi_0 - \cos \psi \right) = \left(n_0 + a \frac{dn}{dh} \right) h \cos \psi + h^2 \frac{dn}{dh} \cos \psi. \tag{6.82}$$

Since the Earth radius is large compared to the vertical extent of a tropospheric path, $h \ll a$, and since dn/dh is expected to be small, the last term on the right-hand side of equation (6.82) can be neglected. For the remaining terms on the right, $\cos\psi \simeq 1$. On the left-hand side, this approximation is too crude since we are subtracting two terms, but $\cos x \simeq 1 - \frac{x^2}{2}$ can be used to give

$$\psi^2 = \psi_0^2 + \frac{2h}{a}\left[1 + \frac{a}{n_0}\frac{dn}{dh}\right]. \tag{6.83}$$

Referring to Figure 6.10, we see that there are two mechanisms for a change in ψ as h is varied. One is the fact that ψ is measured with respect to the horizontal direction that changes as the field propagates around the Earth; the other is the ray curvature due to the nonuniformity of n. The first effect corresponds to the 1 in the bracket of equation (6.83) and the second to the last term in the bracket.

Clearly, the validity of equation (6.83) is unchanged if we write

$$\psi^2 = \psi_0^2 + \frac{2h}{\kappa a}, \tag{6.84}$$

with

$$\kappa = \frac{1}{1 + \frac{a}{n_0}\frac{dn}{dh}}. \tag{6.85}$$

However, if we interpret equation (6.84) as we did in equation (6.83), it appears the same as if a were multiplied by κ and dn/dh were zero; in other words, as if the Earth radius were κa instead of a and rays were straight. This situation is shown in Figure 6.11a. Equations (6.83)–(6.85) give assurance that the dependence of ψ on h is the same for Figures 6.10 and 6.11. Moreover, since the path length along the ray can be written as

$$S = \int dS = \int \frac{dh}{\sin\psi}, \tag{6.86}$$

the path length is also preserved between Figures 6.10 and 6.11a, and these two figures are equivalent representations of the same situation. The quantity κa is often called the *effective Earth radius*, meaning the radius that would produce the same geometrical relationship between the ray and the Earth surface without refractive index variation. In other words, the vertical distance between the Earth and propagating ray is the same at every point on the path for both the real Earth radius-curved ray and the effective Earth radius-straight ray models.

The concept of Figure 6.11a is often useful in theoretical developments, but it is even more useful in graphical constructions for computing the performance of tropospheric paths. Typical values of dn/dh are such that a value of $\kappa = 4/3$ results for "average" conditions.

A second rewriting of equation (6.83) as

$$\psi^2 = \psi_0^2 + \frac{2h}{a}\frac{a}{n_0}\left(\frac{dn}{dh} + \frac{n_0}{a}\right)$$

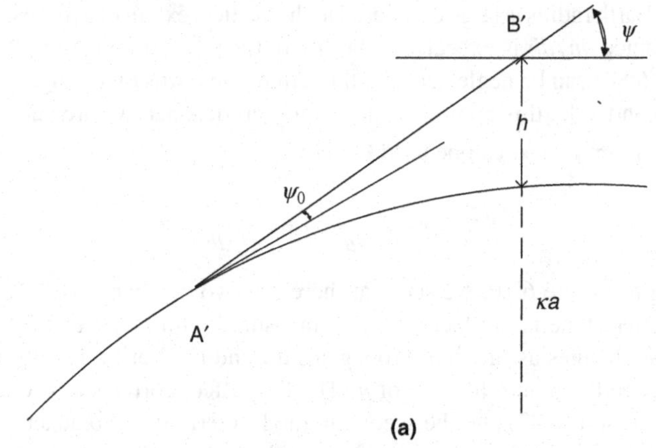

(a)

Straight ray, equivalent Earth radius (κa) model

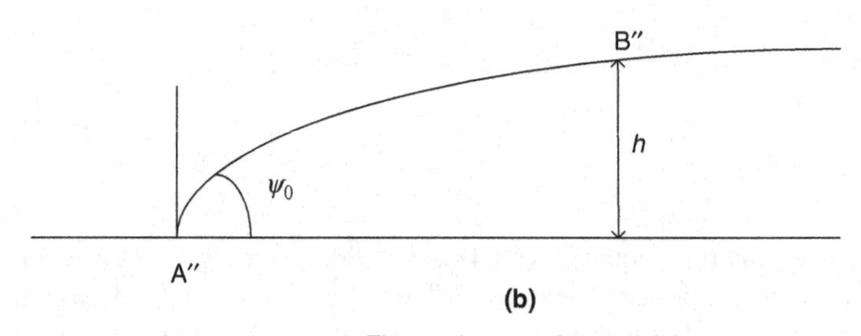

(b)

Flat earth, curved ray model

FIGURE 6.11 Equivalent models for refraction.

$$= \psi_0^2 + \frac{2h}{n_0}\left(\frac{dm}{dh}\right), \tag{6.87}$$

where

$$\frac{dm}{dh} = \frac{dn}{dh} + \frac{n_0}{a} \tag{6.88}$$

leads to a flat Earth-effective ray curvature model, as shown in Figure 6.11b. In this case, the refractivity gradient of the atmosphere is increased by n_0/a (or, by integration, a modified index of refraction m replaces the true index of refraction n) leading to modified ray curvatures above a flat Earth surface that again provide the same height of rays above the Earth surface for a specified location as the real Earth radius and real ray curvature model (the true situation) and the effective Earth radius and straight ray model. The choice of which of these models to use is largely a matter of convenience, so long as the implicit assumptions are satisfied. Again, the assumptions implied in Figure 6.11 are (i) nearly horizontal paths and (ii) n versus h is a linear function over the heights of interest.

6.7 REFRACTION IN THE EARTH'S ATMOSPHERE

The preceding section showed that the effects of a linear change in refractive index with height can be modeled through use of a modified Earth radius κa, where κ is given by

$$\kappa = \frac{1}{1 + \frac{a}{n_0}\frac{dn}{dh}} \tag{6.89}$$

and assuming that the rays travel along straight paths.

A typical value for n at the Earth's surface is $n_0 = 1.000350$, and n usually becomes even closer to unity as the height is increased. The very small deviations in this value from unity make it more convenient to describe the atmospheric refractive index in terms of "N-units," defined as

$$N = (n - 1) \times 10^6, \tag{6.90}$$

so that, for example, $n = 1.000350$ can be represented as 350 N-units. From equation (6.90) follows

$$\frac{dN}{dh} = \frac{dn}{dh} \times 10^6. \tag{6.91}$$

Under standard atmospheric conditions, $dN/dh = -39$ N-units/km, yielding a κ value of 1.33. However, dN/dh can range from much more negative to positive values depending on atmospheric conditions. Figure 6.12 plots κ versus dN/dh, showing the effects of varying refractivity gradients on the effective Earth curvature. For positive

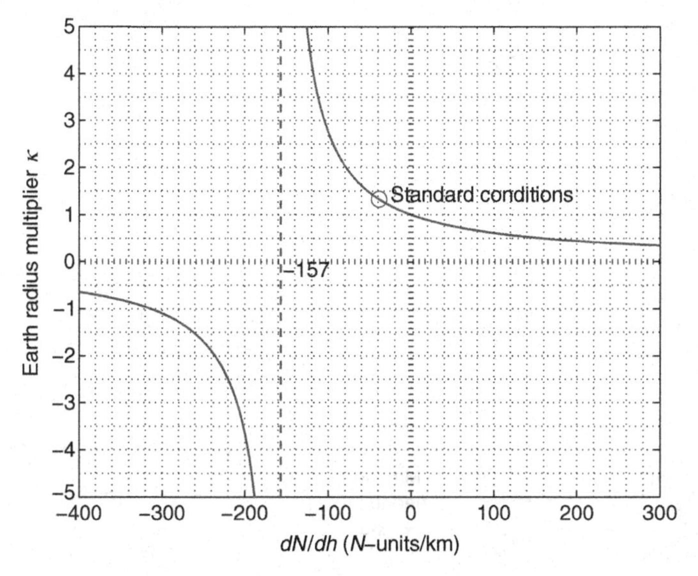

FIGURE 6.12 Variation in κ with dN/dh.

values of dN/dh, known as *subrefractive* conditions, the Earth appears to have a smaller radius (κ values between 0 and 1) with the straight ray model because rays are being bent upward by the atmosphere, causing the ray to move further away from the horizontal than that caused by Earth curvature alone. The $\kappa = 1$ point corresponds to an atmosphere with a constant refractivity profile, so that the effective Earth radius is equal to the true Earth radius a.

For increasingly negative values of dN/dh, the effective Earth radius increases until dN/dh is -157 N-units/km, at which point κ becomes infinite. These effects are due to the fact that rays are being bent downward by the atmosphere, with -157 N-units/km the point at which refractive curvature matches that of the Earth surface, so that rays remain at the same altitude above the surface with no apparent bending, resulting in $\kappa = \infty$. Typical classifications label 0 to -79 N-units/km the *normal region*, with κ values ranging from 1 to 2, while atmospheres with -79 to -157 N-units/km are known as *superrefractive*, with κ values ranging from 2 to ∞. Gradients more negative than -157 N-units/km, known as *trapping gradients*, cause negative values of κ, indicating that rays are bent toward the surface of the Earth more rapidly than the Earth curvature, causing rays to return to the surface, as modeled by a negative Earth radius. Ducting phenomena, which are the result of trapping gradients, are discussed in more detail in the next section. Effects of varying atmospheric conditions are illustrated in Figure 6.13, which plots both the actual curved ray paths using the true Earth radius a and the effective Earth radius model using straight ray paths.

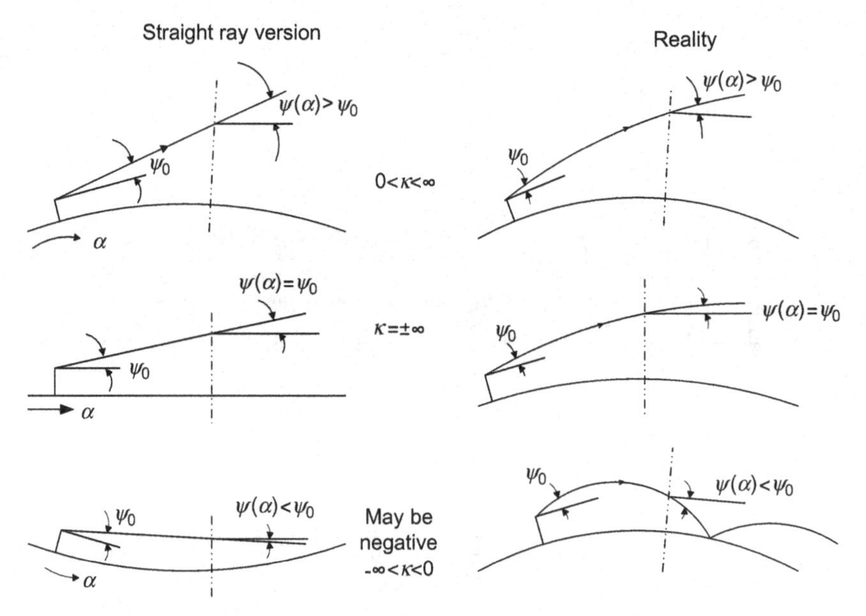

FIGURE 6.13 Use of effective Earth radius model.

In designing communication systems, it is important to be aware of the likely variations in dN/dh for a given location, since changes in refractive index gradient can lead to greatly increased or decreased propagation distances. A database of refractive index information has been compiled by the ITU-R (as described in ITU-R Recommendation P.453 [4]) for sites over the globe, so that this database can be used if no local information can be obtained. Communication systems are often designed with the "worst-case" scenario in mind although typical κ values considered for land paths are usually only $2/3, 4/3$, and ∞. A wider range of refractivity gradients must be considered over sea paths, for which ducting is a much more common phenomenon, as discussed in the next section.

6.8 DUCTING

Trapping gradients are refractive conditions for which $dN/dh < -157$ N-units/km. In this case, κ becomes negative indicating that rays will return to the surface of the Earth. From the definition of κ in equation (6.89)

$$\kappa = \frac{1}{1 + \frac{a}{n_0}\frac{dn}{dh}}, \tag{6.92}$$

it is seen that the -157 N-units/km condition results from

$$\frac{a}{n_0}\frac{dn}{dh} = -1, \tag{6.93}$$

$$\frac{dn}{dh} = -\frac{n_0}{a}, \tag{6.94}$$

$$\frac{dm}{dh} = 0, \tag{6.95}$$

where the last equation uses the "modified refractivity" gradient from equation (6.88). The detection of trapping gradients is simpler by examining gradients of the modified refractivity $m(h)$ as opposed to the true index of refraction $n(h)$. Recalling equations (6.87) and (6.88) therefore leads to the definition of a *modified refractive index*, m, designed to identify trapping gradient situations more clearly as

$$m(h) = n(h) + \frac{h}{a}, \tag{6.96}$$

where the n_0 term has been approximated as unity. Analogous to the N-units, M-units can be defined as

$$M = (m - 1) \times 10^6. \tag{6.97}$$

Equations (6.95) and (6.97) show that ducting becomes possible when

$$\frac{dM}{dh} < 0, \tag{6.98}$$

so that plots of M versus altitude are more useful for identifying trapping gradient regions than plots of N.

Trapping gradients are unlikely to occur throughout the entire height of the atmosphere. They are usually localized between heights h_1 and h_2, with more standard refractive index variations present above and below the trapping gradient. A trapping gradient localized in altitude and that exists over a large horizontal extent is known as a duct. Rays inside the duct are curved back toward the surface of the Earth. For some duct and propagation geometries, rays will return to the Earth surface and be specularly reflected if the Earth surface is sufficiently smooth (as described in Chapter 7). Specularly reflected rays then return to the trapping gradient to be bent again to the Earth surface. In this manner, the duct–Earth surface boundaries can act as boundaries of a waveguide that can allow more efficient propagation than in normal conditions.

Ducts are usually classified into three types: elevated, surface-based, and evaporation ducts, as illustrated in Figure 6.14. Elevated ducts are trapping gradients that are sufficiently high in altitude so that no rays from a surface source can be trapped. These ducts however can trap rays from an elevated source and thus can be important for airborne propagation. For sources above the duct, there are sometimes combinations of ranges and altitudes for which no rays can penetrate, known as radar or radio "holes," as can be observed with the ray-tracing methods described in the next section. Note that the waveguide interpretation of ducting effects indicates that there should be a lower cutoff frequency, but this frequency is usually far below standard radar frequencies, so for radar problems elevated duct effects are usually considered frequency independent. Trapping gradients within elevated ducts are usually caused by a rapid transition between a cool, moist air mass (greater n) below and a warm, dry air mass (smaller n) above. Elevated ducts can occur more than 50% of the time in many areas of the world at altitudes from near 0 to several km. However, by definition they have little effect on surface-based propagation.

Surface-based ducts are ducts that extend down to the Earth surface and are usually formed by the same meteorological mechanisms as elevated ducts. Surface-based ducts can greatly extend propagation ranges for surface-based systems and are also usually assumed to be frequency independent for VHF or higher frequencies. Although use of surface-based ducts in communication systems may seem to be ideal for improving system communication ranges, the general unpredictability of the existence of ducts makes them more often a source of interference and radar clutter than a viable component of systems. Surface-based ducts occur annually about 8% of the time worldwide, ranging from about 1% of the time in the North Atlantic to about 46% in the Persian Gulf. The worldwide average thickness of such ducts is 85 m, but thicknesses up to a few hundred meters are common. Ray-tracing methods with surface-based ducts can be used to determine propagation effects and, in particular, reveal a "skip" zone, where the duct has no influence, near the normal horizon.

Evaporation ducts are created by the extremely rapid decrease in moisture immediately adjacent to the sea surface. For obvious reasons, the water vapor content of air adjacent to the sea surface is very high (approaching 100% relative humidity)

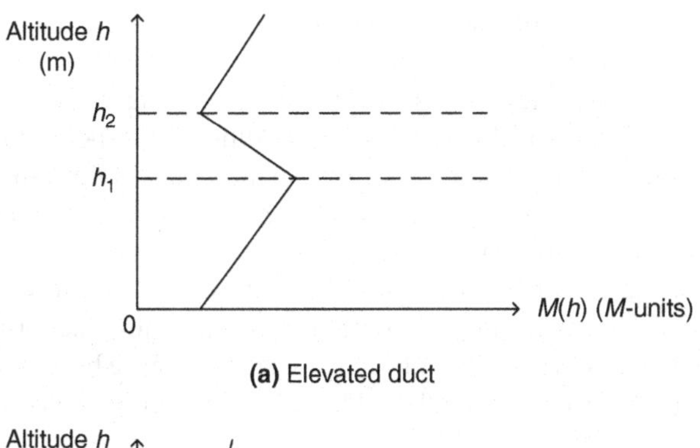

(a) Elevated duct

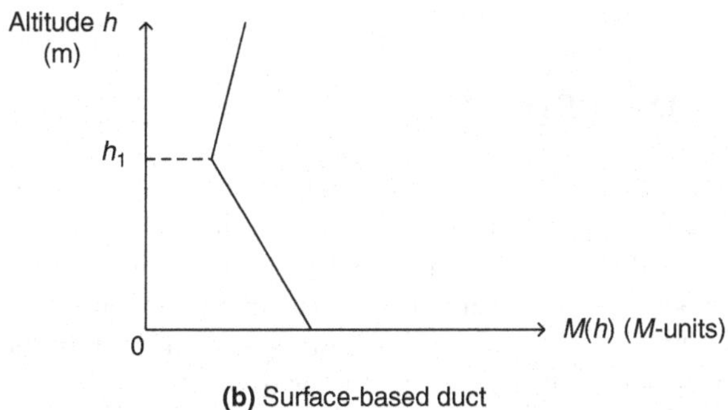

(b) Surface-based duct

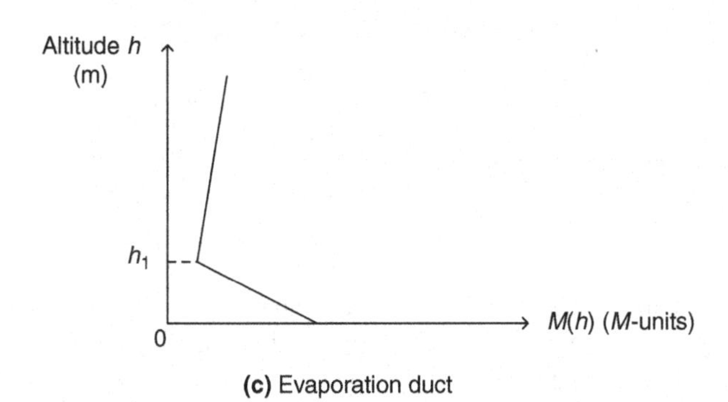

(c) Evaporation duct

FIGURE 6.14 Types of ducts.

and decreases rapidly with altitude to an ambient value that depends on the local meteorological conditions. The rapid decrease in moisture content near the sea surface can create a trapping gradient that gradually weakens with increasing height until a height is reached where $dM/dh = 0$, known as the evaporation duct height. Evaporation duct heights vary between 0 and 40 m, with an average height of 13 m worldwide. In contrast to surface-based and elevated ducts, evaporation duct effects vary strongly with frequency, with frequencies below 2 GHz only slightly affected. For frequencies above 2 GHz and antenna heights near the sea surface, evaporation ducts can provide substantially enhanced signal levels at ranges well beyond the horizon.

For short-range, near-horizon paths, the evaporation duct can be the dominant propagation mechanism over the sea surface.

Information on the long-term statistical occurrence of surface-based ducts and their average thickness, as well as the frequency distribution of evaporation duct heights, is available for many areas of the world. Information on specified surface-based or elevated ducts can be obtained by measuring their refractivity profiles at a given time and location with balloon-borne radiosondes or other airborne instruments. These techniques are not appropriate for evaporation ducts, due to their lower altitudes and more rapid variations, but techniques for relating evaporation duct properties to sea surface air–water temperatures, humidity, and wind speed have been developed. ITU-R Recommendation P.834 [5] provides additional information on ducting effects.

6.9 RAY-TRACING METHODS

Although we have seen that propagation through an atmosphere with a linear refractive index profile can be analyzed through an effective Earth radius or ray curvature model, such simple models do not apply to more general refractivity profiles. In that case, *ray-tracing* methods provide a useful tool for obtaining some insight into the effects of various ducts and refractivity profiles on propagation through the atmosphere. Beginning with the spherical form of Snell's law (6.81)

$$n_0 a \cos\psi_0 = n(h)\,(a+h)\cos\psi, \tag{6.99}$$

we can substitute $n_0 = m_0$ and $n(h) = m(h) - h/a$ to obtain

$$\begin{aligned} m_0 \cos\psi_0 &= (m(h) - h/a)\,(1 + h/a)\cos\psi \\ &= \left[m(h) + \frac{h}{a}\,(m(h) - 1) - \frac{h^2}{a^2}\right]\cos\psi. \end{aligned} \tag{6.100}$$

The first term on the right-hand side dominates, and we can make small-angle approximations for the cosines as well. Upon doing so and rearranging terms as in equations (6.82) and (6.83), we obtain

$$\psi^2 = \psi_0^2 \frac{m_0}{m(h)} + 2\frac{m(h) - m_0}{m(h)}. \tag{6.101}$$

Using $m_0/m(h) \simeq 1$ in the first term (a reasonable approximation since it is assumed that ψ_0 is also small) and $m(h) \simeq 1$ in the denominator of the second term (reasonable since the numerator is small) yields

$$\psi^2 = \psi_0^2 + 2\,(m(h) - m(0))\,, \tag{6.102}$$

where the ray tracing is assumed to start at $h = 0$ with angle ψ_0. Note that equation (6.102) reduces to equation (6.87) for a linear modified refractive index profile.

Consider a ray path $h = f(x)$, where x is the horizontal coordinate along the Earth surface and h is the ray height above that surface. Recognizing that ψ determines the slope of the tangent to $f(x)$, we obtain

$$\frac{df}{dx} = \tan\psi \simeq \psi, \tag{6.103}$$

so that a differential equation for $f(x)$ can be obtained from equation (6.102) as

$$\left(\frac{df}{dx}\right)^2 = \psi_0^2 + 2\left(m(f) - m(0)\right), \tag{6.104}$$

which is a nonlinear equation for f if $m(f)$ is not a constant. A closed-form solution of equation (6.104) would be difficult to obtain and obviously depends on the particular form chosen for $m(h)$. However, a discretized version of (6.104) leads to a straightforward procedure known as "ray tracing" for finding the ray path $f(x)$ and is easily implemented on a computer.

The method is based on approximating the derivative by finite differences

$$\frac{df}{dx}\Big|_{x=x_i} \simeq \frac{f(x_{i+1}) - f(x_i)}{\Delta x}, \tag{6.105}$$

where the horizontal coordinate x is discretized according to

$$x_i = i\Delta x. \tag{6.106}$$

Use of this approximation in equation (6.104) at $x = x_i$ results in

$$\frac{f(x_{i+1}) - f(x_i)}{\Delta x} = \sqrt{\psi_0^2 + 2m(f(x_i)) - 2m(0)}, \tag{6.107}$$

$$f(x_{i+1}) = f(x_i) + \Delta x\sqrt{\psi_0^2 + 2m(f(x_i)) - 2m(0)}. \tag{6.108}$$

Thus, if $f(x_0)$ is set to be the height of a transmitter above the Earth surface, then $f(x_1)$ can be found from equation (6.108) and the process repeated to determine all values of $f(x_i)$. Connection of these points $(x_i, f(x_i))$ with straight lines leads to an approximation for the true curved ray path, with the accuracy increasing as Δx becomes smaller. In some ducting situations there may be problems with equation (6.108) since the term inside the square root may become negative. This situation indicates a total internal reflection of the ray since it is propagating from a more dense to a less dense medium. In this case, equation (6.108) should be modified to obtain

$$f(x_{i+1}) = f(x_i) - \Delta x\sqrt{\psi_0^2 + 2m(f(x_i)) - 2m(0)}. \tag{6.109}$$

This equation should be used until the ray path reaches the ground, upon which equation (6.108) should again be used. Care is required in treating the points where directions are reversed to ensure consistent calculations.

Ray paths computed by ray tracing for a quadratic refractivity profile having a trapping gradient up to 500 m height are illustrated in Figure 6.15. Note the importance of the transmission angle ψ_0 on the effectiveness of coupling to a duct.

The method presented here provides a basic description of a ray-tracing procedure in complex atmospheres, but several more accurate procedures that avoid some of the

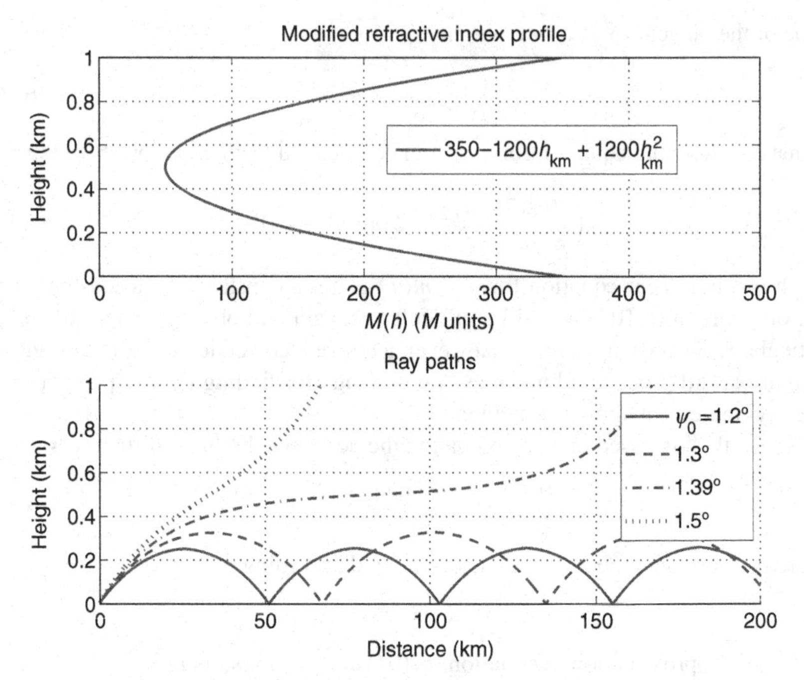

FIGURE 6.15 Ray paths obtained by ray-tracing procedure.

approximations introduced here are also available—see, for example, the "Advanced Refractive Effects Prediction System" (AREPS) [6] produced by the Atmospheric Propagation Branch of the Space and Naval Warfare Systems Command (SPAWAR).

REFERENCES

1. Budden, K. G., *The Propagation of Radio Waves*, Cambridge University Press, 1985.
2. Chew, W. C., *Waves and Fields in Inhomogeneous Media*, IEEE Press, 1999.
3. Kong, J. A., *Electromagnetic Wave Theory*, EMW Publishing, 2008.
4. ITU-R Recommendation P.453-9, "The radio refractive index: its formula and refractivity data," International Telecommunication Union, 2003.
5. ITU-R Recommendation P.834-6, "Effects of tropospheric refraction on radiowave propagation," International Telecommunication Union, 2007.
6. Patterson, W. L., "Advanced Refractive Effects Prediction System," Proc. IEEE Radar Conf., pp. 891–895, 2007.

7

TERRAIN REFLECTION AND DIFFRACTION

7.1 INTRODUCTION

In many applications, line-of-sight communications are influenced by reflections from the surface of the Earth. Transmitting and receiving paths near the surface of the Earth will experience reflections from the ground if sufficient antenna directivity is not available to avoid ground illumination. Examples include VHF and UHF broadcast systems where receivers (such as radio and television antennas) are located near the ground surface, preventing its influence from being removed even with large transmitting antenna heights.

In this chapter, the theory of direct plus reflected wave propagation is presented, along with a graphical method for predicting propagation effects in the presence of irregular terrain. This chapter provides a first discussion of "multipath" interference for the simple case of a single multipath (reflection from the ground surface). Multipath propagation effects are further examined from an empirical and statistical point of view in Chapter 8. In addition, a rule of thumb for siting antennas in point-to-point links to avoid obstruction of the line of sight by terrain is presented. The latter is "site specific" in that the focus is on situations where the terrain elevation is known as a function of distance from the transmitter and the receiver. Empirical models for propagation loss in cases where the terrain is not specified (as for a mobile communication system) are dealt with in Chapter 8. Because the obstruction considered here is the

Radiowave Propagation: Physics and Applications. By Curt A. Levis, Joel T. Johnson, and Fernando L. Teixeira

Earth's surface itself, the discussion applies mainly to nonurban environments; the influence of buildings, vehicles, and other such "clutter" objects is not considered.

7.2 PROPAGATION OVER A PLANE EARTH

For terrain that is relatively smooth on a wavelength scale, the surface of the Earth may be approximated as a dielectric plane at sufficiently short distances where Earth curvature effects are negligible. In this situation, a transmitting antenna with finite directivity illuminates both the receiving antenna and a portion of the ground surface, so that the receiver receives a direct and a specularly reflected signal, as discussed in Chapter 6. Figure 7.1 illustrates the geometry of this situation, from which the following geometrical relationships can be derived:

$$R_2 = R_2' + R_2'', \tag{7.1}$$

$$\tan \psi_2 = (h_1 + h_2)/d, \tag{7.2}$$

$$\tan \psi_1 = (h_1 - h_2)/d, \tag{7.3}$$

$$d = R_1 \cos \psi_1 = R_2 \cos \psi_2, \tag{7.4}$$

where ψ_1 and ψ_2 refer to the declination angles along the direct and reflected paths R_1 and $R_2' + R_2''$, respectively. Note that the reflected wave can be interpreted as arising from an "image" of the true source that is located a distance h_1 below the Earth's surface. The reflected wave contributed by the image source is computed as if the image source were in free space, except that the field is multiplied by the reflection coefficient encountered on the true reflected wave path.

Throughout the following, we will assume that the receiver is in the far field of both the true and image sources and that the horizontal distance d between the two antennas is much larger than either of the antenna heights.

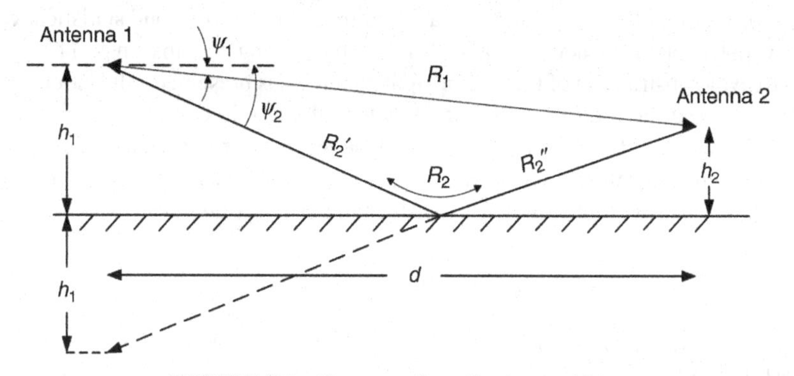

FIGURE 7.1 Geometry for reflection problem.

7.2.1 Field Received Along Path R_1: The Direct Ray

From equation (5.2), assuming impedance and polarization matching and no atmospheric attenuation, the power density of the direct ray, if it existed alone, would be

$$|\overline{S}_{\text{dir}}| = \frac{P_{\text{T}} G_{\text{T}}}{4\pi R_1^2}, \tag{7.5}$$

where P_{T} is the power transmitted and G_{T} is the transmitter gain in the direction of the receiving antenna. From this and

$$|\overline{S}| = \frac{|\underline{E}|^2}{2\eta}, \tag{7.6}$$

where η is the characteristic impedance of free space and peak values are used for the field amplitude, we can obtain

$$|\underline{E}| = \sqrt{2\eta|\overline{S}|}, \tag{7.7}$$

so

$$\begin{aligned} |\underline{E}^{\text{dir}}| &= \frac{\sqrt{\frac{\eta}{2\pi} P_{\text{T}} G_{\text{T}}}}{R_1} \\ &\approx \frac{\sqrt{60 P_{\text{T}} G_{\text{T}}}}{d}, \end{aligned} \tag{7.8}$$

where the latter equation substitutes $\eta \approx 120\pi$ ohms. In equation (7.8), if $|\underline{E}|$ is to be in volts/meter, P_{T} must be in watts and d in meters. In the literature, equation (7.8) is often written as

$$|\underline{E}^{\text{dir}}| = \frac{E_0}{d}, \tag{7.9}$$

where

$$E_0 = \alpha\sqrt{2P_{\text{T}}} \tag{7.10}$$

and

$$\alpha = \sqrt{30 G_{\text{T}}}. \tag{7.11}$$

E_0 is called the "unattenuated field intensity at unit distance" and is usually specified in mV/m at 1 km with the understanding that the distance d is then specified in km. From equation (7.10), E_0 depends only on the transmitted power (usually measured in kW) and α, which in turn depends only on the transmitter antenna gain G_{T}. Using the standard units (mV/m at 1 km for E_0 and kW for P_{T}), values of α can be tabulated; examples for common antennas are given in Table 7.1. Values using mV/m at 1 mile units for E_0 are also provided.

TABLE 7.1 Values for α so That E_0 in mV/m at Specified Distance = $\alpha\sqrt{2P_T}$ with P_T in kW

Antenna Type	α for d in km	α for d in mi
Isotropic	173	108
Short dipole	212	132
$\lambda/2$ dipole	222	138
Short monopole[a]	150	93.2
$\lambda/4$ monopole[a]	157	97.5

This definition is such that the rms field strength for a 1 kW transmitter is α mV/m at the specified distance.

[a]Antennas are fed against ground: to find unattenuated field intensity in presence of ground use $2E_0$.

7.2.2 Field Received Along Path R_2: The Reflected Ray

The magnitude of the reflected ray can be calculated similarly except for the reflection process. The plane wave reflection coefficients from Chapter 6 may be used to obtain

$$|\underline{E}^{\text{ref}}| = |\frac{E_0}{d}\underline{\Gamma}|, \tag{7.12}$$

where $R_2 \approx d$ has been used in the denominator, and it has also been assumed that the antenna gain is identical along the direct and reflect ray paths so that E_0 is the same for both. $\underline{\Gamma}$ is to be taken as either $\underline{\Gamma}_{\|}$ or $\underline{\Gamma}_{\perp}$ depending on the polarization utilized; for ground-to-ground station propagation problems, both reflection coefficients are near -1 due to the small declination angles involved.

7.2.3 Total Field

The total field is obtained by adding the direct and reflected fields in the proper phase relationship,

$$|\underline{E}^{\text{tot}}| = \frac{E_0}{d}\left|e^{-jk_0R_1} + \underline{\Gamma}e^{-jk_0R_2}\right|, \tag{7.13}$$

where $k_0 = \omega\sqrt{\mu_0\epsilon_0}$ is the phase constant in free space. Though correct, this relationship is not in a useful form since k_0R_1 and k_0R_2 are generally many thousands or millions of radians and $\underline{\Gamma}$ is often near -1,[1] therefore great precision would have to be used in evaluating the exponentials. Instead, use

$$\begin{aligned}\left|e^{-jk_0R_1} + \underline{\Gamma}e^{-jk_0R_2}\right| &= \left|e^{-jk_0R_1}\left(1 + \underline{\Gamma}e^{-jk_0(R_2-R_1)}\right)\right| \\ &= \left|1 + \underline{\Gamma}e^{-jk_0(R_2-R_1)}\right| \end{aligned}\tag{7.14}$$

[1]A possible exception is aircraft communication. Another frequently encountered exception is reflection from rough ground.

to obtain

$$|\underline{\overline{E}}^{\text{tot}}| = \frac{E_0}{d}\left|1 + \underline{\Gamma} e^{-jk_0(R_2 - R_1)}\right|. \tag{7.15}$$

From the geometry,

$$R_2 = \left[d^2 + (h_1 + h_2)^2\right]^{1/2}, \tag{7.16}$$

and use of the binomial theorem $(x + y)^{1/2} \approx x^{1/2} + \frac{y}{2x^{1/2}}$ for $x >> y$ gives

$$R_2 \approx d + \frac{1}{2d}(h_1 + h_2)^2. \tag{7.17}$$

Similarly,

$$R_1 \approx d + \frac{1}{2d}(h_1 - h_2)^2 \tag{7.18}$$

so that

$$R_2 - R_1 \approx \frac{2h_1h_2}{d}, \tag{7.19}$$

a generally important and useful relationship for antennas located much further from one another than their heights above the ground.

Use of this in equation (7.15) gives

$$|\underline{\overline{E}}^{\text{tot}}| = \frac{E_0}{d}\left|1 + \underline{\Gamma} e^{-j2k_0h_1h_2/d}\right|. \tag{7.20}$$

This is the relationship that normally should be used to calculate the field. If the heights are small, the reflection coefficient approaches -1 and the exponential may be replaced by the first two terms of its Maclaurin's series expansion, $1 - j2k_0h_1h_2/d$, to give

$$|\underline{\overline{E}}^{\text{tot}}| = \frac{E_0}{d}\left(\frac{4\pi h_1h_2}{\lambda d}\right). \tag{7.21}$$

The factor in parenthesis is dimensionless, and any consistent units may be used in it. Equation (7.21) is applicable only for cases where $2k_0h_1h_2/d$ is less than approximately 0.5; this condition can be rewritten as

$$h_1h_2 < 12\frac{d_{\text{km}}}{f_{\text{GHz}}} \tag{7.22}$$

with h_1 and h_2 in meters. When applicable, equation (7.21) shows that the received field decays as one over the distance squared, as opposed to the one over distance decay in free space. This more rapid loss of field intensity is caused by destructive interference between the direct and reflected rays.

Equation (7.20) provides our first specific example of "multipath" in that the total received field consists of contributions from waves that have different phases due to their differing path lengths. Since the total field amplitude is proportional to the sum

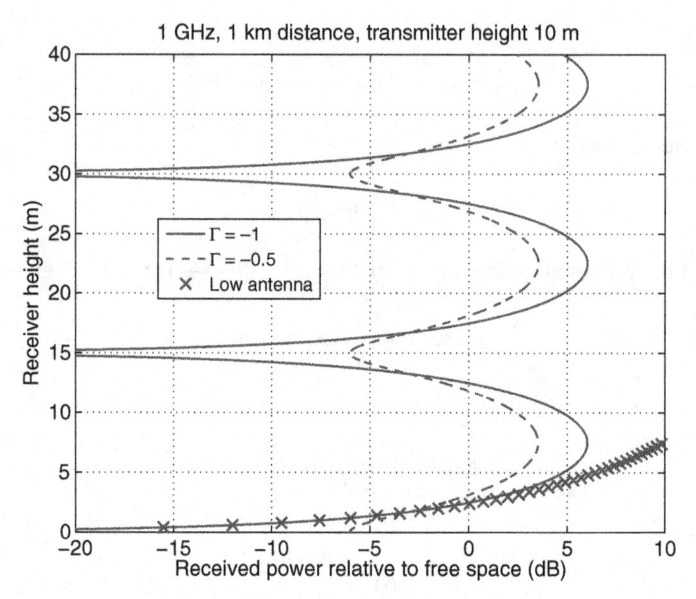

FIGURE 7.2 Height-gain curve example: received power relative to free space in dB as the receiver height is varied: 1 GHz, 1 km distance between antennas, and transmitter height 10 m. Results using $\underline{\Gamma} = -1$, $\underline{\Gamma} = -0.5$, and equation (7.21) are compared.

of two numbers, 1 and $\underline{\Gamma}e^{-j2k_0h_1h_2/d}$, the constructive or destructive interference of these terms depends on their relative phases. In situations where $\underline{\Gamma} \approx -1$, the two terms have identical amplitudes, and the total field amplitude can vary from zero to twice that of the direct ray.

7.2.4 Height-Gain Curves

A quantity of interest in many propagation applications is a "height-gain curve": this is a plot of the received power as either the receiver or the transmitter height is varied at a fixed separation. It is common to scale such plots by the power that would have been received if the direct ray alone were present, in this case the power associated with E_0/d. The quantity $\left|1 + \underline{\Gamma}e^{-j2k_0h_1h_2/d}\right|^2$ is thus the "received power relative to the power that would be received in free space," typically presented in decibels. Figure 7.2 plots this quantity on the horizontal axis versus the receiving antenna height on the vertical axis (a common practice with height-gain curves) for a 1 GHz transmitter frequency, a 1 km separation between the antennas, and a 10 m transmitter height. Curves are included for $\underline{\Gamma} = -1$, $\underline{\Gamma} = -0.5$, and the simplified form in equation (7.21). The results show an oscillation of the received power as a function of receiver height due to the changing phase difference between the direct and reflected rays as the receiver height varies. For $\underline{\Gamma} = -1$, the nulls in the oscillation approach minus infinity decibels since zero field strength can occur in this case, while the maxima are at 6 dB since this represents a doubled field strength. The oscillation is smaller for $\underline{\Gamma} = -0.5$ since complete constructive or destructive interference does not

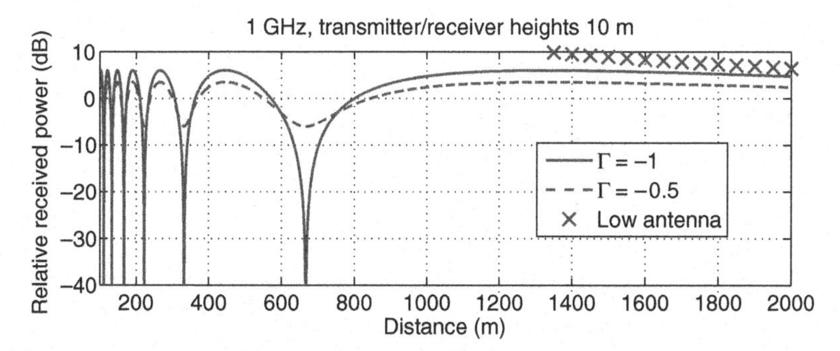

FIGURE 7.3 Received power relative to free space in dB as a function of distance: 1 GHz, transmitter and receiver heights of 10 m. Results using $\underline{\Gamma} = -1, \underline{\Gamma} = -0.5$, and equation (7.21) are compared.

occur in this case. Finally, the linear increase in field amplitudes (quadratic increase in power) with receiver height predicted by equation (7.21) holds only for portions of the height-gain curve significantly below the first maximum. It is of course desirable in establishing point-to-point links to avoid destructive multipath interference; this issue will be discussed further later in the chapter.

Similar effects can be observed in the received power versus range for fixed transmitter and receiver heights (Figure 7.3). Since the phase difference varies inversely with the distance, more rapid oscillations are observed at shorter ranges that eventually resolve into a slow decay as the fourth power of distance when equation (7.21) is applicable. It would be more likely in practice to consider height-gain curves at a fixed distance (as in Figure 7.2) when considering sites for point-to-point systems, while an examination of propagation losses versus range (as in Figure 7.3) would be more appropriate for airborne communications or radar applications.

7.3 FRESNEL ZONES

The discussion of the previous section has considered the direct and specularly reflected rays for propagation over a planar surface. For nonplanar surfaces, it is also of interest to consider the impact of reflections from other surface points. Signals arriving through these reflected rays would be delayed compared to the direct ray. To be delayed by $n \times 180°$, the path via reflection must exceed the direct path length by $n\lambda/2$, where λ is the wavelength.

The set of points at which reflection would produce an excess path length of precisely $n\lambda/2$ is called the nth Fresnel zone. From analytic geometry, it is known that in a plane this collection is an ellipse; in three-dimensional space, it is an ellipsoid of revolution with the direct ray as its axis. The diameter of the ellipsoid increases with increasing n, as sketched in Figure 7.4. While reflections can occur anywhere around the path, most good reflecting surfaces are horizontal, and it is therefore usually sufficient to consider only the vertical plane containing the direct ray.

A signal arriving via reflection from the first Fresnel zone suffers a 180° phase shift at the reflection point (since $\underline{\Gamma}_\perp$ or $\underline{\Gamma}_\parallel \approx -1$ for grazing incidence) and another 180°

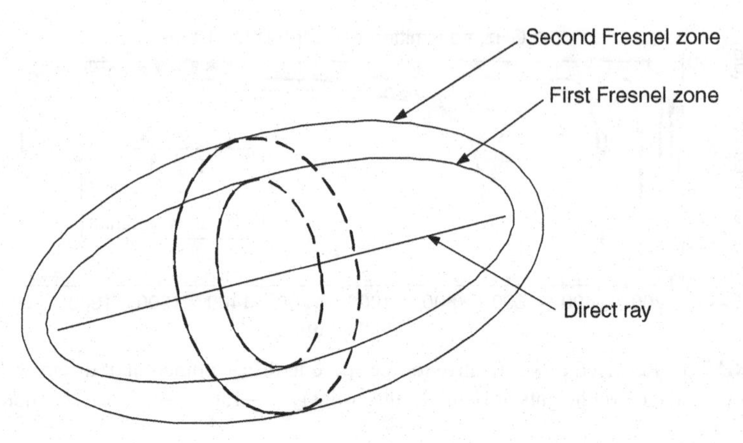

FIGURE 7.4 Fresnel ellipsoids.

phase shift by virtue of its $\lambda/2$ excess path. It will therefore reinforce the direct ray and produce a maximum. A signal arriving via reflection from the second Fresnel zone suffers a 180° phase shift at the reflection point and a 360° phase shift due to excess path; it will therefore cancel the direct ray. In similar manner, all odd Fresnel zone reflections produce maxima and all even Fresnel zone reflections produce minima. A convenient formula for calculating the radius of the nth Fresnel zone at a distance d_1 from one end of the path is

$$F_n = 17.32\sqrt{\frac{n d_{1,\mathrm{km}} d_{2,\mathrm{km}}}{f_{\mathrm{GHz}} d_{\mathrm{km}}}}, \tag{7.23}$$

where F_n is the radius of the nth Fresnel zone in m for a path of length d at a distance d_1 from one end of the path, d_2 from the other end, all in km, at a frequency given by f in GHz. This formula can be derived as follows:

Let AB be the direct path and ACB the path via the reflection point C as shown in Figure 7.5. The condition that will locate C on the nth Fresnel zone is

$$r_1 + r_2 = d_1 + d_2 + n\lambda/2, \tag{7.24}$$

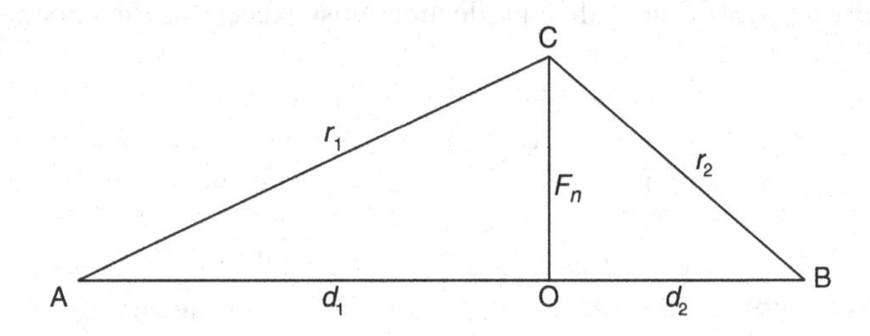

FIGURE 7.5 Triangle for calculation of Fresnel zones.

where λ is the wavelength and the other dimensions are as defined in Figure 7.5. Using the right triangle relations gives

$$\sqrt{d_1^2 + F_n^2} + \sqrt{d_2^2 + F_n^2} = d_1 + d_2 + n\lambda/2. \tag{7.25}$$

The distance AB is on the order tens of km while F_n will turn out to be on the order of tens of meters; therefore, except right near the end points A or B, we can expand the square roots by the binomial theorem and retain only the first two terms since $d_1^2 >> F_n^2, d_2^2 >> F_n^2$:

$$d_1 + \frac{F_n^2}{2d_1} + d_2 + \frac{F_n^2}{2d_2} = d_1 + d_2 + n\lambda/2. \tag{7.26}$$

Therefore,

$$\frac{F_n^2}{2}\left[\frac{1}{d_1} + \frac{1}{d_2}\right] = F_n^2 \frac{d_1 + d_2}{2d_1 d_2} = n\lambda/2. \tag{7.27}$$

Using $d = d_1 + d_2$, we get

$$F_n = \sqrt{\frac{n\lambda d_1 d_2}{d}}, \tag{7.28}$$

where F_n, λ, d_1, d_2, and d are all in the same units. Use of $\lambda = c/f$ and conversion of units gives equation (7.23). Note that the Fresnel zone radii scale directly with $\sqrt{n}$ and with $\sqrt{\lambda}$, and are therefore larger at lower frequencies.

Figure 7.6 is a plot of the Fresnel zone radii at 3.95 GHz versus distance along a 20 km path and shows an increase from zero at the transmitting and receiving locations to a maximum at the middle of the path.

In practice, complete cancelation and reinforcement ($\underline{\Gamma}_\perp$ or $\underline{\Gamma}_\parallel = -1$) occur only for very smooth surfaces such as lakes, tennis courts, paved roads, and so on. For rough surfaces (i.e., surfaces whose roughness is an appreciable part of a wavelength), the effective reflection coefficient amplitude varies between zero and -1.

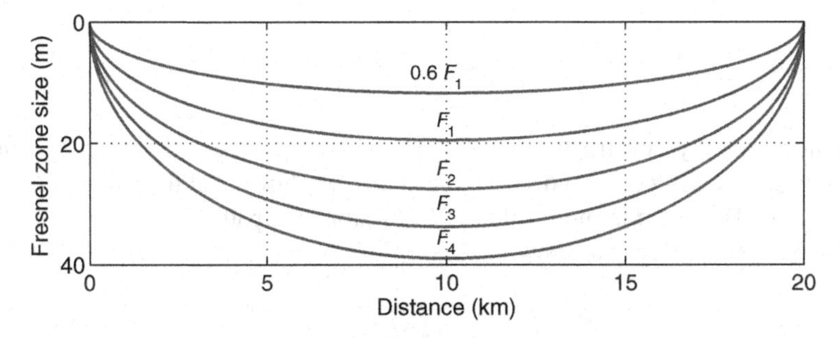

FIGURE 7.6 Variation in 3.95 GHz Fresnel zone radii along a 20 km path.

7.3.1 Propagation Over a Plane Earth Revisited in Terms of Fresnel Zones

Propagation over a plane Earth provides an interesting example for using Fresnel zones to interpret reflections, since there is only a single reflection point from the terrain and the location of this reflection point is known. Referring to Figure 7.1, the reflection point on the planar surface is located at the distance $d_1 = d_\mathrm{r}$, where a line drawn from the image source to the receiver intersects the ground surface (height zero), that is,

$$-h_1 + \frac{h_2 + h_1}{d} d_\mathrm{r} = 0 \tag{7.29}$$

or

$$d_\mathrm{r} = d \frac{h_1}{h_1 + h_2}. \tag{7.30}$$

The direct ray can be represented by a line whose height above the ground surface is

$$h_1 + \frac{h_2 - h_1}{d} d_1, \tag{7.31}$$

which for $d_1 = d_\mathrm{r}$ yields a reflection point clearance h of

$$h = \frac{2h_1 h_2}{h_1 + h_2}. \tag{7.32}$$

It is of interest to compare the height of the direct ray above the reflection point with the Fresnel zone radii at the location of the reflection point. The latter are

$$\begin{aligned} F_n &= \sqrt{\frac{n\lambda d_\mathrm{r}(d - d_\mathrm{r})}{d}} \\ &= \sqrt{n\lambda d \frac{h_1 h_2}{(h_1 + h_2)^2}}. \end{aligned} \tag{7.33}$$

Combining equations (7.32) and (7.33), we obtain

$$\frac{h}{F_1} = 2\sqrt{\frac{h_1 h_2}{\lambda d}}. \tag{7.34}$$

When h/F_1 is unity, constructive interference will occur; when $h/F_1 = \sqrt{2}$, destructive interference will occur, and so on. It can also be shown that the value of h/F_1 in equation (7.34) is the minimum value of h/F_1 along the path.

Furthermore, it is easy to show that

$$\pi \left(\frac{h}{F_1}\right)^2 = 2k_0 h_1 h_2 / d. \tag{7.35}$$

The latter is the phase difference used in combining the direct and reflected rays in equation (7.20). The received power relative to free space can thus be rewritten as

$$\left|1 + \underline{\Gamma} e^{-j\pi\left(\frac{h}{F_1}\right)^2}\right|^2, \tag{7.36}$$

which is a form that can also be used to predict the received power in the presence of a single reflection point on a nonplanar terrain profile, regardless of its location. In this formula, both the path clearance h and the first Fresnel zone radius F_1, as well as $\underline{\Gamma}$, are the values at the location of the reflection point.

One point of interest in height-gain curves is the minimum receiver height for which the received power is equal to that that would be received in free space. This condition implies that

$$\left|1 + \underline{\Gamma} e^{-j\pi\left(\frac{h}{F_1}\right)^2}\right| = 1. \tag{7.37}$$

If $\underline{\Gamma}$ is approximated as -1, the above equation can be solved using $e^{-ja} = \cos a - j \sin a$ to yield

$$\pi (h/F_1)^2 = \cos^{-1}\frac{1}{2}, \tag{7.38}$$

$$h/F_1 \approx 0.577. \tag{7.39}$$

Thus, a received power equal to that received in free space is achieved when the path clears the reflection point by 0.577 times the first Fresnel zone radius at the reflection point.

7.4 EARTH CURVATURE AND PATH PROFILE CONSTRUCTION

A path profile is a sketch showing a signal path and the intervening terrain, with the vertical scale greatly magnified compared to the horizontal. In the actual signal path, both the Earth surface and the ray path are curved due to atmospheric refraction. In most engineering tasks (e.g., locating suitable transmitting and receiving antenna heights for a specific point-to-point microwave link) only the height of the ray above the terrain is of importance. It is therefore often convenient to assume a linear refractive index variation with height and to use the straight ray, modified Earth radius method (Section 6.6) to draw rays as straight lines and scale the Earth radius by the multiplier κ. Alternatively, the flat Earth, modified ray curvature method may be used. The first method is the most convenient when many antenna heights are to be considered for a given refractive index condition. The second is convenient when the effect of varying refractive index conditions is to be explored for specified antenna heights, particularly when a method has been constructed for tracing the curved rays, such as a set of templates when graphical methods are used or an equivalent computer program. To some extent, it is merely a matter of preference; either method works satisfactorily when the basic principle- that any combination that gives the same

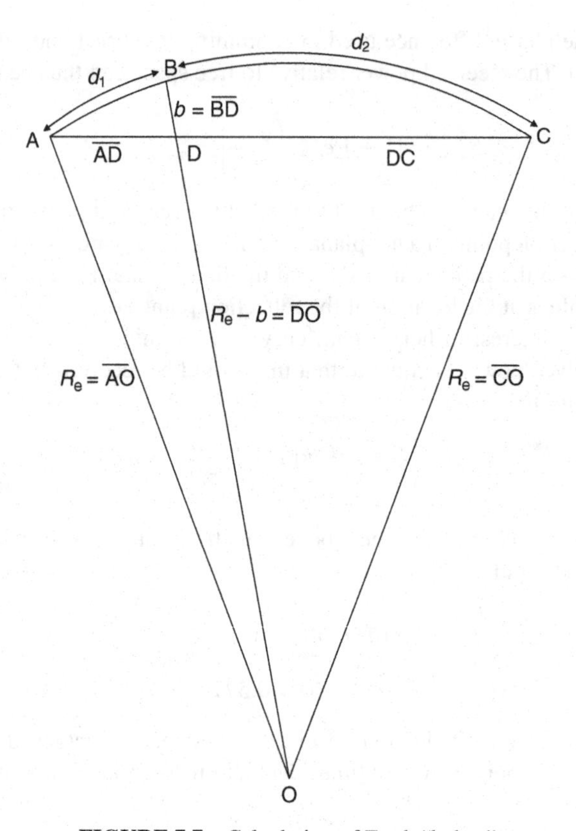

FIGURE 7.7 Calculation of Earth "bulge."

vertical separations from ray to ground all along the path is valid- is understood. The straight ray, modified Earth radius approach will be used in what follows.

Using this approach, the "bulging" of the Earth, that is, the vertical displacement of the spherical surface along the path, can be calculated by means of Figure 7.7. In this figure, ABC is a sector of the effective Earth surface with effective radius $\kappa a = R_e$ and center at O. In practice, the distance AC is on the order of tens of kilometers while R_e is several thousands of kilometers; if drawn to scale, BD (drawn perpendicular to the arc ABC at B) would also be almost precisely perpendicular to the cord AC at D. The distance BD $= b$ is the "bulging" to be calculated; the distance DO therefore is $R_e - b$.

From Figure 7.7, using a theorem that relates the lengths of subsections of intersecting chords in a circle,

$$\overline{AD} \cdot \overline{DC} = \overline{BD} \cdot \left(\overline{DO} + R_e\right), \tag{7.40}$$

where

$$\overline{AD} \approx d_1, \tag{7.41}$$

$$\overline{\mathrm{DC}} \approx d_2, \tag{7.42}$$

$$\overline{\mathrm{BD}} = b, \tag{7.43}$$

$$\overline{\mathrm{DO}} \approx R_e = \kappa a, \tag{7.44}$$

gives

$$d_1 d_2 = (2b)(\kappa a) \tag{7.45}$$

and

$$b = \frac{d_1 d_2}{2\kappa a}. \tag{7.46}$$

In the above equations, b, d_1, d_2, and a are all in the same units. It is convenient to use d_1, d_2, and a in km or miles, but b in meters or feet. Using km and m,

$$b = 1000\frac{d_1 d_2}{2\kappa a} \tag{7.47}$$

meters. The Earth radius is approximately 6370 km (or 3960 miles), giving

$$b = \frac{d_1 d_2}{12.74\kappa} \tag{7.48}$$

meters, with the distances in km. This is a convenient equation from which the Earth "bulging" is plotted; since $d_2 = d - d_1$, the Earth bulge is a quadratic fit to the Earth's curvature as a function of d_1. With miles and feet as units, the corresponding relationship is

$$b = \frac{d_1 d_2}{1.5\kappa} \tag{7.49}$$

feet.

7.5 MICROWAVE LINK DESIGN

Proper point-to-point link design is governed by three considerations: (i) to minimize cost, one would prefer to space the stations far apart; (ii) if they are too far apart, the intervening Earth "bulge" will intercept the beam for some atmospheric conditions; (iii) if the intervening terrain is smooth, it will have a reflection coefficient of nearly -1. In that case, for certain antenna heights (which depend on the value of κ) the direct and ground-reflected signal will add out of phase and produce deep fades or signal loss. The aim of microwave link design is to assure proper path clearance and avoid destructive interference due to reflections; this requires consideration of the range of refractive conditions (i.e., κ values) that may occur at a specific location.

Consider two antennas within the line of sight as shown in Figure 7.8, with one antenna height (at the right-hand side B) variable. This figure represents a 30 km path. The arc near the bottom represents a convenient datum or reference height (such as 150 m AMSL (above mean sea level)) of the effective spherical surface of the Earth and is constructed by the use of equation (7.48). The terrain elevation above this level,

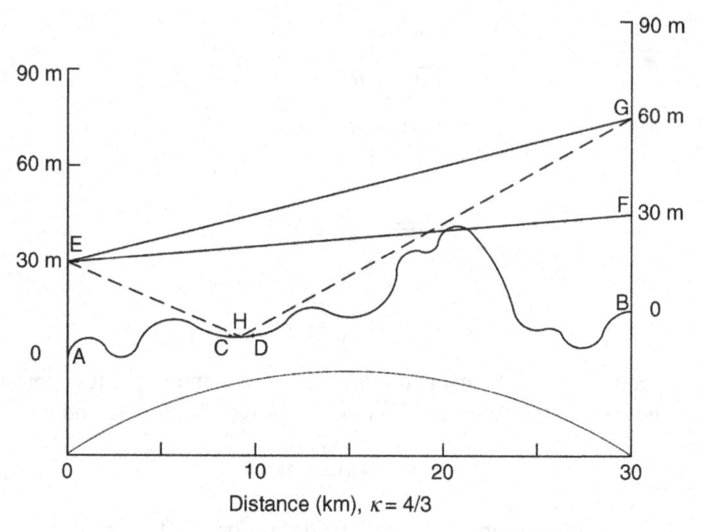

FIGURE 7.8 Sample path profile.

obtained from maps or by measurement, is plotted to the same scale above the arc (i.e., added to the arc) to give the terrain profile. One end of the path appears at the left edge and the other at the right edge of the plot. At each edge a vertical scale, drawn to the same scale as the datum circle and terrain elevations, indicates possible antenna heights above the terrain.

When both antennas are 30 m above their respective terrains (path EF), the path is obstructed and transmission will be poor. If the antenna at B is raised from 30 m (F) to 60 m (G), the path appears to clear the obstacle. However, we should not expect a completely abrupt transition from the "obstructed" to the "clear" path condition. A quantitative understanding of this transition is needed.

We also note that as the antenna is raised, reflections become possible. If the arc CD represents a lake, the reflection EHG is highly likely (it should be noted that reflections often come from places that are not always obvious from the path profile plot because roads, large flat roofs, ballparks, flat pastures, etc. may not be apparent on these scales).

As discussed previously, a plot of the received signal power as one antenna height is varied, all other system parameters remaining fixed, is known as a height-gain curve. For the simple case of Figure 7.8 (one obstruction, one reflection point), the received signal strength relative to the strength that would be received in free space (without the surface below) might look as shown in Figure 7.9. This figure is similar to that observed for the planar Earth case (Figure 7.2) at higher receiver heights but is different at lower receiver heights due to the influence of obstructions caused by the terrain profile. On paths allowing more than one reflection, the height-gain curve is more complicated; nevertheless, the point A, where the field first reaches its free-space (Friis transmission formula) value, can often be identified with the obstruction, while minima such as B or C can be used to identify reflection points.

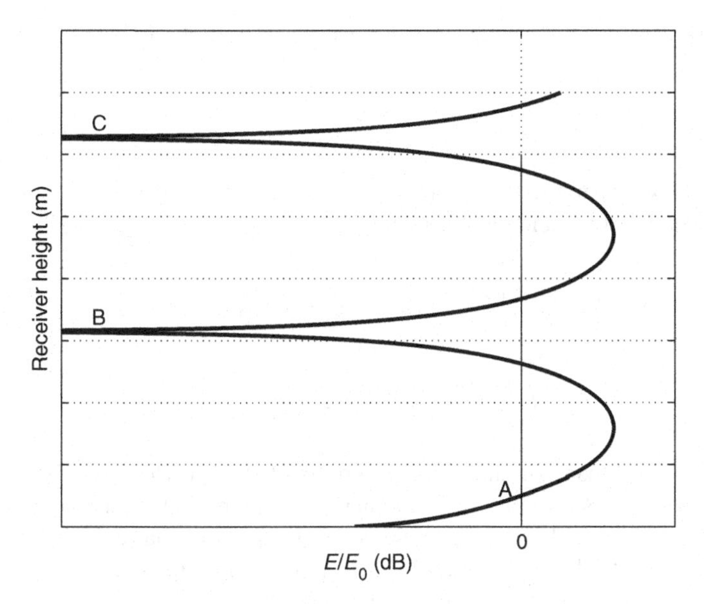

FIGURE 7.9 Sample plot of received power relative to free space versus receiver height: one obstruction, one reflection point.

7.5.1 Distance to the Radio Horizon

The “radio horizon” is the maximum distance by which a transmitter and receiver can be separated (Figure 7.10) and still remain within the radio line of sight, that is, the line of sight when refraction effects are included [1]. The modifier “radio” is often omitted below and in the literature. These concepts are applicable only in the VHF and higher frequency bands where groundwave and ionospheric effects are negligible.

A straight line ray between a transmitter at height h_1 and a receiver at height h_2 for path length d can be written as

$$h_1 + \frac{h_2 - h_1}{d} d_1, \tag{7.50}$$

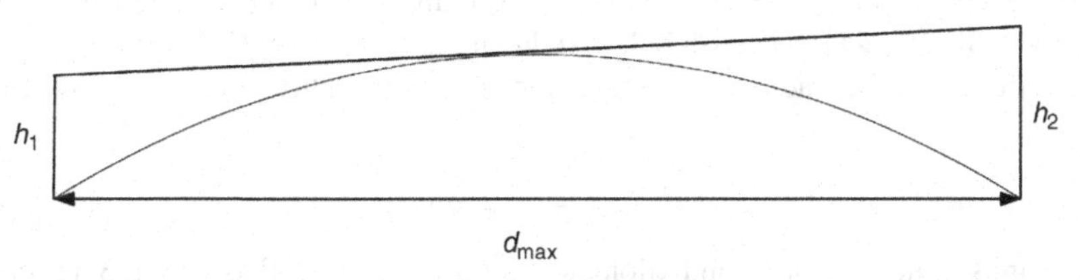

FIGURE 7.10 Applicable geometry for determining the maximum radio line-of-sight distance for level terrain.

where d_1 is the distance along the Earth's surface from the transmitter. The Earth bulge is

$$\frac{d_1 d_2}{2\kappa a}. \tag{7.51}$$

Equating these two expressions determines the points of intersection (i.e., d_1 values) between the direct ray and the Earth bulge:

$$h_1 + \frac{h_2 - h_1}{d} d_1 = \frac{d_1(d - d_1)}{2\kappa a}, \tag{7.52}$$

or

$$d_1^2 + \left(\frac{2\kappa a (h_2 - h_1) - d^2}{d}\right) d_1 + 2\kappa a h_1 = 0. \tag{7.53}$$

This equation is quadratic in d_1; the two roots of this quadratic will be imaginary when the path clears the bulge (no real intersection points) and real otherwise. When the path intersects the bulge in a single point, the quadratic has a double root at the intersection point. This is the situation that determines the radio horizon.

Applying the quadratic formula to equation (7.53) yields

$$d_1 = \frac{d^2 - 2\kappa a (h_2 - h_1)}{2d} \pm \frac{1}{2}\sqrt{\left(\frac{2\kappa a (h_2 - h_1) - d^2}{d}\right)^2 - 8\kappa a h_1}. \tag{7.54}$$

To obtain a double root, the term inside the square root must be zero:

$$\frac{2\kappa a (h_2 - h_1) - d^2}{d} = \pm\sqrt{8\kappa a h_1}, \tag{7.55}$$

which is the quadratic equation to determine $d = d_{\max}$, the maximum distance of the line of sight. Applying the quadratic formula a second time and finding the largest resulting root yields

$$d_{\max} = \sqrt{2\kappa a h_1} + \sqrt{2\kappa a h_2}, \tag{7.56}$$

which is the maximum distance by which antennas at heights h_1 and h_2 can be separated while remaining within each other's radio horizon. This does not necessarily imply that the link between the two will be effective! An adequate path clearance and freedom from destructive reflections are required, as will be discussed later in the Chapter. Note that $d_{\max}$ depends on the Earth radius multiplier, with smaller values of κ (which cause a larger Earth bulge) reducing the line-of-sight distance.

The distance to the radio horizon for a transmitter at height h_1 is obtained by setting h_2 to zero:

$$d_{\text{horizon}} = \sqrt{2\kappa a h_1}. \tag{7.57}$$

Using units of meters for h_1 and kilometers for d_{horizon}, as well as $\kappa = 4/3$, produces

$$d_{\text{horizon,km}} \approx 4.12\sqrt{h_{1,\text{m}}}. \tag{7.58}$$

Using miles and feet, we have

$$d_{\text{horizon,mi}} \approx \sqrt{2h_{1,\text{ft}}}. \tag{7.59}$$

Thus, for example, a transmitter or receiver on an aircraft at 30,000 ft should see the radio horizon at a distance of approximately 245 miles. These results of course assume that the local terrain is sufficiently flat that only the curvature of the Earth need be considered.

7.5.2 Height-Gain Curves in the Obstructed Region

Obstruction by actual terrain may occur even for transmitters and receivers that would be within the radio line of sight if there were no obstructing obstacles. Since no simple mathematical solution is available for the diffraction of a radio wave by an arbitrary obstacle, our understanding of the effect of obstructions is derived from the solution of two canonical problems.

The first of these is the diffraction of an electromagnetic wave around a smooth conducting sphere. The geometry is shown in Figure 7.11 and the solution in Figure 7.12, curve A. Since no universal solution exists for spherical diffraction in the obstructed region, this curve was computed using the method of Chapter 9 for a 146 MHz transmitter at height 94 m and a receiver at distance 80 km whose height is varied to produce the curve illustrated. The path clearance is shown in Figure 7.11 by arrows; it is considered positive when the straight line (as represented on an equivalent Earth radius profile!) path clears the obstruction (Figure 7.11a), negative otherwise (Figure 7.11b). Although in real life h is perpendicular to the Earth surface, it appears vertical in the figure because of the very different scales in the vertical and horizontal dimensions.

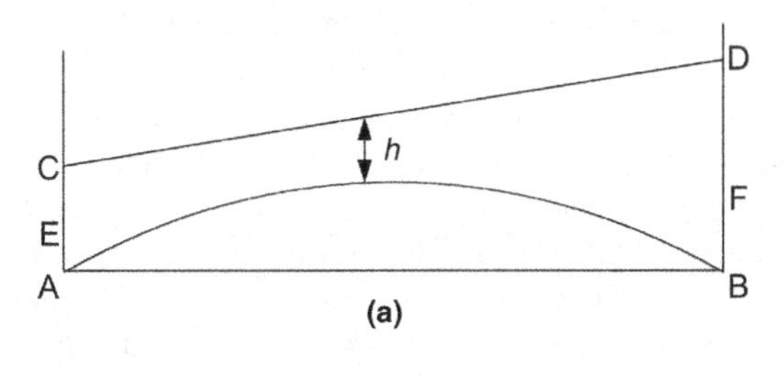

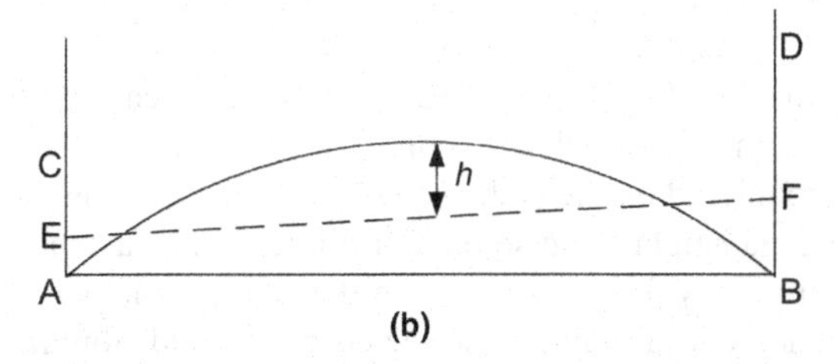

FIGURE 7.11 Calculation of path clearance.

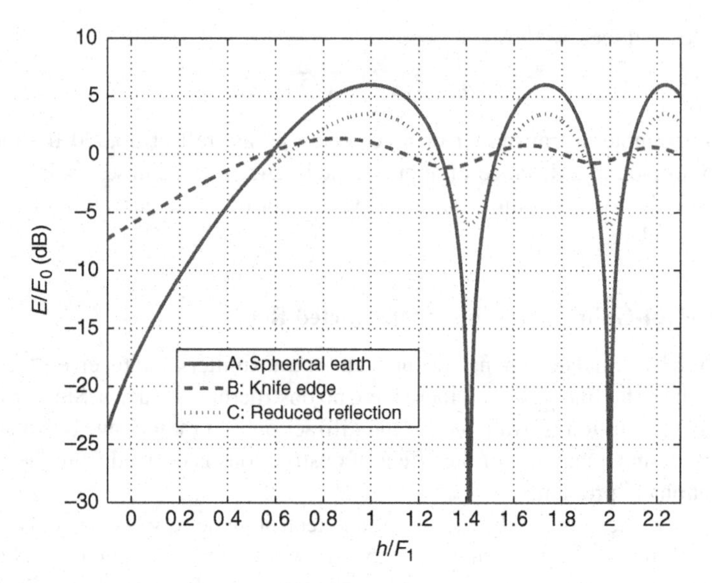

FIGURE 7.12 Received power relative to free space versus normalized path clearance: PEC sphere (146 MHz, 80 km path, 94 m transmitter height, varying receiver height), knife edge, and reduced reflection coefficient interference.

The horizontal axis of Figure 7.12 is expressed in terms of the ratio of the clearance h to the radius F_1 of the first Fresnel zone. In the unobstructed region h is measured, as usual, from the point where the reflection occurs. It can be shown that, for a smooth sphere, this point corresponds to the minimum value of h'/F_1 along each path, where h' is the distance from the path to the surface, perpendicular to the surface. This property was used in calculating Figure 7.12. In Figure 7.12 the abscissa is labeled in h/F_1 (dimensionless) units, and the ordinate is the (also dimensionless) ratio of the actual field strength, $|\overline{E}|$, to what it would be in free space, $|\overline{E_0}|$.

Appreciable signal is present even when the path has negative clearance (e.g., EF in Figure 7.11). After the clearance becomes positive, the signal increases rapidly and attains the free-space value $\left|\overline{E_0}\right|$ at $h \approx 0.6F_1$, that is, when the path is cleared by approximately 0.6 times the radius of the first Fresnel zone at the location of the obstacle. Beyond this point, the curve behaves as we would expect from reflection theory. At $h = F_1$, the reflected ray reinforces the direct ray, causing a field strength $|\overline{E}|/|\overline{E_0}| = 2$, or 6 dB. At $h = \sqrt{2}F_1 = F_2$, the reflected ray cancels the direct ray, at $h = \sqrt{3}F_1 = F_3$, it reinforces, and so on.

The horizontal axis units in Figure 7.12 were chosen because it is found in many diffraction problems that the solution is "universal" in that it depends on the minimum value of h/F_1 encountered on the path and not independently on factors such as the path distance, antenna heights, and so on. For diffraction by a spherical surface, this is approximately true only for paths that clear the obstruction by 0.6 F_1 or more.

The second canonical problem, first solved by Arnold Sommerfeld, is that of diffraction by an infinite half-plane of zero thickness, as shown in Figure 7.13. The

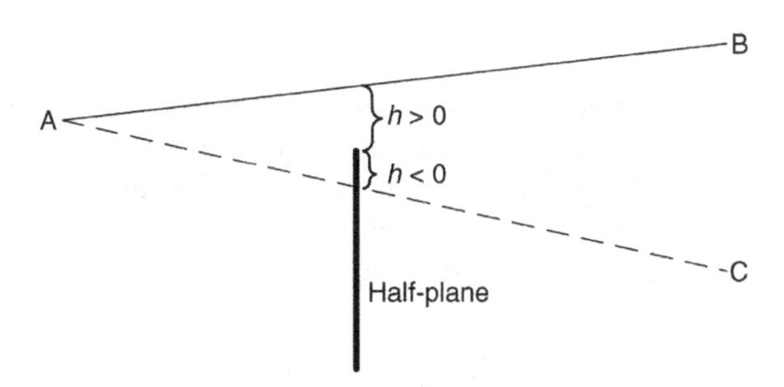

FIGURE 7.13 Canonical knife edge diffraction problem.

path clearance h is taken as positive if the direct ray clears the obstructing screen as for AB and negative if it is obstructed as for path AC. Because of the zero thickness assumption, the problem is sometimes referred to as "diffraction by a knife edge" in the literature; it is valid whenever the obstacle thickness is a small fraction of a wavelength. The solution for the knife edge problem is shown as curve B in Figure 7.12.

It is a curious coincidence, evident from Figure 7.12, that both the smooth sphere diffracted and the knife edge diffracted fields attain their free-space value ($|\underline{E}|/|\underline{E_0}| = 1$) with increasing h first when $h/F_1 \approx 0.6$. This is also consistent with the planar Earth approach, where it was found in equation (7.39) that $h/F_1 \approx 0.577$ provides a free-space propagation level. It is even more astonishing that this relationship generally holds true for actual propagation paths, even when these do not resemble either canonical problem. These results provide a useful "rule of thumb": paths can be considered nonobstructed only when all obstacles are cleared by approximately 0.6 times the first Fresnel zone radius at the obstacle locations.

For obstructed paths, a variety of models have been described in the literature to predict the path loss; some models, such as in ITU-R Recommendation P.526-10 [2], also attempt to include the influence of multiple obstacles along the path. ITU-R Recommendation P.526-10 also provides an approximation to the knife edge curve in Figure 7.12 as $-6.9 - 20\log_{10}\left(\sqrt{(t+0.1)^2+1} - t - 0.1\right)$ decibels, where $t = \sqrt{2}h/F_1$ at the knife edge location; this approximation holds for $h/F_1 < 0.55$. An approximation for diffraction by a spherical Earth is also provided. In addition, ITU-R Recommendation P. 530-12 [3] uses the formula $-10 + 20h/F_1$ decibels as a general approximation for the received power relative to free space for $h/F_1 < -0.25$. However, it has been found that a wide variability in path loss can occur depending on the specific nature of the terrain profile, so producing accurate quantitative predictions through simple analytical means is difficult. Numerical approaches can provide some improvements, as will be discussed later in the chapter. For the moment, we will limit ourselves to classifying whether paths are obstructed or not. The knife edge prediction is also often used as an indicator of the maximum received power that is likely to occur.

7.5.3 Height-Gain Curves in the Reflection Region

When all obstructions are cleared, some minor signal variations due to the diffraction may still exist, but these are usually of little significance (see Figure 7.12 curve B for $h/F_1 > 0.6$). Instead, path behavior when all obstructions are cleared by at least $0.6F_1$ is governed by reflections if any are allowed by the terrain. We have previously discussed (Section 7.3.1) the fact that such effects can be modeled for a single reflection point using

$$\left|1 + \underline{\Gamma} e^{-j\pi\left(\frac{h}{F_1}\right)^2}\right|^2 \tag{7.60}$$

for the received power relative to free space, where h/F_1 is to be taken at the reflection point. Referring to Figure 7.12, curve A was for diffraction by a smooth conducting sphere. When $h/F_1 > 0.6$, the path is no longer obstructed, and curve A is predicted well by equation (7.60) with $\underline{\Gamma} = -1$. For a rough surface, smaller amplitudes of $\underline{\Gamma}$ are applicable; curve C shows the result for $\underline{\Gamma} = -0.5$. As stated previously, it is difficult in general to predict potential reflection points along a path profile without extensive knowledge of the terrain properties that is sometimes not available. In some situations, an analysis of multiple height-gain curves can be utilized to determine the location of reflection points.

For antennas in the reflection region that are separated by an appreciable fraction (30% or more) of the maximum line-of-sight distance, a modified direct plus reflected wave formulation is available that includes the effects of Earth curvature [1]. Two primary effects are important in this modified formulation: the determination of the reflection point location on a smooth curved Earth and a modification of the reflection coefficient $\underline{\Gamma}$ to include a "spherical divergence factor" that models the spreading of the reflected wave's energy by the Earth's curvature. Equation (7.60) remains applicable once these factors are included. This approach assumes that the Earth is a smooth sphere along the path of interest, a situation that is encountered infrequently for overland paths.

7.6 PATH LOSS ANALYSIS EXAMPLES

When a microwave path is to be used heavily (thus justifying high capital investment) and requires high reliability, the path may be tested by erecting temporary antenna towers and obtaining numerous height-gain curves with smaller, and hence more easily handled, antennas than will be used in the final installation. From these, all obstructions and reflective regions are identified. The performance of the path is then calculated for various antenna configurations and κ values, and the final configuration is chosen to maximize reliability for the range of κ values to be expected in that location.

Given a terrain profile, it is possible to make a preliminary prediction of the minimum transmitter or receiver height required in order to avoid obstruction for specified refractive conditions. Figure 7.14 illustrates a 36 km terrain profile to be used in an illustration of this procedure. Consider a 600 MHz transmitter located at the left side

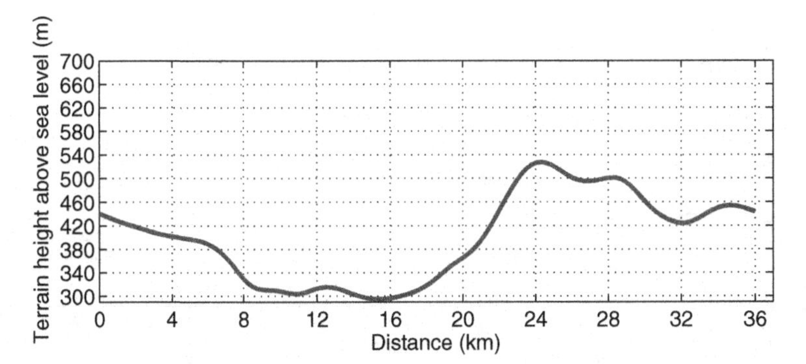

FIGURE 7.14 An example 36 km terrain profile to be used at 600 MHz with a transmitter at height 20 m.

of the profile, at a height of 20 m above the local terrain. To avoid obstruction, a line drawn from this transmitter to a receiver at the right side of the profile must clear all terrain points by the Earth bulge plus 60% of the first Fresnel zone radius. Figure 7.15 plots the original terrain profile (solid curve) as well as the terrain profile plus the Earth bulge (using $\kappa = 4/3$) and $0.6F_1$ (dashed). The minimum receiver height for which a free-space propagation level is predicted can now be determined by finding a line from the transmitter that intersects the dashed curve in only a single location. The line illustrated shows that the required receiver height is around 200 m above the local terrain.

If the location of a reflection point is known *a priori*, a similar process can be used to determine the location of maxima and minima in the height-gain curve. For a reflection point located at distance 24 km, the line tangent to the dotted curve in Figure 7.15 determines the receiver height near 280 m that causes destructive interference because the dotted curve is the terrain height plus the Earth bulge plus the second Fresnel zone radius.

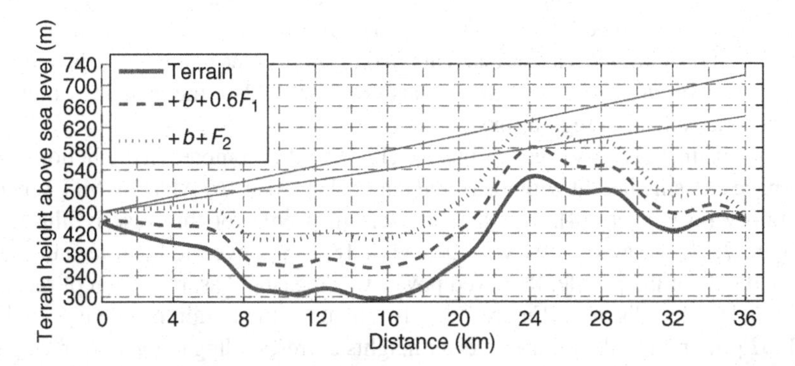

FIGURE 7.15 Determination of the minimum receiver height to achieve a nonobstructed path, and prediction of a receiver height experiencing destructive interference (assuming the hill at distance 24 km is the primary reflection point).

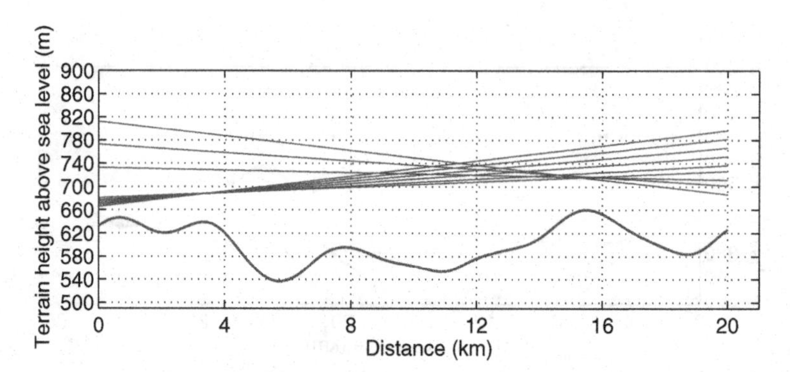

FIGURE 7.16 Paths giving free-space transmission for determination of obstructions.

For actual paths with hills and trees, the points of obstruction are not always obvious. The fact that the field reaches its free-space value, as antenna heights are increased, when the path passes $0.6F_1$ above the obstruction point can often be used to identify the obstruction. In Figure 7.16, a group of propagation paths is shown that were found to give path loss equal to that of free space, that is, the path loss predicted by the Friis transmission formula. It is apparent that some paths are obstructed by the terrain at 16 km from the left end; however, paths that are low at the left end and high on the right end are obstructed at approximately 4 km distance. It would seem that the hill near 4 km is not high enough to cause an obstruction, but the particular cause of an obstruction or reflection may not be evident from the path profile, especially if it has been constructed from a topographic map instead of a careful survey. Buildings, forested areas, farm ponds, driveways, etc. are often not shown on such maps.

The location of the reflection points by similar means is also possible. At heights where the path is not obstructed, the curves will exhibit maxima due to odd Fresnel zone reflections and minima due to even Fresnel zone reflections (Figure 7.9, points B and C). The minima, being sharper, are better suited for locating reflection points. In practice, reflections may be possible over a region, in which case the precise reflection point will be determined by the path geometry, shifting to the left for paths inclined upward to the right, and vice versa. This makes the true crossings less distinct and more easily confused with any extraneous crossings. Also, the crossings multiply when several reflection regions exist.

A final example is presented in Figures 7.17–7.19 that compares a prediction of the free-space crossing height with measured data collected at 167 MHz by MIT Lincoln Laboratory researchers [4]. The 15.2 km terrain profile for this case is illustrated in Figure 7.17; the transmitter was located at 18.3 m above the local terrain at the left-hand side of the profile. Received power was measured as a function of receiver height at 15.2 km distance. Figure 7.17 also includes the terrain profile elevated by $b + 0.6F_1$ in order to determine receiver heights corresponding to the first free-space crossing. Lines are drawn to varying receiver heights until a receiver height is found that results in only a single intersection with the elevated terrain profile; the resulting ray and receiver height determination is shown in the figure. Note in this case that this ray almost intersects the elevated profile in two locations.

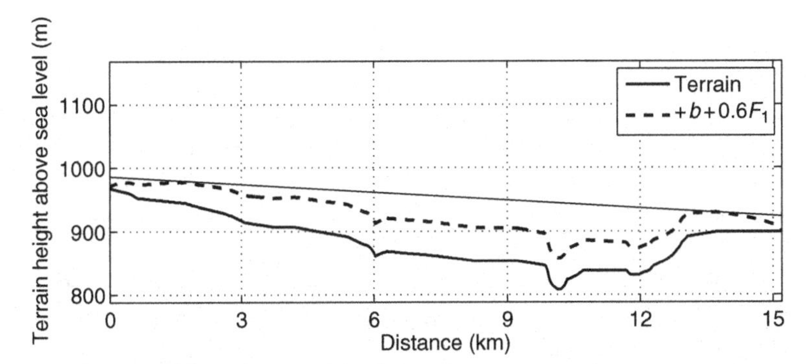

FIGURE 7.17 Beiseker N15 terrain profile for which propagation measurements at 167 MHz were performed by MIT Lincoln Laboratory [4]. Transmitter is located 18.3 m above local terrain at left of the profile. Dashed curve is terrain profile elevated by $b + 0.6F_1$ ($\kappa = 4/3$).

Figure 7.18 compares measured data with the predicted receiver height that obtains a free-space propagation level (marked by an "x"). For this example, the basic rule utilized appears to provide a good indication of the free-space crossing point, although the measured data cross the free-space power level at a slightly lower receiver height.

The height-gain curve in Figure 7.18 also shows significant multipath interference; the graphical methods we have learned can also be utilized to attempt to determine the location of the associated reflection point on the terrain. Given local minima in the measured height-gain curve at receiver heights 71.6 and 113 m, respectively,

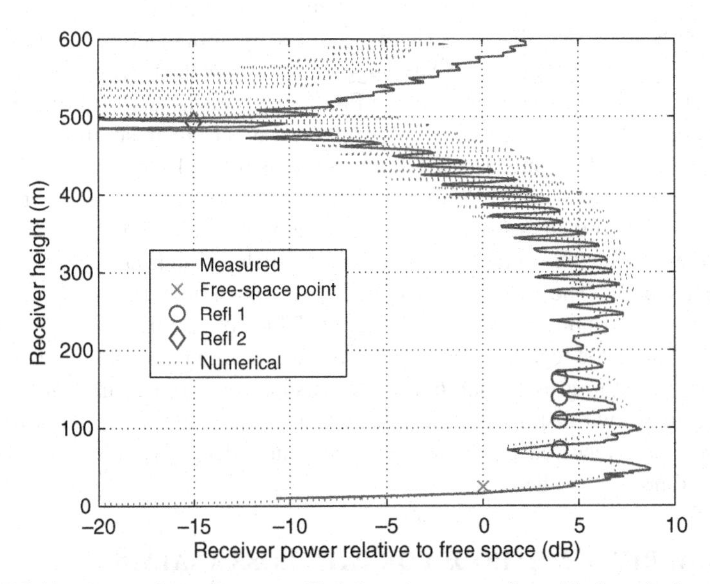

FIGURE 7.18 Comparison of measured height-gain curve and graphical method predictions. Free-space crossing predicted using the rules of this chapter is shown, along with predictions of minima based on determination of two terrain reflection points. Results computed by a numerical method are also included.

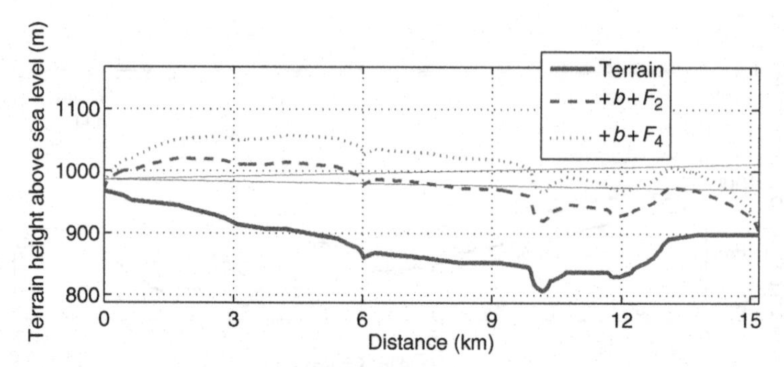

FIGURE 7.19 Determination of reflection point location, using knowledge of two receiver heights (71.6 and 113 m) observing minima in the height-gain curve of Figure 7.18. Common point of intersection is determined for rays to these receiver heights with the terrain elevated by $b + F_2$ and $b + F_4$, respectively.

Figure 7.19 uses the graphical method to assess whether any intersections with the terrain $+b + F_2$ and the terrain $+b + F_4$ occur for these receiver heights at a common terrain location. It is found that a common intersection occurs near 13.2 km distance; the circles in Figure 7.18 mark the predicted minima locations in the height-gain curve, as well as predicted minima for clearance by F_6 and F_8. A reasonable agreement with the measured minima locations is obtained although some small differences remain. Note that the predicted received powers for these minima are assigned arbitrarily, as no knowledge of the local reflection coefficient is available.

It is interesting that even the first minimum here occurs at a power level that is around 1.5 dB larger than that received in free space, while a much stronger minimum occurs near receiver height 490 m. This suggests that more than one reflection may be present. An examination similar to that in Figure 7.19 was used to determine that the minimum near receiver height 490 m apparently results from an additional reflection point that is located at range 0.5 km from the transmitter. Figure 7.18 includes the predicted minimum location for this reflection point, marked with a diamond.

Figure 7.18 also includes the predictions of a numerical method that is found to provide a reasonable match to the measurements; such methods are discussed further in the next section. The path of Figure 7.17 was chosen here, in part, because both measurements and numerical results were available for comparison. It is, in some respects, unusual: the receiver was on a helicopter, enabling the continuous variation of receiver height to greater than 600 m, the results for only one transmitter height were examined, two distinguishable reflection points were present, and diffraction effects produced received power levels more than 6 dB greater than those in free space for some antenna heights.

7.7 NUMERICAL METHODS FOR PATH LOSS ANALYSIS

The ever increasing power of computers is now making numerical predictions of propagation over terrain possible. Instead of the simplified direct plus reflected and canonical diffraction analytical models studied earlier in the chapter, computers can be applied to solve Maxwell's equations with a specified irregular boundary surface (the terrain profile under investigation) and have been shown to yield reasonably accurate predictions when compared with measurements. Of course, the accuracy with which terrain and atmospheric refractivity profiles are known limits the numerical approach, but numerical methods can avoid many of the approximations in the solution of Maxwell's equations used by the analytical approaches, thereby obtaining a more accurate prediction.

Standard numerical techniques for electromagnetics, such as the method of moments, have been applied to propagation predictions, as in Ref. [4]. However, the large size of propagation problems on a wavelength scale often limits these studies because it is usually required in such methods to sample the electric field along the terrain profile at a rate of at least 10 points per electromagnetic wavelength. A more efficient numerical approach can be obtained by making a parabolic approximation to the wave equation that assumes that signals propagate nearly horizontally, which is reasonable for long-range propagation predictions between Earth stations. The parabolic equation (PE) method is also able to model variations in the atmospheric refractivity profile and can thus be used to study coupling effects between ducts and terrain irregularities. The reader is referred to the literature for additional information on such approaches; the "Advanced Refractive Effects Prediction System" (AREPS) produced by the Atmospheric Propagation Branch of the Space and Naval Warfare (SPAWAR) command (mentioned in Chapter 6) also includes a parabolic equation-based code.

The numerical results illustrated in Figure 7.18 were produced using a method of moments algorithm and required substantial computational time to be realized. Figure 7.20 compares these method of moments results (labeled "MOM") with computations of a parabolic equation code (labeled "PE"); the curves are nearly indistinguishable in this example, indicating the generally high accuracy of the PE

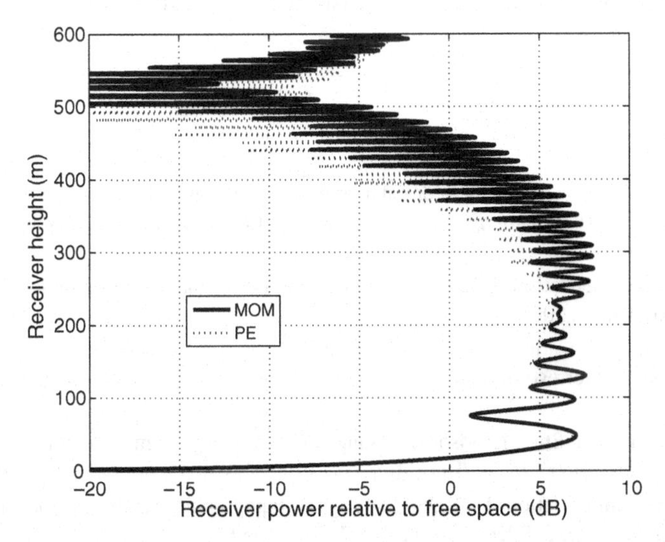

FIGURE 7.20 Comparison of height-gain curves predicted by the method of moments and a parabolic equation code for the Beiseker N15 example of Figure 7.17.

method for propagation problems. The terrain profiles used in these comparisons were obtained from a government mapping agency and contained terrain profile heights recorded at 30 m spacings. While the PE method utilized this profile directly, a linear interpolation between terrain profile points was used to obtain the much finer resolved surface required by the MOM approach. It was also assumed in both MOM and PE computations that the surface was perfectly conducting and that an effective Earth radius multiplier of $\kappa = 4/3$ was applicable.

7.8 CONCLUSION

The methods presented in this chapter have provided insight into the basic physical processes of multipath interference and diffraction by terrain. The path profile analysis rules we have utilized are intended for "point-to-point" applications for which terrain profiles are known, as opposed to "point-to-area" (i.e., broadcast) services or mobile systems. ITU-R Recommendation P.530-12 [3] provides additional information for these applications, including a discussion of diversity methods for overcoming destructive interference. The terrain profile information necessary for the application of these methods is available from several government agencies, including the U.S. Geological Survey; ITU-R Recommendation P.1058 [5] provides additional information on topographic data sets.

Our analyses have focused on a deterministic prediction of path loss properties given a specified value for κ; statistical information on κ can then be utilized to investigate statistics of the path analysis. Generally, it is the worst-case performance that is of interest. In addition, many mechanisms beyond reflections from terrain can cause multipath interference, including contributions from nearby objects (e.g., vehicles, buildings, etc.) Chapter 8 discusses example empirical path loss models for such situations and also introduces basic statistical fading models for predicting propagation losses.

REFERENCES

1. Fishback, William T. "Methods for calculating field strength with standard refraction," in Kerr, Donald E. (ed.), *Propagation of Short Radio Waves*, McGraw-Hill, New York, 1951, pp. 112–140.
2. ITU-R Recommendation P. 526–10, "Propagation by diffraction," International Telecommunication Union, 2007.
3. ITU-R Recommendation P. 530–12, "Propagation data and prediction methods required for the design of terrestrial line-of-sight systems," International Telecommunication Union, 2007.
4. Johnson, J. T., R. T. Shin, J. Eidson, L. Tsang, and J. A. Kong, "A method of moments model for VHF propagation," *IEEE Trans. Antennas Propag.*, vol. 45, pp. 115-125, 1997.
5. ITU-R Recommendation P. 1058–2, "Digital topographic databases for propagation studies," International Telecommunication Union, 1999.

8

EMPIRICAL PATH LOSS AND FADING MODELS

8.1 INTRODUCTION

While the physical models studied in Chapter 7 provide insight into reflection, refraction, and diffraction effects that can occur in propagation between two locations near the Earth's surface, the simplified descriptions of the Earth surface (often modeled as a smooth sphere) and diffraction from obstacles that were used result in only moderate accuracy. The site-specific methods considered in Chapter 7 also require knowledge of terrain along the propagation path; such knowledge sometimes may not be available for fixed transmitters and receivers, or for all possible paths in communication systems involving mobile users or broadcast applications. The methods of Chapter 7 are also oriented primarily to situations involving reflection and/or diffraction from a small number of locations along the propagation path. In many propagation environments, particularly those in urban and suburban areas, there are many possible reflection and diffraction points, resulting in a highly complex propagation environment that is not easily treated using purely physical models.

In complex terrestrial point-to-point propagation problems that exceed the limitations of the methods given in Chapter 7, it is common to apply—in lieu of a physical approach—*empirical* models for predicting propagation losses. This is also true for point-to-area propagation problems as encountered, for example, in predicting cellular communication coverage. As often occurs when empirical data are involved, there are numerous models that have been proposed based on fits to different sets

Radiowave Propagation: Physics and Applications. By Curt A. Levis, Joel T. Johnson, and Fernando L. Teixeira

of measurements. Four popular models for median path loss behaviors at VHF and UHF frequencies are discussed in this chapter: the Okumura–Hata model [1], the COST-231 extension of Okumura–Hata [2], and the Lee model [3], all for macrocells, and the ITU model for indoor picocells [4]. Macrocells refer to wireless links that provide cellular coverage over a radius between 1 km and many tens of kilometers, and where the base station height is larger than the mean height of surrounding buildings and obstacles. Picocells refer to wireless links in indoor environments such as "WiFi" access points. These models have been well accepted in the propagation community, but they should not be applied without careful assessment that the conditions of the underlying experiments are similar to those of the propagation environment in question. One should keep in mind that the main objective of such empirical models is to provide estimates for a wide range of propagation conditions, but this objective is necessarily traded for a reduced accuracy in specific propagation scenarios.

Beyond empirical models for the median path loss in specified environments, it is also possible to develop statistical descriptions of the *variability* of the signal strength found in different environments. We have already encountered statistical descriptions of rain fading in Chapter 5. Here, the underlying probability theory is briefly reviewed before proceeding to a description of some of the statistical fading models used for terrestrial links in urban and suburban environments. Finally, a discussion of the differences between "narrowband" (or flat) fading and "wideband" (or frequency selective) fading is provided along with a summary of particular propagation channel properties that are important for "wideband" systems.

It should be stressed that, although the examples considered in this chapter often focus on terrestrial links (i.e., transmitter and receiver on or near the ground), the concepts have wider applicability. Empirical path loss and signal fading descriptions are ubiquitous in radiowave propagation, and we will encounter further examples in later chapters as well.

8.2 EMPIRICAL PATH LOSS MODELS

For a canonical propagation problem, such as reflection from a planar interface with specified dielectric and conductivity parameters, there is a single correct value for the received field strength. In more complex situations, however, signals may vary in time due to local atmospheric changes, movements of intervening vegetation, nearby traffic changes, movement of the receiver or transmitter (as in mobile systems), and many other possible causes. Such signal changes over a relatively short period of time (often seconds or less, but sometimes lasting for minutes) are called "fading"; they are difficult to predict precisely and will be treated statistically later in this chapter. The empirical models now under discussion attempt to predict an "unfaded" signal strength; that is, the signal strength averaged in some sense over many seemingly similar measurements. Usually, the median value of the field strength (or, equivalently, the median path loss) is described, as opposed to the true average. Even though use of a median value obtained from many measurements eliminates fading effects, the median values that result still represent statistically averaged behaviors because they are computed from propagation measurements in many different situations. For example, given a fixed range, transmitting frequency, and transmitting and receiving antenna heights, models of the median path loss in "urban" environments are to be used to predict propagation losses for a myriad of paths within a single city, or even for many different cities.

Recall in Chapter 5 that we introduced a quantity called the "system loss" to describe the total power loss encountered along a propagation path. The equation for this system loss when specified in decibels was

$$L_{\text{sys}} = 20\log_{10} R_{\text{km}} + 20\log_{10} f_{\text{MHz}} + 32.4 + L_{\text{gas}} + L_{\text{rain}} + L_{\text{pol}} + L_{\text{imp}} + L_{\text{coup}} - G_{\text{Tdb}} - G_{\text{Rdb}}, \tag{8.1}$$

where the subscripts db used in Chapter 5 have been omitted for simplicity. In this chapter, quantities denoted with L will always represent values in decibels.

To simplify the discussion, we focus on the first three terms of equation (8.1) that describe the propagation loss that occurs in the absence of other system and environmental (i.e., L and G) factors. This "path loss" term in free space is therefore

$$L_{\text{p}}^{\text{free}} = 20\log_{10} R_{\text{km}} + 20\log_{10} f_{\text{MHz}} + 32.4, \tag{8.2}$$

and the total system loss is

$$L_{\text{sys}} = L_{\text{p}} + L_{\text{gas}} + L_{\text{rain}} + L_{\text{pol}} + L_{\text{imp}} + L_{\text{coup}} - G_{\text{Tdb}} - G_{\text{Rdb}}. \tag{8.3}$$

If the superscript free is added to L_{sys} and L_{p}, equation (8.3) follows directly from the preceding two equations. The superscripts have been omitted because (8.3) is also used in more general situations, with L_{p} suitably defined, as discussed below.

8.2.1 Review of the Flat Earth Direct plus Reflected Model

A brief review of the direct plus reflected model from Chapter 7 is useful for the later discussions of this section. Recall that we found in Chapter 7 for "low-height" transmitting and receiving antennas located at heights h_1 and h_2, respectively, and separated by a distance d

$$|\underline{E}^{\text{tot}}| = \frac{E_0}{d}\frac{4\pi h_1 h_2}{\lambda d}. \tag{8.4}$$

Given that E_0/d is the field strength in the absence of the ground, the presence of the ground introduces the term

$$\left(\frac{4\pi h_1 h_2}{\lambda d}\right)^2 \tag{8.5}$$

into the power computations of the Friis formula. Doing so results in

$$\begin{aligned} L_{\text{p}}^{\text{flat}} &= 20\log_{10} R_{\text{km}} + 20\log_{10} f_{\text{MHz}} + 32.4 - 20\log_{10}\left(\frac{4\pi h_1 h_2}{\lambda d}\right) \\ &= 120 + 40\log_{10} R_{\text{km}} - 20\log_{10} h_1 - 20\log_{10} h_2, \end{aligned} \tag{8.6}$$

where the distance d used in Chapter 7 here has been replaced with the R_{km} notation of Chapter 5, and the antenna heights are assumed to be in meters.

Equation (8.6) indicates an R^4 loss in the power with range, caused by the interference of the direct and reflected waves when the antenna heights are low. This is a dramatically increased loss with range compared to the R^2 loss encountered in free-space propagation. Furthermore, there is no frequency dependence in equation (8.6) because the frequency dependence of the Friis formula is exactly canceled by the frequency dependence of the term in (8.5). Finally, equation (8.6) states that the dependence of received power on both transmitting and receiving antenna heights is quadratic.

8.2.2 Empirical Model Forms

Empirical studies confirm that for most propagation environments there can be significant variations in the path loss versus range, frequency, and antenna heights that are not predicted by equation (8.6). It is not surprising that the flat Earth model does not achieve a good match to empirical data in most environments, given the model's simplicity and lack of scattering and diffraction phenomena. A specific peculiarity is the absence of frequency dependence in equation (8.6); such behavior is rarely found in practical applications. While the methods of Chapter 7 allow somewhat improved predictions to be made in locations with known terrain profiles, in many applications a more simple and general empirical approach is desirable.

A typical generic form for the median path loss is

$$L_{\mathrm{p}} = L_0 + 10\,\gamma\,\log_{10} R_{\mathrm{km}} + 10\,n\,\log_{10} f_{\mathrm{MHz}} + L_1(h_1) + L_2(h_2), \tag{8.7}$$

with the range and frequency scaling parameters, γ and n, determined empirically, as well as the antenna height factors L_1 and L_2.

The flat Earth model is a special case of equation (8.7) with $L_0 = 120$, $\gamma = 4$, $n = 0$, and $L_1(h) = L_2(h) = -20\log_{10} h$. Most empirical path loss models for complex terrestrial environments specify these parameters according to the specific region type, that is, urban, rural, and so on. Of course, it should be expected that such predictions represent median results that will not necessarily produce highly accurate predictions in a particular location. Nevertheless, predictions should be reasonable if used in a statistical sense for a set of locations within a given region type. Three models for macrocell behaviors (i.e., to be applied for R_{km} from around 1–20 km with a base station antenna at moderate to large heights and a mobile station antenna at heights closer to the ground) are discussed in this section. In particular, these models are very popular for cellular communication system design and are intended for situations in which the base station antenna height in urban situations is above the rooftops of surrounding buildings. One indoor model for picocells is also discussed.

8.2.3 Okumura–Hata Model

The Okumura–Hata model (sometimes simply called Hata model) is a set of empirical curves constructed by Hata [1] to fit data measured by Okumura [5] in Tokyo, Japan, and surrounding areas. The specific form of the Okumura–Hata model described here applies for frequencies between 100 and 1500 MHz and for ranges from 1 to 20 km, and is given by

$$L_0 = 69.55 \quad \text{(urban areas)}, \tag{8.8}$$

$$L_0 = 64.15 - 2\left[\log_{10}(f_{\mathrm{MHz}}/28)\right]^2 \quad \text{(suburban areas)}, \tag{8.9}$$

$$L_0 = 28.61 - 4.78\left[\log_{10} f_{\mathrm{MHz}}\right]^2 + 18.33\log_{10} f_{\mathrm{MHz}} \quad \text{(open rural areas)}, \tag{8.10}$$

and

$$\gamma = 4.49, \tag{8.11}$$

$$n = 2.616, \tag{8.12}$$

$$L_1(h_1) = -\left(13.82 + 6.55\log_{10} R_{\mathrm{km}}\right)\log_{10}(h_1), \tag{8.13}$$

with the antenna heights in meters. It is assumed that h_1 is a base station-type antenna, at heights ranging from 30 to 200 m, while h_2 is a mobile receiver antenna, at heights ranging from 1 to 10 m. For urban areas, $L_2(h_2)$ is specified as follows:

$$L_2(h_2) = 1.56\log_{10} f_{\mathrm{MHz}} - 0.8 + \left(0.7 - 1.1\log_{10} f_{\mathrm{MHz}}\right)h_2 \quad \text{small/medium city}, \tag{8.14}$$

$$L_2(h_2) = 4.97 - 3.2\left[\log_{10}(11.75h_2)\right]^2 \quad \text{large city and } f_{\mathrm{MHz}} \geq 400, \tag{8.15}$$

$$L_2(h_2) = 1.1 - 8.29\left[\log_{10}(1.54h_2)\right]^2 \quad \text{large city and } f_{\mathrm{MHz}} \leq 200. \tag{8.16}$$

Here a "large city" is defined as an area having an average building height greater than 15 m [8]. The original references do not address the form of $L_2(h_2)$ for open rural areas and suburban areas, but some authors have taken $L_2(h_2)$ as either equal to the "small/medium city" case or equal to zero. Some authors for the large city case also replace the conditions $f_{\mathrm{MHz}} \geq 400$ and $f_{\mathrm{MHz}} \leq 200$ in equations (8.15) and (8.16) with $f_{\mathrm{MHz}} \geq 300$ and $f_{\mathrm{MHz}} < 300$, respectively; the original data sets did not include measurements near 300 MHz.

Figure 8.1 illustrates several plots of path loss under the Okumura–Hata model. The upper left plot demonstrates that variations in the dependence on range occur as the transmitter height is varied (note the loss for the higher antenna has been increased by 15 dB to facilitate comparison of the dependencies on range), while the upper right plot is a similar illustration of the varying frequency dependencies that can occur as the receiver height is changed. Variations in frequency dependence also occur for the "suburban" and "rural" areas. Similar comparisons are illustrated in the lower plots for the dependencies on transmitting and receiving antenna heights; the lower right plot includes curves versus receiver height in the "small" and "large" city cases. The slower decrease in loss for the "large" city case as the receiver height is increased can be explained by the increased level of obstructions in the large city environment.

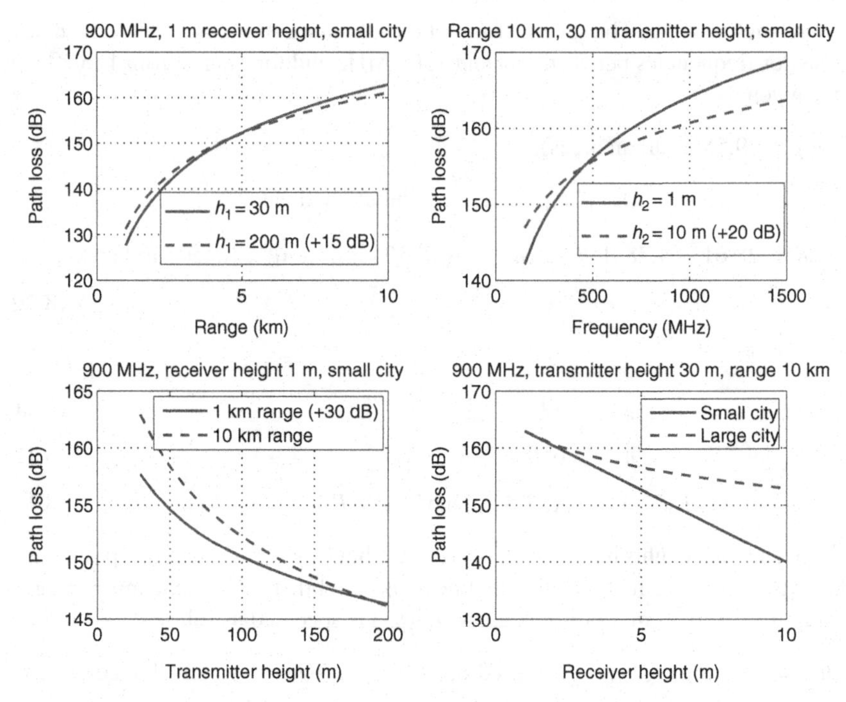

FIGURE 8.1 Sample plots of path loss under the Okumura–Hata model.

8.2.4 COST-231/Hata Model

Since the Okumura–Hata model is applicable only in the 100–1500 MHz range, and given the interest in personal communication systems operating near 1.9 GHz, the intergovernmental agency COST (European Cooperation in Science and Technology) performed propagation measurements to extend it to 2 GHz [2], where the original Okumura–Hata model is known to underestimate the path loss. The parameters of the COST-231/Hata model are

$$L_0 = 46.3, \tag{8.17}$$

$$\gamma = 4.49, \tag{8.18}$$

$$n = 3.39, \tag{8.19}$$

and the L_1 and L_2 functions are identical to those of Okumura–Hata. The above parameters are used in "medium-sized cities" and suburban areas; L_0 is increased to 49.3 in dense urban areas. This empirical COST-231/Hata model is applicable for frequencies from 1.5–2 GHz, with other limitations identical to those of Okumura–Hata. In particular, neither model should be used to predict path loss in scenarios where the base station height is smaller than the surrounding buildings or for transmission distances less than 1 km.

8.2.5 Lee Model

The (area-to-area) Lee model [3] has also been widely used in prediction of median path losses in macrocell applications, particularly for systems operating near 900 MHz and ranges greater than 1.6 km. The Lee model specifies distinct parameters for varying region types. The simplest specification is

$$n = 2 \qquad f_{\text{MHz}} < 850, \tag{8.20}$$

$$n = 3 \qquad f_{\text{MHz}} > 850. \tag{8.21}$$

The antenna height factors are also simple:

$$L_1(h_1) = -20\log_{10}(h_1/30), \tag{8.22}$$

$$L_2(h_2) = -(10)(\chi)\log_{10}(h_2/3), \tag{8.23}$$

where again the heights are given in meters, and it is assumed that h_1 is at a moderate to large height (greater than 30 m) while h_2 is around 3 m. In the above, $\chi = 2$ for $h_2 > 3$ m, and $\chi = 1$ for $h_2 < 3$ m. For the Lee model, the L_0 parameter is written as

$$L_0 = 50.3 + P_0 - 10\,\gamma\,\log_{10}(1.61) - 10\,n\,\log_{10}(900), \tag{8.24}$$

where P_0 and γ are determined empirically through a set of measurements in a given location. The following results from such measurements are available [3]:

$$P_0 = 49 \qquad \text{open terrain}, \tag{8.25}$$

$$P_0 = 62 \qquad \text{suburban areas}, \tag{8.26}$$

$$P_0 = 70 \qquad \text{Philadelphia}, \tag{8.27}$$

$$P_0 = 64 \qquad \text{Newark, NJ}, \tag{8.28}$$

$$P_0 = 77 \qquad \text{New York City}, \tag{8.29}$$

$$P_0 = 84 \qquad \text{Tokyo}, \tag{8.30}$$

and

$$\gamma = 4.35 \qquad \text{open terrain}, \tag{8.31}$$

$$\gamma = 3.84 \qquad \text{suburban areas}, \tag{8.32}$$

$$\gamma = 3.68 \qquad \text{Philadelphia}, \tag{8.33}$$

$$\gamma = 4.31 \qquad \text{Newark, NJ}, \tag{8.34}$$

$$\gamma = 4.80 \qquad \text{New York City}, \tag{8.35}$$

$$\gamma = 3.05 \qquad \text{Tokyo}. \tag{8.36}$$

The variability in the above empirical parameters for seemingly similar cities emphasizes the importance of empirical determination of L_0 and γ whenever possible. The Lee model should be expected to give increasingly less accurate predictions as the conditions of operation deviate away from the standard measurements used in its compilation, that is, $f = 900$ MHz, $d = 1.6$ km, $h_1 = 30$ m, and $h_2 = 3$ m.

8.2.6 Site-General ITU Indoor Model

The preceding models are applicable for macrocells only. For indoor propagation channels in which both transmitter and receiver are situated in the same building, other empirical models for the path loss should be applied. A popular, site-general indoor path loss model in ITU-R Recommendation P.1238 [4] has

$$L_{\mathrm{p}} = -28 + 10\gamma \log_{10} R_{\mathrm{m}} + 20\log_{10} f_{\mathrm{MHz}} + L_{\mathrm{f}}(n_h) \tag{8.37}$$

in decibels, where R_{m} is the separation between the transmitter and receiver in meters, and L_{f} is a floor separation loss factor that depends on the number n_h of floors penetrated between the transmitter and the receiver. The parameters γ and $L_{\mathrm{f}}(n_h)$ depend on the type of building and also on frequency. For the 1.8–2.0 GHz frequency range, they are $\gamma = 2.8$ and $L_{\mathrm{f}}(n_h) = 4n_h$ in residential buildings, $\gamma = 3.0$ and $L_{\mathrm{f}}(n_h) = 15 + 4(n_{\mathrm{h}} - 1)$ in office buildings, and $\gamma = 2.2$ and $L_{\mathrm{f}}(n_h) = 6 + 3(n_h - 1)$ in commercial buildings. Parameters for other frequency ranges can be found in ITU-R Recommendation P.1238 [4] and in Refs. [6,7].

8.2.7 Other Models for Complex Terrain

The models that have been considered are examples among numerous other empirical models for complex terrestrial environments that have been described in the literature for macrocell applications as well as for microcells (smaller range cellular systems operating in urban environments) and picocells (i.e., indoor or very short-range systems) [2,5–9]. The simplest forms of all of these models employ path loss equations of the type described in Section 8.2.2 and require only minimal knowledge of a specific location. As more information on a location becomes available, more sophisticated models can be considered, ranging from *site-specific* empirical methods to *physics-based* models such as ray-tracing [10] or parabolic equation [10,11] models in a completely characterized environment. These physical models are more accurate but computationally intensive and therefore less widely applied.

One approach incorporating site-specific information in an empirical theory for urban environments is the COST231-Walfish–Ikegami model [2]. For line-of-sight cases, this model retains a form similar to that of Section 8.2.2. However, when no line of sight from transmitter to receiver is available, the model expresses the received power in terms of parameters involving average street widths and building heights in the urban environment. This model is applicable to frequencies from 800 to 2000 MHz and to distances from 20 m to 5 km.

8.2.8 An Example of Empirical Path Loss Model Usage

Problem: Consider a cellular communication system operating at 900 MHz in a Columbus, OH, suburb; we will classify the terrain type as "suburban" for this location. The transmit power is 25 W, and we will assume that all receivers and transmitters are polarization and impedance matched. For a base station antenna of gain

9 dBi located at 30 m above the ground, and a receiver antenna gain of 2 dBi located at 2 m above the ground and 3 km away from the transmitter, compare the predicted power received under the direct plus ground reflected (Section 8.2.1), Okumura–Hata, and Lee models.

Solution: From the definition of system loss discussed previously,

$$P_{\mathrm{R,dbW}} = P_{\mathrm{T,dbW}} - L_{\mathrm{sys}}, \tag{8.38}$$

and from equation (8.3), neglecting all mismatch and polarization losses as well as atmospheric gas and rain attenuation, we obtain

$$P_{\mathrm{R,dbW}} = P_{\mathrm{T,dbW}} + G_{\mathrm{T,db}} + G_{\mathrm{R,db}} - L_{\mathrm{p}} = 25 - L_{\mathrm{p}}. \tag{8.39}$$

When the plane-Earth direct-plus-reflected-ray model for low antennas (equation (8.6)) is used, the result is

$$\begin{aligned} L_{\mathrm{p}} &= 120 + 40\log_{10} R_{\mathrm{km}} - 20\log_{10} h_1 - 20\log_{10} h_2 \\ &= 120 + 40\log_{10} 3 - 20\log_{10} 30 - 20\log_{10} 2 \\ &= 103.5 \text{ dB}. \end{aligned} \tag{8.40}$$

The Okumura–Hata model gives

$$\begin{aligned} L_{\mathrm{p}} &= 64.15 - 2\left[\log_{10}(f_{\mathrm{MHz}}/28)\right]^2 + 44.9\log_{10} R_{\mathrm{km}} + 26.16\log_{10} f_{\mathrm{MHz}} \\ &\quad + \log_{10}(h_1)\left(-13.82 - 6.55\log_{10} R_{\mathrm{km}}\right) \\ &= 133.3 \text{ dB}, \end{aligned} \tag{8.41}$$

where $L_2(h_2)$ was taken as zero; use of the "small city" $L_2(h_2)$ reduces the path loss by 1.3 dB.

The Lee model results in

$$\begin{aligned} L_{\mathrm{p}} &= 50.3 + 62 + 38.4\log_{10}\frac{R_{\mathrm{km}}}{1.609} + 30\log_{10}\frac{f_{\mathrm{MHz}}}{900} \\ &\quad -20\log_{10} h_1/30 - 10\chi\log_{10} h_2/3 \\ &= 124.5 \text{ dB}. \end{aligned} \tag{8.42}$$

These path loss predictions result in received power predictions of −78.5, −108.3, and −99.5 dBW for the direct plus reflected model, Okumura–Hata model, and Lee model, respectively.

The flat-Earth, direct-plus-reflected ray, low-antenna theoretical model was chosen because it has the same generic form as the empirical models, but it is not really appropriate for this calculation. It is easy to verify that neglecting the Earth bulge for this short distance is not a problem, but equation (7.22) shows that the low-antenna assumption is violated, and Figure 7.2 (although not drawn for the same conditions) shows that the height-gain functions will be overestimated seriously when the low-antenna approximation is used beyond its appropriate range. Also it must be remembered that the very purpose of the empirical methods is to include the effects of obstructions and reflections in this suburban environment in a statistical sense; this is also missing in the flat-Earth low-antenna theoretical model. The difference between the Lee and Okumura-Hata results shows the variability that can be encountered with empirical model estimates; in this case, the Lee model may be recommended as somewhat more reliable, given the Okumura-Hata model's focus on urban areas, but no empirical model should be assumed to have a high accuracy.

8.3 SIGNAL FADING

The path loss models presented do not include any information on variations in the system loss to be expected with time or, in the case of a moving transmitter or receiver, location. In general, signal strengths do vary over time due to changes in the propagation channel (in the case of a moving vehicle, the signal may change as a function of location, but the changes will be experienced as functions of time). For example, the transmission between two microwave towers may vary due to moving tree branches; the transmission between a ground station and a satellite is affected by atmospheric refractivity variations, clouds, rain, and intervening ionospheric conditions; a mobile receiver will see signal strength fluctuations due to changes in the path geometry or the relative locations of nearby objects as it moves. These variations are called "fading."

Fading is of great importance in predicting the reliability of a propagation link, and it is therefore of great importance to the system designer. Compensating for deep fading can be expensive, especially if transmitter power must be increased to achieve the desired reliability. Since fading cannot be described in a deterministic way, it is necessary in system design to calculate the *probability* that a given fade level will be exceeded. Before going into this in detail, a brief discussion of general fading characteristics is in order. The fading experienced by a mobile receiver in a suburban environment will be used as an example.

It is customary to classify fading into two basic types: *slow fading* and *fast fading*; some authors also use the terms "large-scale" and "small-scale" fading for these effects. Slow fading occurs, for example, when a mobile receiver moves in the environment and signal variability is experienced due to shadowing/diffraction effects from nearby buildings/obstacles. The associated signal fluctuations are "slow" because shadowing will be effective over relatively large distances determined by the size of obstacles. Alternatively, fading can be caused by interference among signals propagating over many different paths to the receiver (e.g. from multiple reflections). This interference can be constructive or destructive, depending on the relative phase of the multipath signals. Since the relative phase among the interfering signals depends on the wavelength, the resulting signal strength can change rapidly in space (especially for VHF frequencies and above); hence, this is denoted as "fast" fading. Even though slow and fast fading in general coexist simultaneously, there is often a clear "separation of scales" so that they can be treated independently.

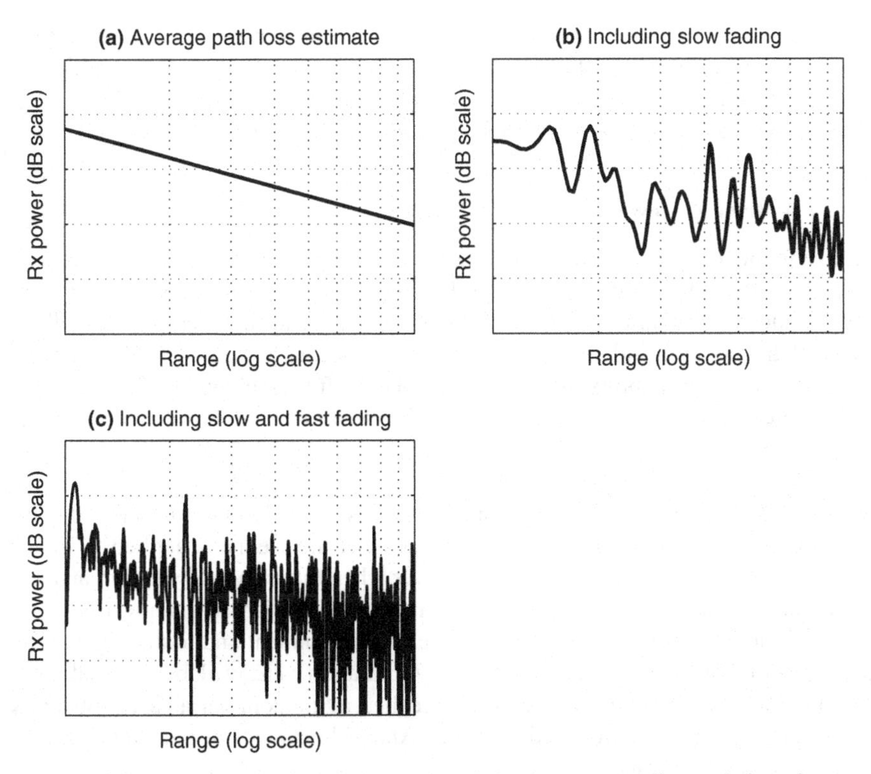

FIGURE 8.2 Illustrations of typical path loss behaviors: (a) median path loss from an empirical model, (b) including effects of slow fading, and (c) including effects of both slow and fast fading.

Figure 8.2 provides an illustration of the typical behavior of these fading mechanisms for a mobile receiver. These curves represent functions of time as the distance from the transmitter is increased. Predictions of the empirical path loss models agree with Figure 8.2a in showing a smooth and slowly increasing median path loss versus range. However, it should be expected that particular obstacles encountered along the path cause variable shadowing effects and give rise to variations in the path loss versus range. An illustration of this slow fading is included in Figure 8.2b. Finally, in environments where many paths exist by which transmitted power can reach the receiver, interference among the multipath receptions will cause the signal to vary rapidly over distances comparable to one wavelength. Figure 8.2c illustrates the rapid signal power variations that occur when fast fading is included.

A typical example of fast fading, that the reader is likely to have experienced when stopping for a traffic light, is an abrupt change in the quality of received FM radio signals as the automobile moves a few meters or less. FM broadcast frequencies correspond to wavelengths on the order of a few meters, and destructive interference from reflections by nearby objects can occur on this scale. It is also interesting to note that a mobile receiver moving at 60 miles per hour travels nearly 27 m/s; for a cellular phone operating at 900 MHz, a 27 m distance equals about 80 wavelengths. Hence, it is not unreasonable to expect that on the order of 100 fades can be encountered within 1 s for such an user—this can surely be denoted as "fast" fading!

8.3.1 A Brief Review of Probability Theory

Similar to the treatment of the median path loss, fading is treated by the measurement of fading effects over many paths and then approximating the statistics of the observed fades with an empirical probability model that fits the data reasonably well. To establish the notation and make the following discussions reasonably self-contained, a brief review of probability theory is included next. It is assumed that the reader has encountered these concepts before, so a detailed treatment is left to other references [12,13].

8.3.1.1 Random Variables Probability theory is applied to situations that involve experiments with unknown outcomes. In our case, the theory is applied to model either the path loss or the final received power when including slow fading and fast fading along a specific path at a specific time; this is taken as unknown due to a lack of complete knowledge of the propagation environment and to the difficulty in performing precise calculations even if such information were available. The unknown outcome of a propagation measurement is called a "random variable"; probability theory provides a means for predicting the expected behavior of a random variable over many measurements.

For concreteness, let us choose a specific random variable for further consideration: the path loss contribution due to slow fading, denoted here as L_S. The slow fading loss is represented in decibels and is added to our empirical median path loss L_P to estimate the received power in a given measurement. Fast fading loss will also be modeled in terms of random variables, as explained later.

8.3.1.2 Probability Density Function and Cumulative Distribution Function A random variable is specified through its *probability density function* (pdf). This function indicates the likelihood that the random variable will assume an outcome within a particular set (interval) of values. In this text, we are concerned only with random variables having continuous (as opposed to discrete) outcomes. For a continuous random variable L_S, we can define the probability density function $f_{L_\mathrm{s}}(l_\mathrm{S})$ by

$$f_{L_\mathrm{s}}(l_\mathrm{S}) = \lim_{\Delta l_\mathrm{S} \to 0} \frac{\mathcal{P}(|L_\mathrm{S} - l_\mathrm{S}| < \Delta l_\mathrm{S})}{\Delta l_\mathrm{S}}, \quad (8.43)$$

where the $\mathcal{P}$ symbol represents the probability of obtaining the experiment outcome specified in the argument of $\mathcal{P}$. Here, the outcome specified is a measured slow fading loss L_S being within a small range Δl_S of a specified value l_S. A similar definition can be made for the fast fading loss. The probability $\mathcal{P}$ is always a number between 0 (i.e., the outcome never occurs) and 1 (i.e., the outcome always occurs), with higher values of $\mathcal{P}$ indicating that an event is more likely to occur. Note that pdfs are never negative because this would require negative values of $\mathcal{P}$. However, pdfs can exceed unity since it is only their integral over an interval, not their value, that gives the probability of occurrence within that interval.

Frequently, physical considerations can be used to propose a pdf for a random variable. However, if no clear physical model of a quantity of interest is available, one method for creating a pdf is to perform a large number of measurements, followed by attempts to find a mathematical model that reasonably fits the observed data. This process is sometimes called "density estimation." The models to be described here have been verified through such a process.

The *cumulative distribution function* (cdf) of the slow fading loss describes the probability that the slow fading loss is less than or equal to a specified value. Such information is useful for predicting link margins, as performed for rain fades in Chapter 5. Given a pdf $f_{L_s}(l_s)$, we can define the cdf $F_{L_s}(l_s)$ as

$$F_{L_s}(l_s) = \mathcal{P}(L_s \leq l_s) = \int_{-\infty}^{l_s} dl\; f_{L_s}(l). \tag{8.44}$$

Because pdfs are always nonnegative, cdfs are never decreasing functions of their argument. It is also clear that cdfs must take on the value 0 for argument $-\infty$ and the value 1 for argument ∞. The latter provides a normalization condition that any pdf must satisfy:

$$\int_{-\infty}^{\infty} f_{L_s}(l_s) dl_s = 1. \tag{8.45}$$

Given the fact that the cdf is obtained from an integration of the pdf, we can find

$$f_{L_s}(l_s) = \frac{d}{dl_s} F_{L_s}(l_s), \tag{8.46}$$

so knowledge of either the pdf or the cdf is sufficient to determine the other.

8.3.1.3 Expected Value and Standard Deviation If the pdf of a random variable is known, it is possible to compute the *average* of this random variable. Average values are also called "mean values" or "expected values." It is clear that the average value of a path loss should involve a summation of path loss values multiplied by the fraction of experiments in which these values occur. Given the definition of the pdf, the mean slow fading loss μ_{L_s} is given by

$$\mu_{L_s} = E\left[L_s\right] = \int_{-\infty}^{\infty} dl\; l\; f_{L_s}(l), \tag{8.47}$$

where $E[\cdot]$ is called the *expected value* operator. The expected value operator can be defined more generally for any function g of the random variable L_s through

$$E\left[g(L_s)\right] = \int_{-\infty}^{\infty} dl\; g(l)\; f_{L_s}(l). \tag{8.48}$$

The expected value of L_s turns out to be 0 dB because fading was eliminated in the definition of the median empirical path loss L_p.[1] Even though the expected value of a slow fade is 0 dB, slow fades still produce variations in the signal power level in a given measurement. A second quantity, called the *variance*, is useful in characterizing the level of variation of a random variable. The variance $\sigma_{L_s}^2$ of a random variable L_s is defined through

$$\sigma_{L_s}^2 = E\left[(L_s - E[L_s])^2\right]. \tag{8.49}$$

The *standard deviation* σ_{L_s} is defined as the square root of the variance. Both the variance and the standard deviation are nonnegative quantities; larger variances indicate that a random variable exhibits greater variability. Since slow fading loss is specified in dB, its standard deviation is also expressed in dB.

8.3.2 Statistical Characterization of Slow Fading

The slow fading loss expressed in decibels can be approximated as a *Gaussian* random variable. The pdf of a Gaussian random variable is

$$f_{L_s}(l_s) = \frac{1}{\sigma_{L_s}\sqrt{2\pi}} e^{-\frac{(l_s-\mu_{L_s})^2}{2\sigma_{L_s}^2}}, \tag{8.50}$$

where, as before, $\sigma_{L_s}^2$ is the variance and $\mu_{L_s} = E[L_s]$ is the expected value of the Gaussian variable. As mentioned above, $\mu_{L_s} = 0$ dB for the slow fading loss due to our separation of the slow fading loss from the median path loss term L_p. Many references report a σ_{L_s} range of 4–12 dB as applicable in typical point-to-point ground-based propagation problems. For mobile communications the value of σ_{L_s} tends to increase with frequency because shadowing effects are more pronounced at higher frequencies. Typically, σ_{L_s} also tends to be larger in suburban environments than in most urban environments due to the larger variability of the nearby clutter in the former. An empirical formula for σ_{L_s} in urban and suburban environments determined from the Okumura–Hata measurements is

$$\sigma_{L_s} = \sigma_0 - 1.3\log_{10} f_{\mathrm{MHz}} + 0.65\left(\log_{10} f_{\mathrm{MHz}}\right)^2, \tag{8.51}$$

where σ_{L_s} is expressed in dB, and with $\sigma_0 = 5.2$ dB used in urban environments and $\sigma_0 = 6.6$ dB in suburban environments. This equation is applicable for the conditions defined for the Okumura–Hata model.

A Gaussian random variable is also called a "normal" random variable. The corresponding cdf is

$$F_{L_s}(l_s) = \frac{1}{2}\left[1 + \mathrm{erf}\left(\frac{l_s - \mu_{L_s}}{\sqrt{2}\sigma_{L_s}}\right)\right], \tag{8.52}$$

[1]More precisely, this is generally true only if the *mean* path loss is considered. However, since the pdf of L_s is symmetric as discussed below, this distinction can safely be ignored here.

TABLE 8.1 Percent of L_s Outcomes Exceeding a Specified Argument Y for a Gaussian Random Variable L_s

$Y = \frac{L_s - \mu_{L_s}}{\sigma_{L_s}}$	$100 \times \left(1 - F_{L_s}(Y)\right)$
1.0000	15.87 %
1.2816	10.00 %
1.6449	5.00 %
2.0000	2.28 %
2.3263	1.00 %
3.0000	0.13 %
3.0902	0.10 %
4.0000	3.167×10^{-3} %

The value Y is specified in terms of number of standard deviations from the mean of L_s.

where

$$\text{erf}(\beta) = \frac{2}{\sqrt{\pi}} \int_0^{\beta} dt\, e^{-t^2} \tag{8.53}$$

is the "error function". Routines and tables for computing this special function are widely available.

One useful operation with the Gaussian cdf involves computation of the percent of L_s outcomes that are expected to exceed a specified value. This percent is given by $100 \times \left(1 - F_{L_s}(l_s)\right)$, where l_s is the specified value. Table 8.1 shows these percentages for the Gaussian cdf.

Assuming that the only significant fading contribution is due to the slow fading loss, it is clear that the received signal measured *in decibels*, $P_{\text{R,dBW}} = M$, also has a normal distribution, that is,

$$f_M(m) = \frac{1}{\sigma_M \sqrt{2\pi}} e^{-\frac{(m-\mu_M)^2}{2\sigma_M^2}}, \tag{8.54}$$

where μ_{M} is the mean received signal[2] in dBW, which can be predicted through use of the empirical path loss models considered previously in this chapter, and $\sigma_M = \sigma_{L_s}$.

If we convert back to a linear representation of power in watts, we have $10 \log_{10}(S) = P_{\text{R,dBW}} = M$, where S is the received power relative to 1 W; this can also be rewritten as $\frac{10}{\ln 10} \ln S = M$. The pdf of S can be found as the derivative of the cdf of S, which is

$$\begin{aligned} F_S(s) &= \mathcal{P}(S \leq s) \\ &= \mathcal{P}\left(M \leq 10 \log_{10} s\right) \\ &= F_M\left(\frac{10}{\ln 10} \ln s\right). \end{aligned} \tag{8.55}$$

[2]Also, the median received signal in this case.

The pdf of S is then

$$f_S(s) = \frac{d}{ds} F_M\left(\frac{10}{\ln 10}\ln s\right)$$

$$= f_M\left(\frac{10}{\ln 10}\ln s\right)\frac{d}{ds}\left(\frac{10}{\ln 10}\ln s\right), \tag{8.56}$$

which is a *lognormal distribution*:

$$f_S(s) = \left(\frac{1}{\sigma'\sqrt{2\pi}}\right)\frac{1}{s}e^{-\frac{(\ln s-\mu')^2}{2\sigma'^2}}, \tag{8.57}$$

where

$$\mu' = \frac{\ln 10}{10}\mu_M, \tag{8.58}$$

$$\sigma' = \frac{\ln 10}{10}\sigma_M, \tag{8.59}$$

with $(\ln 10)/10 \approx 0.2303$. It can be shown that the expected value and the variance of S are

$$\mu_S = e^{\mu'+\sigma'^2/2}, \tag{8.60}$$

$$\sigma_S^2 = e^{2\mu'+\sigma'^2}\left(e^{\sigma'^2}-1\right). \tag{8.61}$$

Note that the mean of S is not determined solely by the mean of M; this is because averaging quantities in decibels (μ_M) is different from averaging the corresponding linear values (μ_S).

In our chosen example—a mobile receiver in a suburban environment—slow fading variations correspond to diffraction effects that take place on characteristic scales on the order of tens to hundreds of meters at VHF and UHF frequency bands. However, slow fading is also observed in many other frequency bands and propagation scenarios, and is often characterized with the same probability distribution.

8.3.3 Statistical Characterization of Narrowband Fast Fading

Fast fading results when there is more than one ray path from the transmitter to the receiver and the relative phases of the signals vary in time. The multipath components interfere with each other at the receiver location and can cause great variability in the strength of the resulting signal. Two probability models are commonly used for propagation channels with fast fading. One is the *Rayleigh model*, which corresponds to situations where no line-of-sight (NLOS) path is present between transmitter and receiver. The other is the *Rician model*, which corresponds to situations in which a "line-of-sight"' (LOS) path exists between transmitter and receiver. For the present we assume that all frequency components of the signal experience the same fading magnitude, which characterizes *flat* or "narrowband" fading. *Frequency-selective* or "broadband" fading will be considered later in the chapter.

8.3.3.1 The Rayleigh Channel We begin by writing the time-harmonic electric field at the receiver as a superposition of fields from the various ray paths (multipath components) reaching the receiver. If there are N significant paths, the time-harmonic voltage $\underline{V}$ received can be written as

$$\underline{V} = \sum_{n=1}^{N} w_n e^{-j\psi_n}, \tag{8.62}$$

where w_n and ψ_n are the amplitude and phase, respectively, of each path's contribution. The above can be also written as

$$\underline{V} = \sum_{n=1}^{N} (a_n + jb_n) = \sum_{n=1}^{N} a_n + j\sum_{n=1}^{N} b_n = A + jB, \tag{8.63}$$

where a_n and b_n are the real and imaginary parts of the voltage contributed by each path. For a Rayleigh channel, N is assumed to be a large number, and a_n and b_n are assumed to be identically distributed random variables (this is an assumption because, in practice, N is not necessarily large and the contributions are not necessarily identically distributed). The sums A and B therefore both approach a Gaussian distribution by the central limit theorem of probability theory. The Gaussian distributions of A and B are independent and have identical standard deviations $\sigma_A = \sigma_B$ (volts) and zero mean because it is assumed that the ray phases in equation (8.62) correspond to changes in path length that are large compared to the wavelength. The latter makes the phases equally likely to take on a large range of values, and the resulting interference produces a zero mean received complex voltage when averaged over many measurements.

Even though the average complex voltage is zero, the *magnitude* of the voltage $|\underline{V}| = V = \sqrt{A^2 + B^2}$ has a positive mean value. It can be shown that the pdf of V is a *Rayleigh distribution* [14], given by

$$f_V(v) = \frac{v}{\sigma_A^2} e^{-v^2/2\sigma_A^2}, \tag{8.64}$$

in V^{-1}. The mean and standard deviation of V are

$$\mu_V = E[V] = \sigma_A \sqrt{\frac{\pi}{2}}, \tag{8.65}$$

$$\sigma_V = \sigma_A \sqrt{\frac{4 - \pi}{2}}, \tag{8.66}$$

both expressed in volts.

For the Rayleigh channel, it is also possible to determine the pdf of the power at the receiver P in watts. For this purpose, we write $P = V^2/R_\mathrm{R}$ (using rms values for V), where R_R is the receiver input resistance, and find the pdf of P by first finding the cdf of P and then differentiating. Following this process,

$$\begin{aligned} F_P(p) &= \mathcal{P}(P \leq p) \\ &= \mathcal{P}\left(V^2/R_R \leq p\right) \\ &= \mathcal{P}\left(V \leq \sqrt{pR_R}\right) \\ &= F_V\left(\sqrt{pR_R}\right). \end{aligned} \tag{8.67}$$

Now differentiate to obtain

$$\begin{aligned} f_P(p) &= \frac{d}{dp} F_V\left(\sqrt{pR_R}\right) \\ &= f_V\left(\sqrt{pR_R}\right) \frac{d}{dp}\sqrt{pR_R} \\ &= f_V\left(\sqrt{pR_R}\right) \frac{1}{2}\sqrt{\frac{R_R}{p}}, \end{aligned} \tag{8.68}$$

which can be simplified to

$$f_P(p) = \frac{1}{\mu_P} e^{-p/\mu_P}, \tag{8.69}$$

expressed in W^{-1}, with $\mu_P = 2\sigma_A^2/R_R$ and for $p > 0$. This pdf corresponds to an *exponential distribution*. It is easy to show that the mean received power is $\mu_P = 2\sigma_A^2/R_R$ watts and the standard deviation of the received power σ_P is equal to μ_P. Therefore, the Rayleigh channel clearly has a high degree of variability. Furthermore, the cdf also follows easily and is given by

$$F_P(p) = 1 - e^{-p/\mu_P}. \tag{8.70}$$

At first sight, it may seem curious that the pdf of the Rayleigh voltage vanishes for zero voltage, while the pdf of the Rayleigh power is maximum for zero power. This apparent contradiction is resolved when it is recognized that it is only the integral of the pdf over an interval, and not its value at a point, that determines the probability for that interval.

The Rayleigh channel model approximates propagation behaviors reasonably for VHF and higher frequencies in large cities under NLOS conditions. It can also be used to approximate fading in some tropospheric scatter or ionospheric channels where the intervening refractive index or electron density is highly irregular, fading in some ship-to-ship radio links, and in other applications. Additional properties of the Rayleigh channel will be discussed subsequent to the consideration of another commonly used fading model.

8.3.3.2 The Rician Channel We now turn our attention to the case where there is a LOS path in the presence of a large number N of multipath contributions, the latter all of the same order.[3] In this case, the pdf of the voltage magnitude V at the receiver in the limit of large N is given in terms of the Rice distribution [14,15]

$$f_V(v) = \frac{v}{\sigma_A^2} e^{-\frac{V_m^2+v^2}{2\sigma_A^2}} I_0\left(\frac{V_m v}{\sigma_A^2}\right) \tag{8.71}$$

expressed in V^{-1} and with $v > 0$. The parameter V_m corresponds to the voltage magnitude at the receiver due to the LOS path in the absence of all other multipath components. As before, σ_A is the standard deviation of the real (or the imaginary, as they coincide) part of the time-harmonic multipath component (note that $\sigma_V \neq \sigma_A$). The quantity I_0 represents the modified Bessel function of zeroth order: routines and tables for computing this function are widely available. A series expansion for $I_0(x)$ is

$$I_0(x) = \sum_{k=0}^{\infty} \frac{(x/2)^{2k}}{(k!)^2}, \tag{8.72}$$

and a useful approximation for large argument is

$$I_0(x) \approx \frac{e^x}{\sqrt{2\pi x}} \left(1 + \frac{1}{8x}\right). \tag{8.73}$$

It can be verified that the Rice distribution recovers the Rayleigh distribution for $V_m = 0$, as should be expected since $V_m = 0$ corresponds to a NLOS situation.

Alternatively the Rician channel can be characterized in terms of the pdf of the received power P_{rec} in watts. The procedure is to find the cdf of $P_{rec} = V^2/R_R$, and then differentiating, as in the Rayleigh case. There it was found that

$$f_P(p) = f_V\left(\sqrt{pR_R}\right) \frac{1}{2}\sqrt{\frac{R_R}{p}}, \tag{8.74}$$

which can be simplified for a Rician channel to

$$f_P(p) = \frac{1}{P_f} e^{-\frac{(P_m+p)}{P_f}} I_0\left(\frac{2\sqrt{P_m p}}{P_f}\right), \tag{8.75}$$

expressed in W^{-1} and valid for $p > 0$, and where the subscript "rec" has been omitted. Here, $P_m = V_m^2/R_R$ is the power associated with the LOS path in the absence of fading, and $P_f = 2\sigma_A^2/R_R$ is the average power of the fading components.

It can be shown that the mean received power in a Rician channel is

$$P_{rec,mean} = P_f + P_m = P_f(1 + K), \tag{8.76}$$

[3]The Rician channel does not require a "line of sight" in the sense that the receiver needs to be visible from the transmitter. Generally speaking, this model is also applicable to scenarios where one strongly dominant path exists; sometimes this dominant path may be subject to a specular reflection, for example.

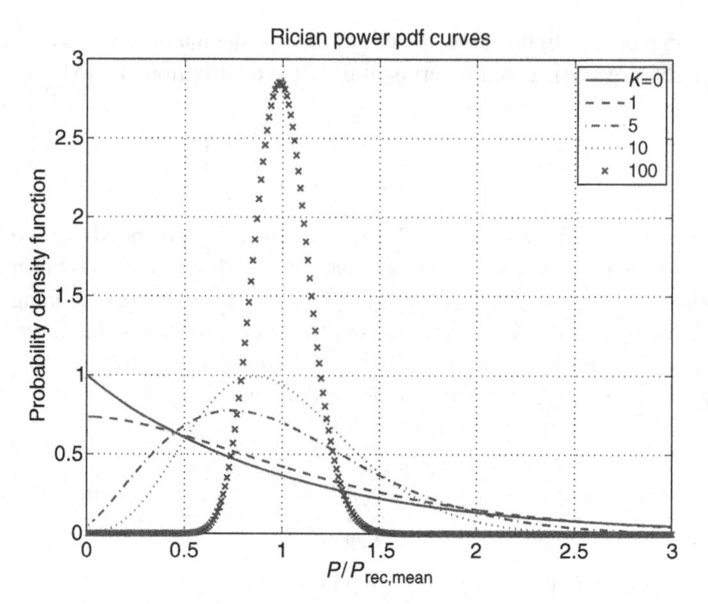

FIGURE 8.3 Rice normalized power (i.e., $P/P_{\mathrm{rec,mean}}$) pdfs with K indicated in the legend.

again in watts, where the "Rician K-factor" is $K = P_{\mathrm{m}}/P_{\mathrm{f}}$. The standard deviation of the received power can be found to be

$$\sigma_{\mathrm{P}} = P_{\mathrm{f}}\sqrt{1+2K} \tag{8.77}$$

in watts.

The cdf of the Rician power pdf is given by

$$F_{\mathrm{P}}(p) = 1 - Q_1\left(\sqrt{2K}, \sqrt{\frac{2p}{P_{\mathrm{f}}}}\right), \tag{8.78}$$

where $Q_1(a, b)$ is called the Marcum function, for which routines and tables are also available.

Figure 8.3 illustrates Rician pdfs for the quantity $P/P_{\mathrm{rec,mean}}$, which eliminates the dependence on P_{f} and makes the resulting pdfs unitless. The $K = 0$ case is identical to a normalized Rayleigh pdf and shows a wide spread of received power levels. As K increases, the pdfs become more centered about unity and show smaller relative variations from the mean. It can also be shown that the Rician pdf approaches a Gaussian pdf in the limit of very large K.

Figure 8.4 illustrates values taken from the Rician cdf, represented as the percent of time that a given received power is expected to fall below the mean by a specified number of standard deviations. The value of K is indicated in the legend, and the curve axes are again scaled so that the plots are independent of P_{f}. It is clear in these plots that the Rician power pdf approaches a Gaussian behavior for small fades as K becomes large. The rapid falloff in the curves for small K values occurs due to the small mean values of P_{m} in these cases. Since the received power cannot fall below

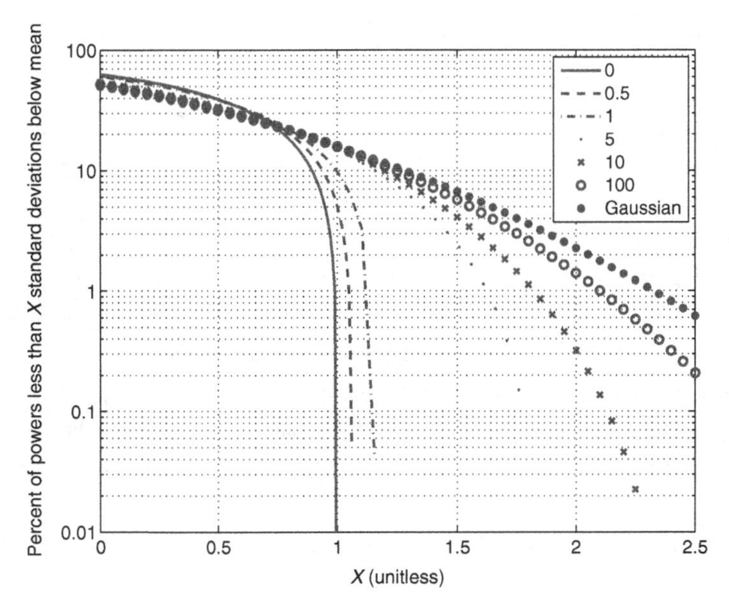

FIGURE 8.4 Rice power cdfs, legend indicates the ratio $K = P_m/P_f$.

zero, it is not possible for these curves to fall a large number of standard deviations below the small mean value obtained.

Note that the probability of deep fades due to strong destructive interference increases as the factor K is reduced. This means that the Rayleigh channel $K = 0$ is a more challenging propagation scenario than the Rice channel. This makes good physical sense since the Rician channel has at least one LOS path, while the Rayleigh signal is due entirely to multipath.

8.3.3.3 Link Reliability and Outage Probability In many applications, the interest lies in the probability that the received power will fall below a specified level (or, equivalently, the probability that the output SNR will fall below a threshold level). This is sometimes called the *outage probability*. Recalling that the cdf represents $\mathcal{P}(P_{rec} \leq p)$, the cdf is precisely the function to be used for such questions, with p the specified minimum power level in watts. It is also common to represent such information in terms of the probability that the received power falls more than a specified number of dB below the mean received power. Figure 8.5 illustrates such curves for Rician channels of varying K values; the case $K = 0$ corresponds to a Rayleigh channel. Larger amplitude fades are greatly reduced as K is increased.

In the case of Rayleigh fading, if the values of p are small compared to the mean received power μ_P, the cdf can be simplified through a power series expansion of the exponential distribution to

$$F_P(p) \approx 1 - \left(1 - \frac{p}{P_f}\right) = \frac{p}{P_f}. \tag{8.79}$$

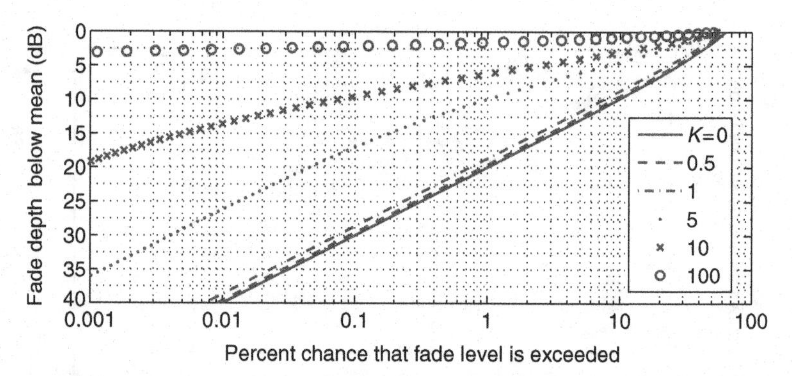

FIGURE 8.5 Percent of time (horizontal axis) that a specified fade depth (vertical axis) below the mean received power is exceeded for a Rician channel, legend indicates the ratio $K = P_m/P_f$.

If the received power falls L dB below the mean power, we have $p = 10^{-L/10}P_f$ and

$$F_P(10^{-L/10}P_f) \approx 10^{-L/10}, \tag{8.80}$$

which accounts for the linear nature of the $K = 0$ curve in Figure 8.5. This simplification can also be used to establish that the power level that will be exceeded q percent of the time in a Rayleigh channel is

$$\left(1 - \frac{q}{100}\right) P_f, \tag{8.81}$$

which is applicable for q greater than approximately 90%.

8.3.3.4 ***Other Fading Distributions*** Numerous other fading distributions have been proposed and used in propagation analyses. Because both the Rayleigh and Rician fast fading models are based on the assumption of a large number of received multipath signals, it is not surprising that there are limits on the range of problems to which these models can be applied. Other useful distributions include the Suzuki distribution (used for modeling a mixture of fast and slow fading in mobile radio links and forest environments), the Nakagami-m distribution (used for modeling fast fading in some land and indoor mobile channels, and scintillating ionospheric radio links—see Chapter 11),[4] the Nakagami-q or Hoyt distribution (used for fading in some satellite links), and the Weibull distribution (also used for mobile radio channels). Many of these distributions are based on curve fits to empirical data, and again, the interpretation of such empirical results should be done with caution to avoid overgeneralizations. It should also be noted that differences in many of these distributions in the domain of high occurrence (large values for the pdfs) are often very

[4]The Rice distribution is sometimes referred to as "Nakagami–Rice" or "Nakagami-n" distribution, which should not be confused with the Nakagami-m or Nakagami-q distributions.

small; major discrepancies are present primarily in the "tails" of the pdfs. The reader is referred to the literature [16] for further discussion of these distributions. ITU-R Recommendation P.1057 also contains a detailed discussion of probability distributions relevant in propagation modeling [17].

8.3.4 Example Fading Analyses

Slow Fading Problem: Assume that an empirical path loss model estimates the mean path loss in a given propagation scenario to be 122 dB with an additional slow fading loss of 4 dB standard deviation. Find the probability that the system loss excluding fast fading exceeds 130 dB.

Solution: We assume that the additional slow fading loss when measured in decibels is distributed normally with zero mean and 4 dB standard deviation as specified. Since the mean path loss is 122 dB, a system loss excluding fast fading of 130 dB means that the slow fading loss exceeds 8 dB, which is two standard deviations above the mean. In Table 8.1, this is a Y value of 2, and $100 \times (1 - F_X(2)) = 2.28\%$ is the applicable entry. Thus, we conclude that 2.28% of measurements would experience a path loss greater than 130 dB.

Fast Fading Problem (Rician): Assume a propagation scenario where a dominant LOS path exists between transmitter and receiver, and an empirical path loss model estimates the mean received power (including slow fading but neglecting fast fading contributions) to be -69 dBW at a particular location. It is also known that the received fast fading power is -79 dBW. Find the probability that the total received power falls below -76 dBW.

Solution: We can use two alternate approaches to answer this question. In the first, we transform the values specified in decibels (relative to 1 W) into watts. The values of interest are the mean power without fast fading

$$P_{\mathrm{m}} = 10^{-69/10} = 0.1259\ \mu\mathrm{W}, \tag{8.82}$$

and the fast fading power

$$P_{\mathrm{f}} = 10^{-79/10} = 0.01259\ \mu\mathrm{W}, \tag{8.83}$$

so $K = P_{\mathrm{m}}/P_{\mathrm{f}} = 10$. Given these parameters, we can find from equation (8.76) that the mean power received is

$$\mu_{\mathrm{P}} = P_{\mathrm{m}} + P_{\mathrm{f}} = 0.1385\ \mu\mathrm{W}, \tag{8.84}$$

and the standard deviation of the received power is

$$\sigma_{\mathrm{P}} = P_{\mathrm{f}}\sqrt{1+2K} = 0.0577\ \mu\mathrm{W}. \tag{8.85}$$

We are asked to find the probability that the received power falls below $10^{-76/10} =$ 0.02512 μW. This power value is equal to

$$\frac{0.1385 - 0.02512}{0.0577} = 1.97 \tag{8.86}$$

standard deviations below the mean value. Reading Figure 8.4 with $K = 10$, we can approximate this probability as around 0.4%. A more precise evaluation using equation (8.78) yields 0.42%.

The alternate approach uses the $K = 10$ curve in Figure 8.5. Using the mean power level computed above, we are interested in the probability that the received power falls 7.4 dB or more below the mean received power. An examination of Figure 8.5 again yields an approximate probability around 0.4%.

Fast Fading Problem (Rayleigh): Assume that the receiver in the previous problem is moved so that no line-of-sight path is available. It is still known that the received fast fading power is −79 dBW. Find the power level that will be exceeded 99% of the time.

Solution: For Rayleigh fading, we know that power levels exceeded more than 90% of the time can be determined through equation (8.81). First, we need the mean received power level in watts, which is 0.01259 μW. The power level exceeded 99% of the time ($q = 99$) is now

$$\left(1 - \frac{q}{100}\right) P_{\mathrm{f}} = P_{\mathrm{f}}/100 \tag{8.87}$$

or 0.1259 nW. This is equivalent to −99 dBW, or a 20 dB fade from the mean power. Similar results can be obtained using Figure 8.5.

8.4 NARROWBAND FADING MITIGATION USING DIVERSITY SCHEMES

Aside from increasing the transmit power—which is usually an expensive or impractical solution—the reliability of propagation links experiencing narrowband fading can be increased by employing *diversity schemes*. We have already encountered an example of a diversity scheme in Chapter 5, when we discussed site (spatial) diversity to mitigate rain attenuation.

The basic idea of diversity schemes is to transmit multiple copies of the same signal over different propagation channels. This is done so that during a deep fade at one channel, the other channel(s) will be unlikely to experience a simultaneous deep fade. In order for this to happen, the multiple channels need to be sufficiently independent to experience different fading behaviors. A *diversity combiner* is then employed either to select the best signal among all the copies (typically the highest amplitude signal) or to extract the signal using some combination of the multiple received signals. Independent (or nearly independent) propagation channels can be sought by transmitting the same signal at different times, frequencies, or polarizations, or by using multiple transmitting and receiving antennas that are spatially separated (a "multiple input, multiple output" (MIMO) system). Empirical models for the degree to which such signals must be separated in time or frequency, or to which antennas must be separated in space, have been developed; see, for example, ITU-R Recommendation P.530-12.

In a diversity scheme employing m (say, $m = 4$) independent propagation channels where each channel has a probability of a specified fade level equal to q (say, $q =$ 10%), the probability that the same deep fade occurs in all channels simultaneously will be q^m (i.e., only 0.01%). In practice, the channels may not be truly independent, and the probability of a simultaneous deep fade may be higher.

8.5 WIDEBAND CHANNELS

Up to this point, we have been considering the path loss and/or received power computed as applicable to a single frequency. Of course, any system utilized for transmitting information will require a finite bandwidth of frequencies to be transmitted and received. Treating propagation loss and fading effects as independent of frequency is reasonable only if the bandwidths of interest are sufficiently small so that fading and loss effects are reasonably constant within these bandwidths (flat fading). "Narrowband" communication channels fall into this category. Both the signal bandwidth and properties of the propagation environment and its variation with frequency determine whether a particular transmission is narrowband for a given propagation channel.

In "wideband" channels, the bandwidth is sufficiently large so that significantly different path loss and fading effects can occur for individual component frequencies within the transmission band. Fading effects in this category are referred to as "frequency selective fading". Such variations in the propagation environment with frequency result in distortion of the transmitted information and can cause significant errors upon reception.

While specific system design techniques and analyses used for wideband systems are beyond the scope of the present discussion (the reader is referred to books on digital communication systems), the basic idea involves attempting to perform channel "sounding" (i.e., estimation of the impulse response on a given transmitter-to-receiver link through transmission of a test signal), followed by correction of subsequent data by knowledge obtained from the channel sounding, that is, *channel equalization*. These techniques have enabled moderate increases in the information bandwidth that can be transmitted, but difficulties still remain as the desired bandwidth is further increased. Estimation and correction of the channel response is additionally complicated in mobile systems by Doppler shifting and delay spread effects caused by relative motion of the user or different time delays from multipath sources that require adaptive equalization [18].

8.5.1 Coherence Bandwidth and Delay Spread

In the description of propagation properties as a function of frequency for a wideband channel, the propagation behaviors of interest should be treated as *stochastic processes* in frequency. Stochastic processes generalize the concept of random variables, that represent measurements of a single quantity, into collections of random variables that represent measurements of *functions*, for example, the path loss versus frequency. Path loss and/or fading phenomena can be treated as slowly varying in frequency at least within a small frequency band. As the bandwidth of interest increases, eventually the separation among the constituent frequencies becomes sufficiently large so that propagation effects at one frequency are decorrelated with those at a second, sufficiently different frequency (this decorrelation is exploited in frequency diversity schemes). A stochastic process description of the propagation channel allows these relationships to be quantified and treated mathematically, but this is beyond the scope of this text.

One particular parameter that results from a stochastic process description and enables insight into basic channel behaviors versus frequency is the *correlation bandwidth* (or *coherence bandwidth*) [19]. This parameter can be interpreted as the maximum separation in frequency over which the propagation channel can be considered "flat." For example, in a channel with coherence bandwidth 1 MHz, received signals at 900 and 900.2 MHz are likely to experience similar propagation effects, while those at 900 and 905 MHz are likely to experience different propagation effects. The coherence bandwidth therefore allows systems to be identified in a simple manner as wideband (bandwidth somewhat larger or much larger than the coherence bandwidth) or narrowband (bandwidth less than the coherence bandwidth). As will be seen in Chapter 11, ionospheric propagation links have notoriously small coherence bandwidths, especially in the HF band. This is the reason for the voice distortion familiar to amateur radio operators and short-wave radio listeners.

The coherence bandwidth of the channel is also related to the channel's *delay spread* that describes the range of delays over which a transmitted impulse is received at the receiver. In urban environments with strong multipath, for example, the various multipath components may arise due to reflections from obstacles at many different locations, and hence arrive at the receiver at different times. The rms delay spread, τ_{rms}, refers to the standard deviation of the multipath component time delays weighted by the relative power of each [19]. The coherence bandwidth is inversely proportional to the delay spread. In macrocellular environments, the rms delay spread is typically on the order of 100–10,000 ns. In indoor environments, it ranges over 20–500 ns for links employing omnidirectional antennas[5] [6]. This range of delays can cause errors upon reception if the receiver is not designed to accommodate delay spread effects. For example, in a digital communication system, delay spread can be a source of intersymbol interference.

8.5.2 Coherence Time and Doppler Spread

Coherence bandwidth and delay spread characterize the frequency dispersive nature of a multipath propagation channel, but do not account for any time variations in

[5]The use of directional antennas tends to reduced the rms delay spread because it suppresses multipath components that are not aligned with the antenna main beam.

the channel itself. In both narrowband and wideband systems, we may expect the propagation channel to change in time, as the transmitter, receiver, and/or surrounding objects move or simply as the environment itself changes. As in the discussion of the coherence bandwidth, the propagation environment determines the timescales over which these variations occur. For systems operating or transmitting information over timescales short compared to the temporal evolution of the channel, the channel's time variations may not be an important factor. On the other hand, propagation channels that vary rapidly within the time of interest for transmission, that is, *time-varying channels*, require a stochastic process description of these time variations. Again, such descriptions are beyond the scope of the discussion here, but enable examination of properties such as the expected duration of fades.

A simple parameter useful for characterization of channel time variations is the *coherence time*, τ_c , which indicates roughly the time interval within which the propagation channel can be considered to remain relatively stable. Received signals separated by a time interval larger than τ_c will experience a channel with different properties. The coherence time is inversely proportional to the *Doppler spread* bandwidth. In mobile communications, Doppler spread is typically a consequence of the *Doppler shift*, that is the change of the carrier frequency of a signal received by a mobile unit caused by the relative motion of the latter [7,9]. Doppler shift causes Doppler spread because different multipath components arrive at the receiver from different directions, and each of these directions will have its own Doppler shift. The net effect is a broadening of the spectrum. Methods for mitigating Doppler spread effects have also been developed and are described in the digital communications literature [16,18].

8.6 CONCLUSION

This chapter has considered some basic empirical path loss and fading models that are used in practical propagation analyses. Due the empirical nature of the path loss models and simplifying assumptions implicit in the fading models, the reader is again reminded to perform a careful review of the underlying propagation conditions before selecting a particular model. When possible, it may be useful to perform propagation measurements for the set of conditions under which a system is to be implemented and then to match a fading model to these measurements. The models presented in this chapter are useful for initial consideration. Fading predictions from a probabilistic model also should not be expected to be highly accurate, especially near the tails of the pdf. Further discussion of fading and other propagation effects (including delay and Doppler spreads) and system designs for mitigating these terms are available in books on digital communication systems, for example, Refs [16,18].

REFERENCES

1. Hata, M., "Empirical formula for propagation loss in land mobile radio services," *IEEE Trans. Veh. Technol.*, vol. 29, pp. 317–325, 1980.

2. Kurner, T., "Propagation models for macrocells," Section 4.4.1 of COST Action 231, "Digital mobile radio toward future generation systems: final report," European Commission Technical Report EUR 18957, 1999.
3. Lee, W. C. Y., *Mobile Communications Engineering,* second edition, McGraw Hill, 1998.
4. ITU-R Recommendation P.1238-5, "Propagation data and prediction methods for the planning of indoor radio communications systems and radio local area networks in the frequency range 900 MHz to 100GHz," International Telecommunication Union, 2007.
5. Okumura, Y., E. Ohmori, T. Kawano, and K. Fukuda, "Field strength and its variability in VHF and UHF land mobile service," *Rev. Electr. Commun. Lab.*, vol. 16, no. 9-10, pp. 825–873, 1968.
6. Barclay, L., "Indoor Propagation," Section 9.3 of *Propagation of Radiowaves*, L. Barclay (ed.), second edition, IEE Press, 2003.
7. Seybold, J. S., *Introduction to RF Propagation*, Wiley, 2005.
8. Siwiak, K., and Y. Bahreini, *Radiowave Propagation and Antennas for Personal Communications*, Artech, 2007.
9. Saunders, S. R., and A. Aragon-Zavala, *Antennas and Propagation for Wireless Systems*, Wiley, 2007.
10. Constantinou, C. C., "Numerically intensive propagation prediction methods," in *Propagation of Radiowaves*, second edition (L. Barclay, ed.), pp. 163–184, IEE, 2003.
11. Barrios, A. E., "A terrain parabolic equation model for propagation in the troposphere," *IEEE Trans. Antennas Propag.*, vol. 42, pp. 90–98, 1994.
12. Papoulis, A., *Probability and Statistics,* Prentice-Hall, 1989.
13. Papoulis, A. and S. Pillai, *Probability, Random Variables, and Stochastic Processes,* McGraw-Hill, 2002.
14. Rice, S. O., "Mathematical analysis of random noise," *Bell Syst. Techn.*, vol. 23, pp. 282–332, 1944.
15. Rice, S. O., "Mathematical analysis of random noise," *Bell Syst. Techn.*, vol. 24, pp. 46–156, 1945.
16. Simon, M. N., and M.-S. Alouini, *Digital Communication over Fading Channels*, second edition, Wiley–Interscience, 2005.
17. ITU-R Recommendation P.1057-2, "Probability distributions relevant to radiowave propagation modelling," International Telecommunication Union, 2007.
18. Rappaport, T.S., *Wireless Communications: Principles and Practice*, second edition, Prentice-Hall, 2002.
19. ITU-R Recommendation P.1407-3, "Multipath propagation and parametrization of its characteristics," International Telecommunication Union, 2007.

9

GROUNDWAVE PROPAGATION

9.1 INTRODUCTION

For antennas located above a perfectly conducting plane, image theory can be used to find the total field intensity at any location as that of the original antenna in free space plus the image contribution. For a lossy ground medium, the image term is multiplied by a reflection coefficient (see Chapter 7), but an additional term must also be included to satisfy the boundary conditions. This term is called the "groundwave" because it is strongest near the ground and decreases with distance from the ground. For transmitting and receiving antennas at heights above the Earth surface that are very small relative to the wavelength (as will be quantified later), direct and Earth reflected signals cancel (since the path lengths become equal and the reflection coefficient approaches -1), leaving only the groundwave component of the received field. Attenuation of the groundwave along the interface is partly due to power loss through the conductivity of the Earth's surface, so more poorly conducting grounds lead to more rapidly attenuating groundwaves.

Because of these properties, groundwave contributions to received field intensities are important when both transmitting and receiving antennas are sufficiently low (so that the direct and reflected rays cancel) and when the imaginary part of the ground dielectric constant is large. These conditions are typically met at lower frequencies (around a few MHz or less) since raising antennas above the hundreds of meter wavelengths at these frequencies is not practical in most situations. However,

Radiowave Propagation: Physics and Applications. By Curt A. Levis, Joel T. Johnson, and Fernando L. Teixeira

at these frequencies the ionosphere often propagates signals much more efficiently, so the groundwave mechanism should only be considered when ionospheric propagation can be neglected. In most situations, consideration of the ionosphere shows that groundwave contributions dominate daytime reception of signals in the HF and lower frequency bands, while ionospheric propagation dominates for long-distance paths at night. At frequencies higher than HF, antennas must be very near the ground for the groundwave mechanism to be appreciable.

The groundwave theory was developed by Sommerfeld [1–2] early in the twentieth century for a planar Earth. At larger propagation distances, Earth curvature effects must be considered, and the planar Earth Sommerfeld groundwave model no longer applies. Practical solutions for the fields of an antenna above a spherical, conducting Earth were first provided by Van der Pol and Bremmer [3–5], and enable groundwave propagation predictions to be extended beyond the radio horizon. Comparisons of the planar and spherical Earth theories show that the planar theory is valid for horizontal distances d of up to approximately $80/f_{\text{MHz}}^{1/3}$ km.

Unfortunately, the electromagnetic theory of groundwave propagation is very complex, so we will be able to present only the resulting calculation methods here. Interested students are referred to the references given at the end of this chapter for a more detailed discussion of the planar and spherical Earth groundwave problems.

Because the calculation of groundwave fields can be quite complex, computer programs are frequently used, especially for repeated calculations. The GRWAVE package, available from the ITU, is such a program. A set of groundwave attenuation curves, as a function of distance and frequency for various ground dielectric constants and conductivities, has also been published by the ITU. Some of these curves are reproduced in Figures 9.11–9.15. Interpolation based on such curves often yields acceptable accuracy. Nevertheless, the calculation procedures developed in this chapter are often useful. First, they provide a feel for the physics of the propagation effects, for example, the concept of "numerical distance" that governs attenuation, and the effects of varying antenna heights. Also, in many cases simple approximations give satisfactory results. Section 9.4 is a valuable guide for this approach.

9.2 PLANAR EARTH GROUNDWAVE PREDICTION[1]

In Chapter 7, we modeled the complex field amplitude radiated by an antenna located above a planar interface as

$$\underline{E}^{\text{tot}} = \frac{E_0}{d}\left(e^{-jk_0R_1} + \underline{\Gamma}e^{-jk_0R_2}\right), \tag{9.1}$$

where R_1 and R_2 represented distances to the source and image locations, respectively, and Γ is the Fresnel reflection coefficient for the appropriate polarization. To derive this equation, we approximated $\frac{1}{R_1} = \frac{\cos\psi_1}{d}$ and $\frac{1}{R_2} = \frac{\cos\psi_2}{d}$ as $\frac{1}{d}$, where d is the horizontal distance between the two antennas, and also neglected any variations in

[1] Section 9.2 is based on the work of K. A. Norton [6].

the antenna pattern as a function of angle. A more accurate version of equation (9.1) for the z component of the radiated electric field that removes these assumptions is

$$\underline{E}_z^{\mathrm{tot}} = \frac{E_0}{d}\left(\cos^3\psi_1 e^{-jk_0R_1} + \underline{\Gamma}\cos^3\psi_2 e^{-jk_0R_2}\right), \tag{9.2}$$

where a directivity pattern of $\cos^2\psi$ for the z component fields radiated by the source has been assumed; this pattern corresponds to that of a short vertical dipole antenna as would be encountered for the lower frequency ranges where groundwave calculations are appropriate.

As mentioned in the introduction, the total electric field obtained in equation (9.2) will vanish as the source and receiving antenna heights above the interface become small compared to the wavelength. This is because the phase difference related to $R_1 - R_2$ becomes small for small antenna heights (as discussed in Chapter 7) and the Fresnel reflection coefficients approach -1. However, in this case an appreciable "groundwave" field is still observed. A modified version of equation (9.2) that includes groundwave contributions is

$$\underline{E}_z^{\mathrm{tot}} = \frac{E_0}{d}\left(\cos^3\psi_1\, e^{-jk_0R_1} + \underline{\Gamma}\cos^3\psi_2\, e^{-jk_0R_2} + (1-\underline{\Gamma})\cos^2\psi_2\, e^{-jk_0R_2}\underline{F}_1(R_2,\underline{\epsilon}_r)\right). \tag{9.3}$$

The final term in the above equation represents the groundwave; note its form appears similar to that of the reflected wave but an additional complex factor $\underline{F}_1(R_2,\underline{\epsilon}_r)$ is included that can significantly modify both the amplitude and the phase of the groundwave relative to the reflected wave. This factor depends on both the distance from the image R_2 and the relative complex permittivity $\underline{\epsilon}_r$ (for simplicity, in this chapter $\underline{\epsilon}_r$ is used to denote the complex relative dielectric constant $= \underline{\epsilon}^e/\epsilon_0$) of the medium over which the groundwave propagates. In practice, usually $(h_1 + h_2) < 0.1R_2$, so that the powers of $\cos\psi_2$ can be replaced by unity.

When transmitting and receiving antennas are very close to the ground, only the groundwave term is required and the received field becomes

$$\underline{E}_z^{\mathrm{tot}} \approx \frac{2E_0}{R_2}\underline{F}_1(R_2,\underline{\epsilon}_r)e^{-jk_0R_2}. \tag{9.4}$$

The field amplitude thus varies with distance as $\underline{F}_1(R_2,\underline{\epsilon}_r)/R_2$.

When transmitting and receiving antennas are close to the ground, studies of $\underline{F}_1(R_2,\underline{\epsilon}_r)$ show that it depends in fact on a single variable $\underline{p}$, which is a function of R_2 and $\underline{\epsilon}_r$, and not on R_2 and $\underline{\epsilon}_r$ independently. Thus, $\underline{F}_1(R_2,\underline{\epsilon}_r)$ can be written as $\underline{F}(\underline{p})$, and description of groundwave propagation is greatly simplified if considered in terms of the "numerical distance" $\underline{p}$. A useful expression for numerical work is equation (9.18), used later in this chapter in a sample calculation. The equations that define $\underline{p}$ for vertically and horizontally polarized sources, respectively, are

$$\underline{p} = -j\frac{k_0R_2}{2\underline{\epsilon}_r}\left(\frac{\underline{\epsilon}_r - \cos^2\psi_2}{\underline{\epsilon}_r}\right) \quad \text{vertical polarization}, \tag{9.5}$$

$$\underline{p} = -j\frac{k_0R_2}{2}(\underline{\epsilon}_r - \cos^2\psi_2) \quad \text{horizontal polarization}. \tag{9.6}$$

The numerical distance is thus related to the actual distance R_2 divided by the electromagnetic wavelength (since $k_0 = 2\pi/\lambda$) scaled by a function of $\underline{\epsilon}_r$. For a fixed actual distance R_2, the amplitude of $\underline{p}$ will increase as frequency increases. Also, since values of $\underline{\epsilon}_r$ will typically have very large, negative imaginary parts when groundwave effects are important, the quantity in parenthesis in equation (9.5) is approximately unity, so $\underline{p}$ is approximately inversely proportional to $\underline{\epsilon}_r$ for vertical polarization. For a fixed distance and frequency, the vertical polarization amplitude of $\underline{p}$ will therefore decrease as the ground becomes more conducting. For a fixed distance and frequency, values of $\underline{p}$ for horizontal polarization will always be much larger than those for vertical polarization, since the inverse dependence on $\underline{\epsilon}_r$ does not occur in horizontal polarization. These dependencies (on frequency, $\underline{\epsilon}_r$, and polarization) are very important for understanding groundwave propagation.

The amplitude of $\underline{p}$ is a function of both R_2 and $\underline{\epsilon}_r$, but it can be shown that the phase of $\underline{p}$ depends only on $\underline{\epsilon}_r$. A commonly used notation defines the phase of $\underline{p}$ as b through

$$\underline{p} = |\underline{p}|e^{-jb}, \tag{9.7}$$

where b depends only on $\underline{\epsilon}_r$. This definition allows plots of the magnitude or phase of $\underline{F}(\underline{p})$ (which are applicable to both vertical and horizontal polarizations) to be generated versus $|\underline{p}|$ with the phase b in degrees as a parameter. For vertical polarization, b is found to be between zero and 90° and to approach 0° for highly conducting grounds. For horizontal polarization, b ranges between 90° and 180° and approaches 180° for highly conducting grounds.

The function $\underline{F}(\underline{p})$ that determines the groundwave dependence on distance (other than the $1/R_2$ already included in equation (9.4)) does not have a simple form. A series expansion that is useful for $|\underline{p}|$ less than approximately 10 is

$$\underline{F}(\underline{p}) = 1 - j\sqrt{\pi \underline{p}}e^{-\underline{p}} - 2\underline{p} + \frac{(2\underline{p})^2}{1 \cdot 3} - \frac{(2\underline{p})^3}{1 \cdot 3 \cdot 5} + \cdots, \tag{9.8}$$

where additional terms in the series follow the pattern of the final two terms above. For larger values of $|\underline{p}|$, another expansion is more appropriate:

$$\underline{F}(\underline{p}) = -\frac{1}{2\underline{p}} - \frac{1 \cdot 3}{(2\underline{p})^2} - \frac{1 \cdot 3 \cdot 5}{(2\underline{p})^3} - \cdots, \tag{9.9}$$

again with additional terms following the pattern above. Note the series expansion of equation (9.9) is an asymptotic series that does not yield a uniform convergence rate; additional terms in the series should not be added if their successive amplitudes begin to increase rather than decrease. For $|\underline{p}| > 20$, a single term in equation (9.9) will provide an accuracy of better than 10%.

A plot of $|\underline{F}(\underline{p})|$ versus $|\underline{p}|$ can help to illustrate the behavior of groundwave propagation. However, since final groundwave field amplitudes involve the product of $1/R_2$ and $|\underline{F}(\underline{p})|$, a plot of $|\underline{F}(\underline{p})|$ versus $|\underline{p}|$ actually indicates the excess loss from $1/R_2$ as a function of distance. The resulting plot should then be representative of additional loss obtained in groundwave field propagation compared to free-space

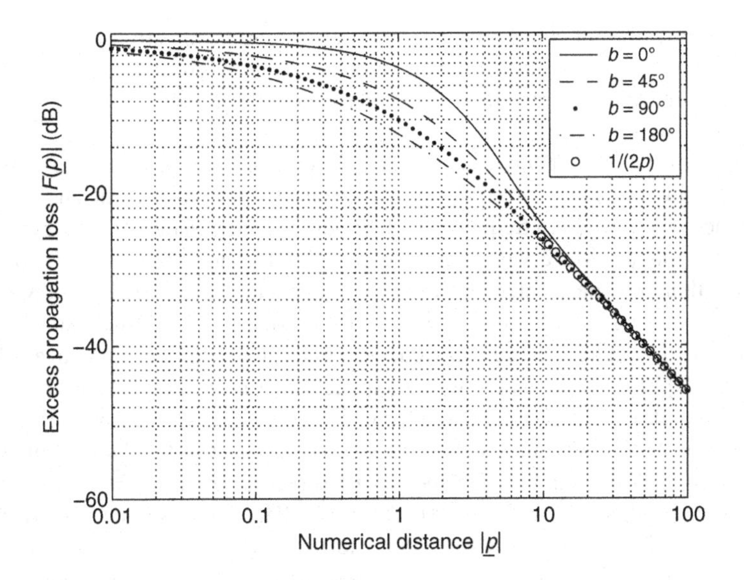

FIGURE 9.1 Excess groundwave propagation loss versus $|\underline{p}|$.

propagation. Figure 9.1 plots $|\underline{F}(\underline{p})|$ in decibels versus $|\underline{p}|$ with b as a parameter and also includes a plot of a $|1/(2\underline{p})|$ dependency for $|\underline{p}| > 10$ to show that the first term of equation (9.9) suffices for large $|\underline{p}|$. Figure 9.2 plots the phase of $\underline{F}(\underline{p})$. Groundwave excess loss in Figure 9.1 is observed to transition between a near-constant behavior for small values of $\underline{p}$ to a $|1/(2\underline{p})|$ (or one-over-distance-squared dependence for final field amplitudes) behavior at large values of $\underline{p}$. The transition between these two trends occurs for $|\underline{p}|$ between 0.1 and 10, depending on the value of b. Smaller values of b

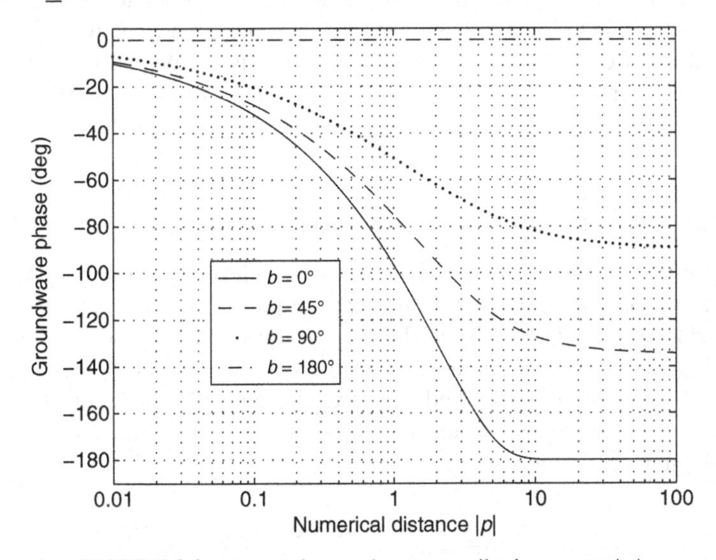

FIGURE 9.2 Phase of groundwave contribution versus $|\underline{p}|$.

show a more constant behavior for larger distances, indicating stronger groundwave propagation for vertical polarization and more conducting grounds. Larger values of b produce a more rapid transition to the one-over-distance-squared dependence for final field values.

Because the parameter $\underline{p}$ is a scaled version of the distance R_2, the actual distance at which fields transition from an inverse distance to an inverse distance squared dependence will depend on frequency and ground dielectric parameters. For example, consider a ground surface with dielectric constant 15 and conductivity 0.003 S/m. Assuming that these parameters are valid for frequencies from 100 kHz to 1 MHz, we can find for vertical polarization (using $\cos^2\psi_2 \approx 1$) that $|\underline{p}| = 1$ at a distance of approximately 515 km for the 100 kHz frequency but at approximately 5.4 km at 1 MHz. Thus, the 100 kHz ground wave will attenuate much more slowly than the 1 MHz ground wave. Propagation over terrain for which conductivity values are much larger (such as seawater) shows much slower rates of increase in $\underline{p}$ as the distance increases. In horizontal polarization, the corresponding distance at which $|\underline{p}| = 1$ for the example above is less than 2 m for both 100 kHz and 1 MHz. This result demonstrates that horizontally polarized groundwaves attenuate very rapidly and have little practical use. Our further groundwave studies will thus concentrate on vertical polarization: practical antennas for generating ground waves are generally vertically polarized.

Our consideration here of fields that vary as one-over-distance-squared may raise questions about the absence of similar terms in the direct and reflected waves of equation (9.3). However, the "near-field" $1/R^2$ and $1/R^3$ terms of the direct and reflected waves will almost always be found negligible compared to the ground wave in the $|1/(2\underline{p})|$ region, since distances for which $|\underline{p}|$ is large typically have much larger values of R_1 and R_2 (due to the $1/\underline{\epsilon}_r$ factor in $\underline{p}$). For smaller values of $|\underline{p}|$ but with R_1 and R_2 values sufficiently large (i.e., $kR_1 \gg 1$, $kR_2 \gg 1$), again the "near-field" terms are negligible compared to the ground wave, so these terms can generally be neglected outside the near-field region of the antenna, even when the ground wave is included. See Ref. [7] for further discussion of this issue. We must also remember that the planar Earth theory applies only for distances up to approximately $80/f_{\mathrm{MHz}}^{1/3}$ km, so the curves of Figure 9.1 become invalid for a $|\underline{p}|$ value that depends on both the frequency and the complex permittivity of the ground $\underline{\epsilon}_r$.

9.2.1 Elevated Antennas: Planar Earth Theory

For transmitting and receiving antennas sufficiently close to the ground surface, the groundwave contribution dominates the received field, $(1 - \underline{\Gamma})$ is approximately 2, and the field amplitude is $\frac{2E_0}{d}|\underline{F}(\underline{p})|$. However, as transmitter and receiver heights are increased, direct and reflected wave contributions must be included, and the groundwave calculation is slightly modified. For vertical antennas "sufficiently close" to the ground, the physical heights of the antennas, h_1 and h_2, are replaced by "numerical antenna heights" q_1 and q_2

$$q_{1,2} = k_0 h_{1,2}\sqrt{\frac{|\underline{\epsilon}_r - \cos^2\psi_2|}{|\underline{\epsilon}_r^2|}}, \tag{9.10}$$

where h_1 and h_2 represent the transmitter and receiver heights, respectively. Numerical antenna heights are defined similarly for horizontal antennas, but the above definition of q is multiplied by $|\underline{\epsilon}_r|$. Again, the numerical antenna heights depend on both frequency and $\underline{\epsilon}_r$ and are scaled versions of the actual heights. Antenna elevation effects usually can be neglected for cases with $q_1 + q_2 < 0.01$.

For elevated antennas, spherical Earth theory shows that the planar Earth theory considered in equation (9.3) becomes invalid if either actual antenna height $h_{1,2}$ becomes larger than approximately $610/f_{\mathrm{MHz}}^{2/3}$ m. For antenna heights within these limits and for distances less than $80/f_{\mathrm{MHz}}^{1/3}$ km, the planar Earth theory remains valid, but calculation of groundwave contributions is slightly modified. Direct, reflected, and groundwave contributions are added in the proper phase relationships in equation (9.3), but the groundwave function $\underline{F}$ is calculated as $\underline{F}(\underline{P})$, where $\underline{P} = 4\underline{p}/(1 - \underline{\Gamma})^2$. Computation of this function versus distance can be quite tedious, so use of computers is recommended. An approximation for the total field is available for cases where $p > 20$, $p > 10q_1q_2$, and $p > 100(q_1 + q_2)$. In this case, the resulting total field amplitude is simply that of the ground wave alone (assuming source and receiving antennas on the ground) multiplied by "height-gain" functions $f(q_1)$ and $f(q_2)$, where

$$f(q) = \left[1 + q^2 - 2q\cos\left(\frac{\pi}{4} + \frac{b}{2}\right)\right]^{1/2}. \tag{9.11}$$

This equation considerably simplifies calculations for elevated antennas that fall into the applicable region. A common procedure is to use (9.18) and to multiply the result by $f(q_1)f(q_2)$. Figure 9.3 plots $f(q)$ versus q for different values of b; note that

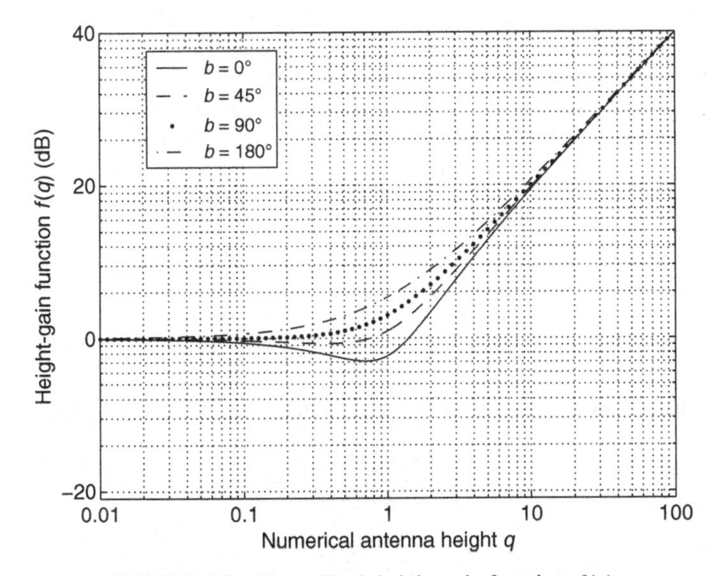

FIGURE 9.3 Planar Earth height-gain function $f(q)$.

increases in antenna height typically result in increases in received power, although there are some combinations of q and b for which reduced power is observed. This behavior occurs because increasing power in the direct and reflected waves as antenna heights increase is compensated by decreased power in the groundwave field, particularly for smaller values of b.

9.3 SPHERICAL EARTH GROUNDWAVE PREDICTION

When distances between the source and the receiver exceed $80/f_{\text{MHz}}^{1/3}$ km or when transmitting or receiving antenna heights are increased above $610/f_{\text{MHz}}^{2/3}$ m, the effect of Earth curvature must be taken into account. The theory of groundwave propagation over a spherical Earth has been considered by many researchers, but the resulting expressions (which are obtained from a "Watson"-type transformation of a spherical eigenfunction solution) remain more complicated than those for the planar Earth model. The outcome of the theory is essentially that groundwave field amplitudes decay more rapidly than one-over-distance-squared at large distances. Refraction effects in the Earth atmosphere are approximately included through the use of the effective Earth radius $a_{\text{eff}} = \kappa a$ described in Chapter 6.

The distance d in the spherical Earth theory now refers to the distance measured along the arc length of the Earth. Groundwave predictions in this case are most conveniently evaluated in terms of a new scaled distance parameter x, defined as

$$\begin{aligned} x &= \left(\frac{k_0 a_{\text{eff}}}{2}\right)^{1/3} \left(\frac{d}{a_{\text{eff}}}\right) \\ &= \left(\frac{d}{\lambda}\right) \sqrt[3]{\pi \left(\lambda / a_{\text{eff}}\right)^2}. \end{aligned} \tag{9.12}$$

The scaled distance x is related to the frequency and the effective Earth radius, but not to the dielectric properties of the ground. Note that for a fixed distance d, as frequency is increased, x will increase only as the cube root of the frequency. New scaled antenna heights $y_{1,2}$ are defined through

$$\begin{aligned} y_{1,2} &= k_0 h_{1,2} \left(\frac{2}{k_0 a_{\text{eff}}}\right)^{1/3} \\ &= \frac{2\, h_{1,2}}{\lambda} \sqrt[3]{\pi^2 \frac{\lambda}{a_{\text{eff}}}} \end{aligned} \tag{9.13}$$

and a new descriptor of the ground dielectric parameters $\underline{\tau}$ is also defined:

$$\begin{aligned} \underline{\tau} &= -j \left(\frac{k_0 a_{\text{eff}}}{2}\right)^{1/3} \frac{\sqrt{\underline{\epsilon}_r - 1}}{\underline{\epsilon}_r} \\ &= -j \sqrt[3]{\pi \frac{a_{\text{eff}}}{\lambda}} \frac{\sqrt{\underline{\epsilon}_r - 1}}{\underline{\epsilon}_r}. \end{aligned} \tag{9.14}$$

Note that $\underline{\tau}$ is related to the frequency in a fashion similar to x, except for variations in $\underline{\epsilon}_r$ that occur with frequency.

Field complex amplitudes for a vertically polarized source at distances beyond the planar Earth theory limit are expressed as

$$\underline{E}_z^{\text{tot}} = \frac{2E_0}{d} e^{-jk_0 d} \sqrt{\frac{\pi x}{j}} \sum_{s=1,2,3,\ldots}^{\infty} \frac{\underline{G}_s(y_1)\underline{G}_s(y_2)\exp\left(-jx\underline{t}_s\right)}{\left(\underline{t}_s - \underline{\tau}^2\right)}, \tag{9.15}$$

where the sum is over an index s to a set of complex "roots" $\underline{t}_s$, and where the $\underline{G}_s$ function is a spherical Earth height-gain function that also depends on s. Although the complexity of this expression makes obtaining insight difficult, values of $\underline{t}_s$ are found to have negative imaginary parts so that each term in the series represents an exponential decay as the distance, and therefore x, increases. Also, because the imaginary parts of the roots $\underline{t}_s$ become more negative as s increases, the relative amplitudes of each term decrease with s so that the number of terms that must be included in the series for a given accuracy is reduced as the distance is increased. For distances $d > 36\sqrt[3]{\lambda}$ km (so that $x > 1.2$ when a 4/3 Earth radius multiplier is used), neglecting all but the first series term typically results in errors no larger than approximately 10%. It should be noted that the above expressions are most effective for regions outside the planar Earth region and for antenna heights at which a direct line of sight is not obtained. For high antennas within the line of sight, a geometrical optics theory of reflection from a spherical Earth can be applied [8], but it is not considered here.

The values of $\underline{t}_s$ in equation (9.15) are obtained from another set of series expansions described in the appendix to this chapter. The spherical Earth height-gain functions $\underline{G}_s(y_1)$ and $\underline{G}_s(y_2)$ are also complicated and are again described in detail in the appendix. The approximate forms given in the appendix show that $\underline{G}_s(y)$ is approximately linear in y for small antenna heights, but increases rapidly for larger heights. For consideration of cases where the groundwave mechanism is applicable (i.e., low frequencies), the elevated antenna procedures for the planar Earth model are usually reasonable, and the planar Earth height-gain functions can often be applied. Thus, approximating $\left|\underline{G}_s(y)\right|$ by $f(q)$ when the conditions for using $f(q)$ are satisfied will typically yield only minor errors.

Figure 9.4 plots the excess propagation loss (determined from the field amplitude divided by $2E_0/d$) from the spherical Earth theory with both antennas on the ground and for three values of the ground parameter $\underline{\tau}$. Note that $\underline{\tau} = 0$ corresponds to perfectly conducting ground, while increasing values of $\underline{\tau}$ for a fixed frequency indicate more poorly conducting grounds. The spherical Earth theory shows excess loss similar to that of the planar Earth theory (although the horizontal axes are scaled differently) for smaller distances, but exponential attenuation (which appears as a curved rather than linear dependency on a log–log scale) is evident at larger distances. The distance at which the exponential attenuation becomes predominant is again observed to depend significantly on the conductivity of the Earth. A plot of the spherical Earth height-gain function $\underline{G}_1(y)$ for the $s = 1$ term is illustrated in Figure 9.5 for three values of $\underline{\tau}$. The plot shows similar behaviors to the planar Earth height-gain function

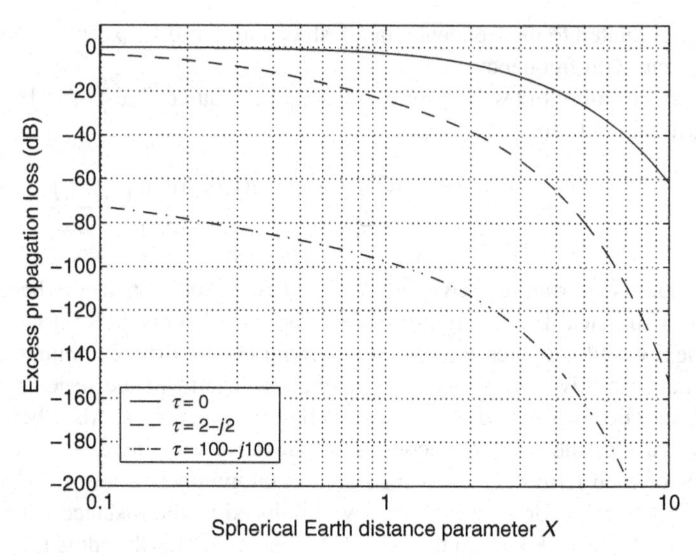

FIGURE 9.4 Spherical Earth excess propagation loss.

in that increases in antenna height can initially decrease field strength before dramatic increases are observed at larger heights. Height-gain functions show larger values for the poorer conducting grounds because field strengths at the surface are much smaller for this case.

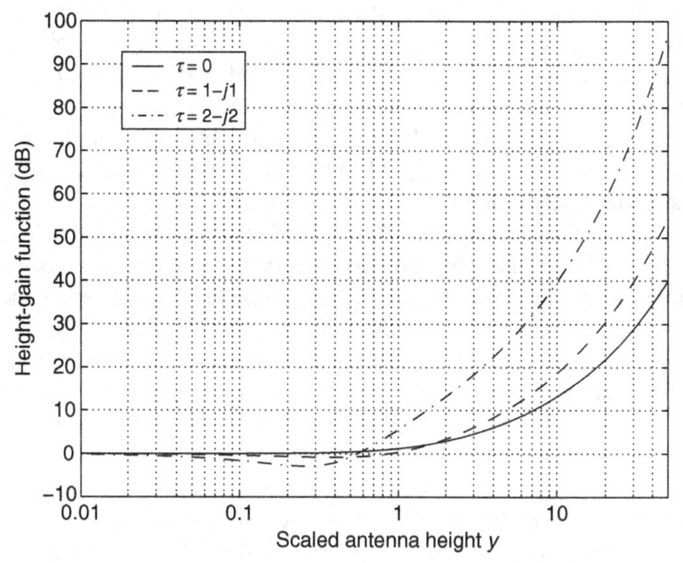

FIGURE 9.5 Spherical Earth height-gain function $d > 36\sqrt[3]{\lambda}$ km.

9.4 METHODS FOR APPROXIMATE CALCULATIONS

The complexity of the equations for groundwave computations can make predictions somewhat difficult. However, approximations can help with this process in many cases. For the planar Earth region with both antennas "on" the ground, the following steps can be useful:

- Compute the distance to which the planar Earth theory is valid: $d = 80/f_{\text{MHz}}^{1/3}$ km.
- Evaluate the relationship between $\underline{p}$ and distance d using equations (9.5) (vertical antennas) or (9.6) (horizontal antennas).
- Values of the excess propagation loss in the planar Earth region can then be read approximately from Figure 9.1.
- For $|\underline{p}| > 20$, the approximation $|\underline{F}(\underline{p})| = \frac{1}{2|\underline{p}|}$ yields acceptable accuracy. An inverse-distance squared dependence in field amplitudes will be obtained for all larger distances within the planar Earth theory limits.
- If more accuracy is needed, an empirical approximation to $|\underline{F}(\underline{p})|$ is described in ref. [9] as

$$|\underline{F}(\underline{p})| = \exp\left(-0.43|\underline{p}| + 0.01|\underline{p}|^2\right) - \sqrt{|\underline{p}|/2}\exp\left(-5|\underline{p}|/8\right)\sin b \quad \text{for } |\underline{p}| < 4.5$$

$$= \frac{1}{2|\underline{p}| - 3.7} - \sqrt{|\underline{p}|/2}\exp\left(-5|\underline{p}|/8\right)\sin b \quad \text{for } |\underline{p}| > 4.5, \tag{9.16}$$

  which is valid to within 3 dB with improved accuracy for b less than 35°.

- If even more accuracy is needed, the series forms of equations (9.8) and (9.9) can be computed and used.

For distances beyond the planar Earth theory limit with both antennas "on" the ground,

- Determine the relationship between the distance and the spherical Earth distance parameter x from equation (9.12).
- For distances $d > 36\sqrt[3]{\lambda}$ km, retaining only a single term in the series (9.15) should yield acceptable accuracy.
- When only a single term is needed, equation (9.15) can be simplified for field amplitudes to

$$\left|E_z^{\text{tot}}\right| = \frac{2E_0}{d}\frac{\sqrt{\pi}}{\left|\underline{t}_1 - \underline{\tau}^2\right|}\sqrt{x}\exp\left[x\,\text{Im}\{\underline{t}_1\}\right]$$

$$= \frac{2E_0}{d}\,Q\,\sqrt{x}\exp\left[x\,\text{Im}\{\underline{t}_1\}\right], \tag{9.17}$$

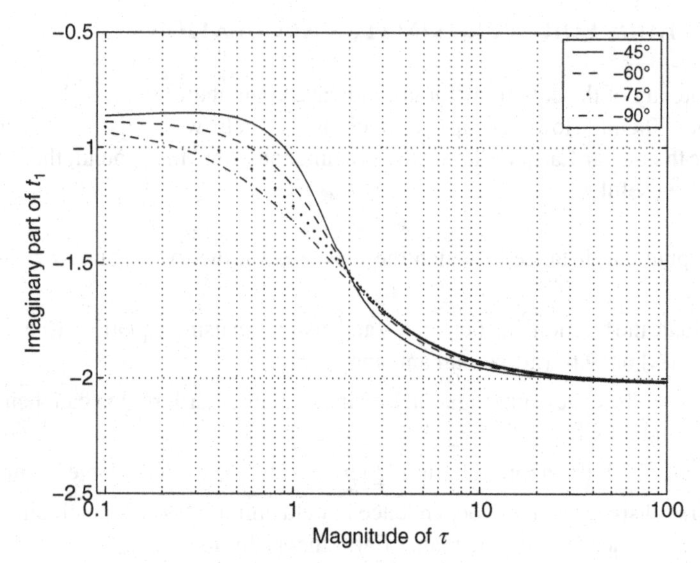

FIGURE 9.6 Spherical Earth attenuation factor $\mathrm{Im}\{\underline{t}_1\}$ versus $|\underline{\tau}|$, with phase of $\underline{\tau}$ as a parameter.

where Q is defined as a field scaling factor and is a function only of the ground dielectric descriptor $\underline{\tau}$.

- Figure 9.6 plots $\mathrm{Im}\{\underline{t}_1\}$ versus $|\underline{\tau}|$, while Figure 9.7 plots Q versus $|\underline{\tau}|$. The phase of $\underline{\tau}$ is included as a parameter in the plots. Values read from these plots can be used in equation (9.17) to predict field strength for $d > 36\sqrt[3]{\lambda}$ km. Exponential attenuation is obtained because the imaginary part of $\underline{t}_1$ from Figure 9.6 is negative.
- For distances between $d = 80/f_{\mathrm{MHz}}^{1/3}$ km and $36\sqrt[3]{\lambda}$ km, the above methods do not apply. However, because the transition in field behavior from the planar Earth to spherical Earth regions is typically very gradual, a simple smooth curve drawn to connect fields in the two regions will often yield reasonable accuracy.

9.5 A 1 MHz SAMPLE CALCULATION

Consider a 1 MHz vertically polarized transmitter transmitting 1 kW of power over "wet ground" with dielectric constant 30 and conductivity 0.01 S/m. For this case, the resulting complex dielectric constant is $30 - j179.8$, and the planar Earth theory should apply for distances to approximately 80 km. If both transmitting and receiving antennas are located close to the ground, the received field is entirely due to the ground

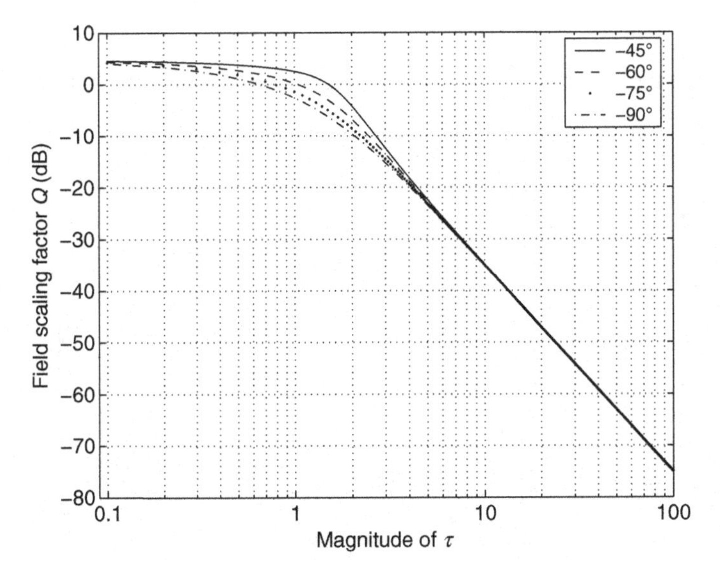

FIGURE 9.7 Spherical Earth field scaling factor Q versus $|\underline{\tau}|$, with phase of $\underline{\tau}$ as a parameter.

wave, and in the planar Earth region is

$$\underline{E}_z^{\text{tot}} \approx \frac{2E_0}{d}\underline{F}(\underline{p})e^{-jk_0 d}, \tag{9.18}$$

where E_0 for a short monopole antenna is 150 mV rms if d is measured in km in the above equation (i.e., the field strength produced by a short monopole antenna transmitting 1 kW of power above a perfectly conducting plane is 300 mV rms at distance 1 km).

Computation of the planar Earth "numerical distance" $\underline{p}$ (using $\cos^2\psi_2 \approx 1$) provides

$$\underline{p} = -j\frac{k_0 R_2}{2\underline{\epsilon}_r}\left(\frac{\underline{\epsilon}_r - 1}{\underline{\epsilon}_r}\right) = \left(5.74 \times 10^{-2}\right) d\ \exp(-j0.17) \tag{9.19}$$

with d in km. Thus, at the distance at which the planar Earth theory becomes invalid, $\underline{p}$ will obtain an amplitude of 4.6. The phase parameter b in this case is approximately 9.8°, indicating the relatively large conductivity of the ground. Due to the moderately small amplitudes of $\underline{p}$ that will be obtained for this problem in the planar Earth region, the series expansion of equation (9.8) can be used for all calculations. Figure 9.8 illustrates the resulting groundwave amplitude assuming that both antennas are located on the ground. Calculations are also included for distances greater than 80 km to investigate the inaccuracy of the planar Earth theory at larger distances. The transition from a one-over-distance to a one-over-distance-squared dependence is evident in the range from 10 to 100 km. Approximate predictions using equation (9.16) are valid for this case to within 0.4 dB at all the distances illustrated in Figure 9.8.

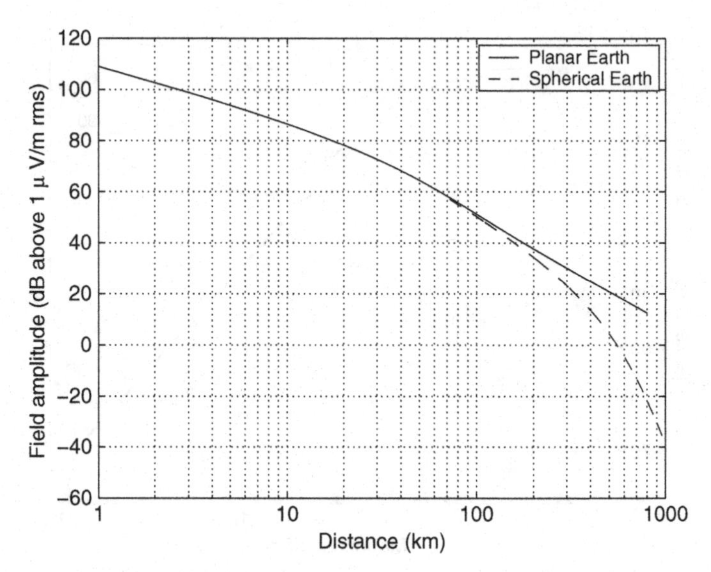

FIGURE 9.8 Sample calculation at 1 MHz.

Spherical Earth theory results are also included in Figure 9.8 using equation (9.15) and the equations in the appendix assuming a 4/3 Earth radius multiplier. The spherical Earth distance x is related to the actual distance d through

$$x = \left(\frac{d}{\lambda}\right) \sqrt[3]{\pi\left(\lambda/a_{\mathrm{eff}}\right)^2} = \left(5.25 \times 10^{-3}\right) d \tag{9.20}$$

with d in km. Note that the spherical Earth distance parameter x is 0.42 at distance 80 km, where spherical Earth results begin to be required. The spherical Earth dielectric parameter $\underline{\tau}$ for this case is approximately $2.13 - j2.53 = 3.3e^{-j0.870}$, indicating again that the ground is a relatively good conductor. The results show a good match between the two theories even up to distances of about 100 km, but the spherical Earth exponential attenuation is clearly observed for larger distances. The approximate spherical Earth method described in equation (9.17) can be computed for this case by finding $\mathrm{Im}\{\underline{t}_1\} = -1.77$ from Figure 9.6 and $Q = -14.6$ dB from Figure 9.7. Results from this approximation are within 0.5 dB of the complete series solution for distances greater than $36\sqrt[3]{\lambda} = 241$ km. Behavior of the exact solution for distances between 80 and 241 km is indeed observed to be a smooth function of distance, so a simple curve drawn to connect the two regions would yield reasonable accuracy.

At this relatively low-frequency, practical antenna elevations are likely to be very small compared to the 300 m wavelength, so elevated antenna effects are not likely to be observed.

9.6 A 10 MHz SAMPLE CALCULATION

Next, consider a 10 MHz vertically polarized transmitter transmitting 1 kW of power over ground with dielectric constant 15 and conductivity 0.003 S/m. For this case, the resulting complex dielectric constant is $15 - j5.4$, and the planar Earth theory should apply for distances out to approximately 37 km.

Computation of the planar Earth "numerical distance" $\underline{p}$ (using $\cos^2\psi_2 \approx 1$) provides

$$\underline{p} = -j\frac{k_0 R_2}{2\underline{\epsilon}_r}\left(\frac{\underline{\epsilon}_r - 1}{\underline{\epsilon}_r}\right) = (6.18)\,d\,\exp(-j1.24) \tag{9.21}$$

with d in km. Thus, at the distance at which the planar Earth theory becomes invalid, $\underline{p}$ will obtain an amplitude of 229, much larger than in the 1 MHz example. The phase parameter b in this case is approximately 71.5°, indicating the relatively low conductivity of the ground. Due to the typically large amplitudes of $\underline{p}$ that will be obtained for this problem in the planar Earth region, the series expansion of equation (9.9) can be used for almost all calculations. Figure 9.9 illustrates the resulting groundwave amplitude assuming that both antennas are located on the ground. Calculations are also included for distances greater than 37 km to investigate the inaccuracy of the planar Earth theory at larger distances. The relatively high frequency considered results in an inverse distance squared dependence for all the ranges illustrated in the figure. Approximate predictions using equation (9.16) are valid for this case to within 2.5 dB at all the distances illustrated in Figure 9.9;

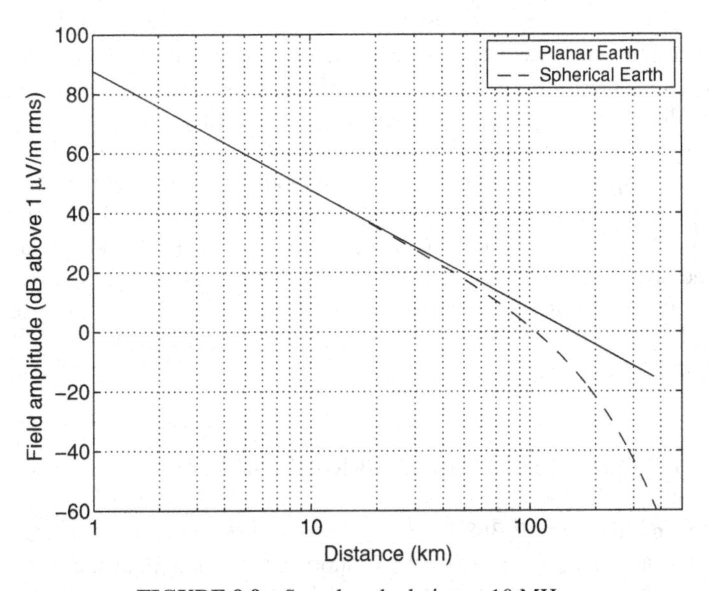

FIGURE 9.9 Sample calculation at 10 MHz.

decreased accuracy is observed compared to the 1 MHz example due to the larger value of b.

Spherical Earth theory results are also included in Figure 9.9 using equation (9.15) and the equations in the appendix assuming a 4/3 Earth radius multiplier. The spherical Earth distance x is related to the actual distance d through

$$x = \left(\frac{d}{\lambda}\right)\sqrt[3]{\pi\left(\lambda/a_{\text{eff}}\right)^2} = \left(1.13 \times 10^{-2}\right)d \tag{9.22}$$

with d in km. Again, the spherical Earth distance parameter x is 0.42 at distance 37 km where spherical Earth results begin to be required. The spherical Earth dielectric parameter $\underline{\tau}$ for this case is approximately $3.75 - j23.0 = 23.3e^{-j1.41}$, indicating again that the ground is a relatively poor conductor. The results show a good match between the two theories even up to distances of about 40 km, but the spherical Earth exponential attenuation is clearly observed for larger distances. The approximate spherical Earth method described in equation (9.17) can be computed for this case by finding $\text{Im}\{\underline{t}_1\} = -1.98$ from Figure 9.6 and $Q = -49.8$ dB from Figure 9.7. Results from this approximation are within 0.5 dB of the complete series solution for distances greater than $36\sqrt[3]{\lambda} = 112$ km. Again, behavior of the exact solution for distances between 40 and 112 km is observed to be a smooth function of distance.

For this higher frequency, reasonable antenna elevations can have a more significant effect. Consider elevated transmitting and receiving antennas with the transmitter at height 80 m and the receiver at height 10 m. In the planar Earth theory, the numerical antenna heights q_1 and q_2 at distances greater than 0.5 km are 4.07 and 0.509, respectively, so a significant influence of the elevated antenna heights should be expected. The simplified planar Earth height-gain function of equation (9.11) is applicable only for distances at which $p > 20$, $p > 10q_1q_2$, and $p > 100(q_1 + q_2)$; the final condition implies distances greater than the planar Earth theory boundary of 80 km. Thus, planar Earth elevated antenna effects must be calculated from equation (9.3) including direct, reflected, and groundwave contributions (using $\underline{F}(\underline{P})$ as described in Section 9.2.1) appropriately. This is most easily accomplished through use of a computer. Spherical Earth results are also modified through use of the spherical Earth height-gain functions described in the appendix. Figure 9.10 plots the modification of field amplitudes obtained with these antenna heights, including planar Earth (solid) and spherical Earth (dashed) results. The larger antenna heights considered here result in an increase in received power of approximately 12.5 dB. A test of the approximate planar Earth height-gain functions for distances greater than 80 km also yields a value of 12.5 dB.

9.7 ITU INFORMATION AND OTHER RESOURCES

ITU-R Recommendation P. 368-9 [10] provides several curves for the prediction of groundwave intensity produced by a 1 kW short vertical monopole antenna at ground level. Curves are presented for varying ground dielectric constants, conductivities, and

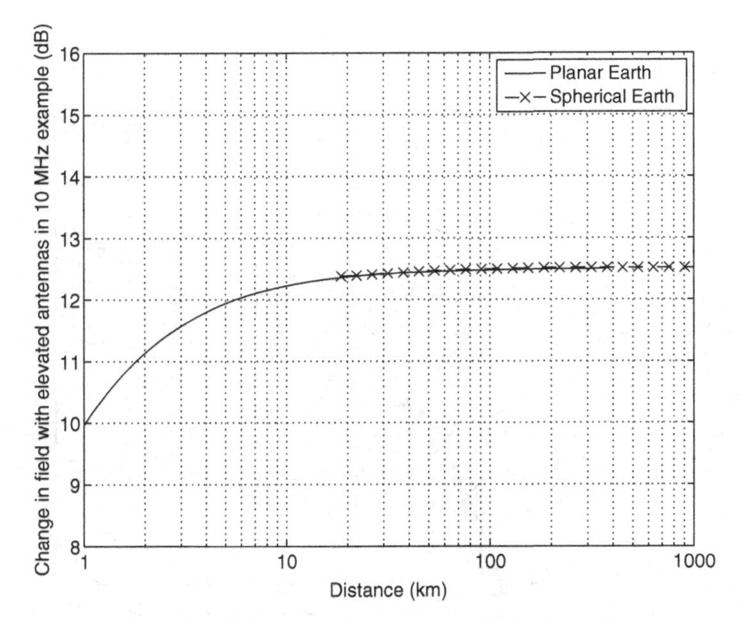

FIGURE 9.10 Change in received power with elevated antennas for sample calculation at 10 MHz.

frequencies, enabling a time-consuming computation to be avoided for many cases. However, the curves apply only for both transmitting and receiving antennas near the ground; for elevated antennas, the height-gain functions must still be applied. These curves were produced using an exponential model for the atmospheric refractive index, which should be more accurate than the linear refractive index model implicit in the modified Earth radius assumption, because the exponential model avoids the exaggerated decrease in atmospheric refractivity with altitude of the linear model. Five sets of curves resulting from the ITU model are presented in Figures 9.11–9.15. It should be noted that the results presented in this section and the models described in this chapter apply for groundwave propagation predictions on average; for a given measurement, site-specific terrain effects may cause significant deviations from these predictions.

Methods for predicting groundwave propagation over mixed paths, that is, paths that are part ground and part sea, for example, or over layered ground surfaces (such as ice-covered seawater) have also been developed; these issues are described in Ref. [7]. In Ref. [11] an empirical model for groundwave propagation in urban areas is described.

9.8 SUMMARY

The groundwave propagation mechanism is important when both transmitting and receiving antennas are relatively close to the ground in terms of the electromagnetic

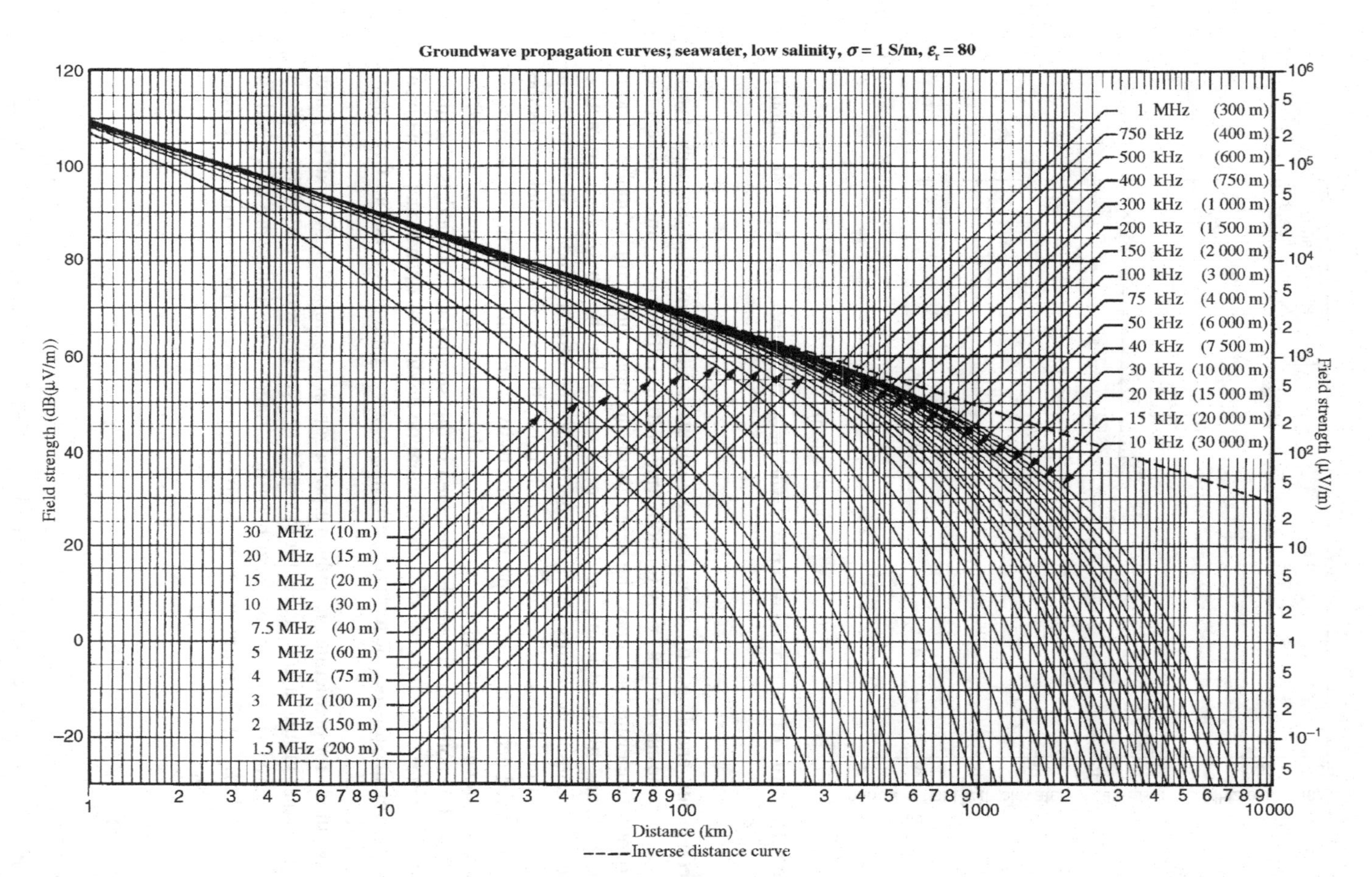

FIGURE 9.11 Groundwave curves for "seawater, low salinity" with relative dielectric constant 80, conductivity 1 S/m. (*Source:* ITU-R Recommendation P. 368-9, used with permission.)

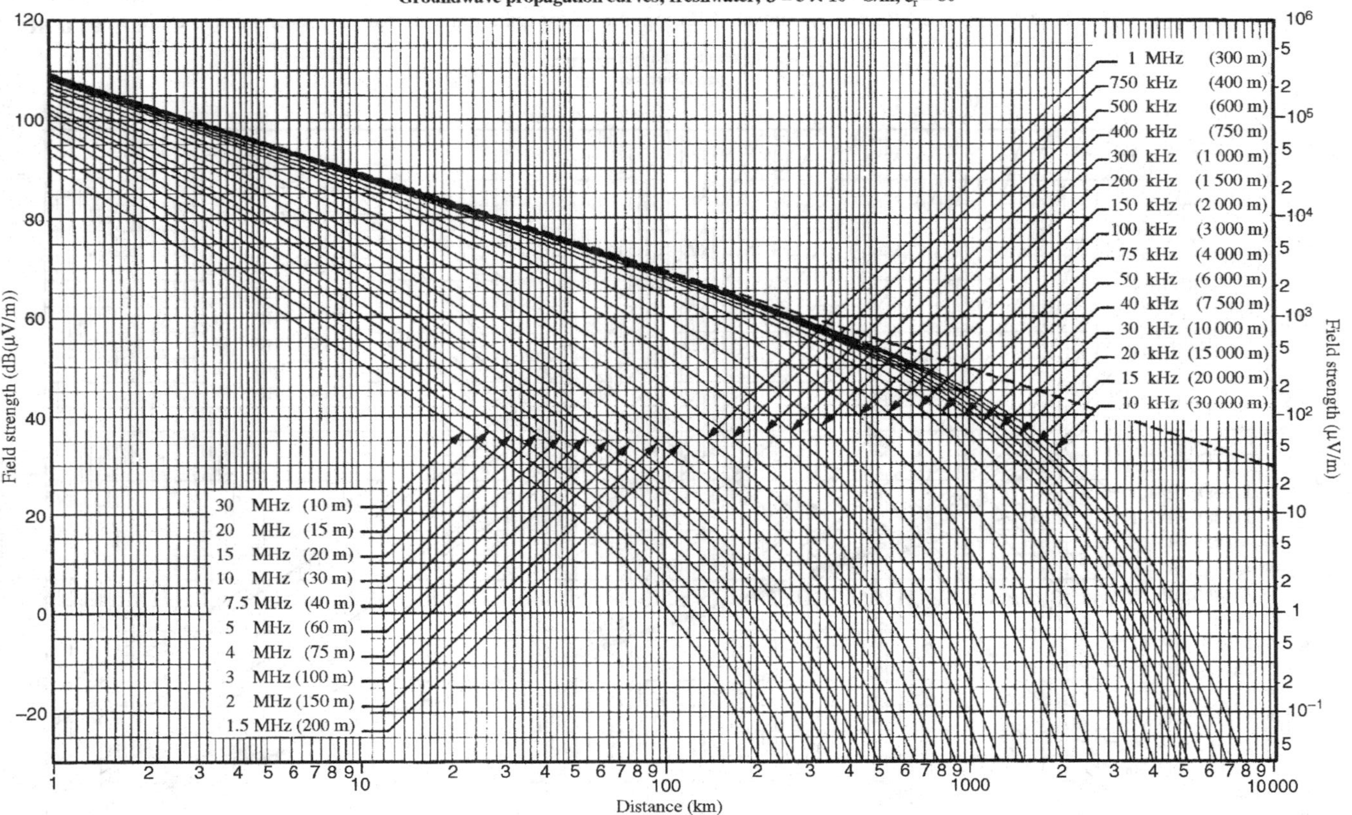

FIGURE 9.12 Groundwave curves for "freshwater" with relative dielectric constant 80, conductivity 0.003 S/m. (*Source:* ITU-R Recommendation P. 368-9, used with permission.)

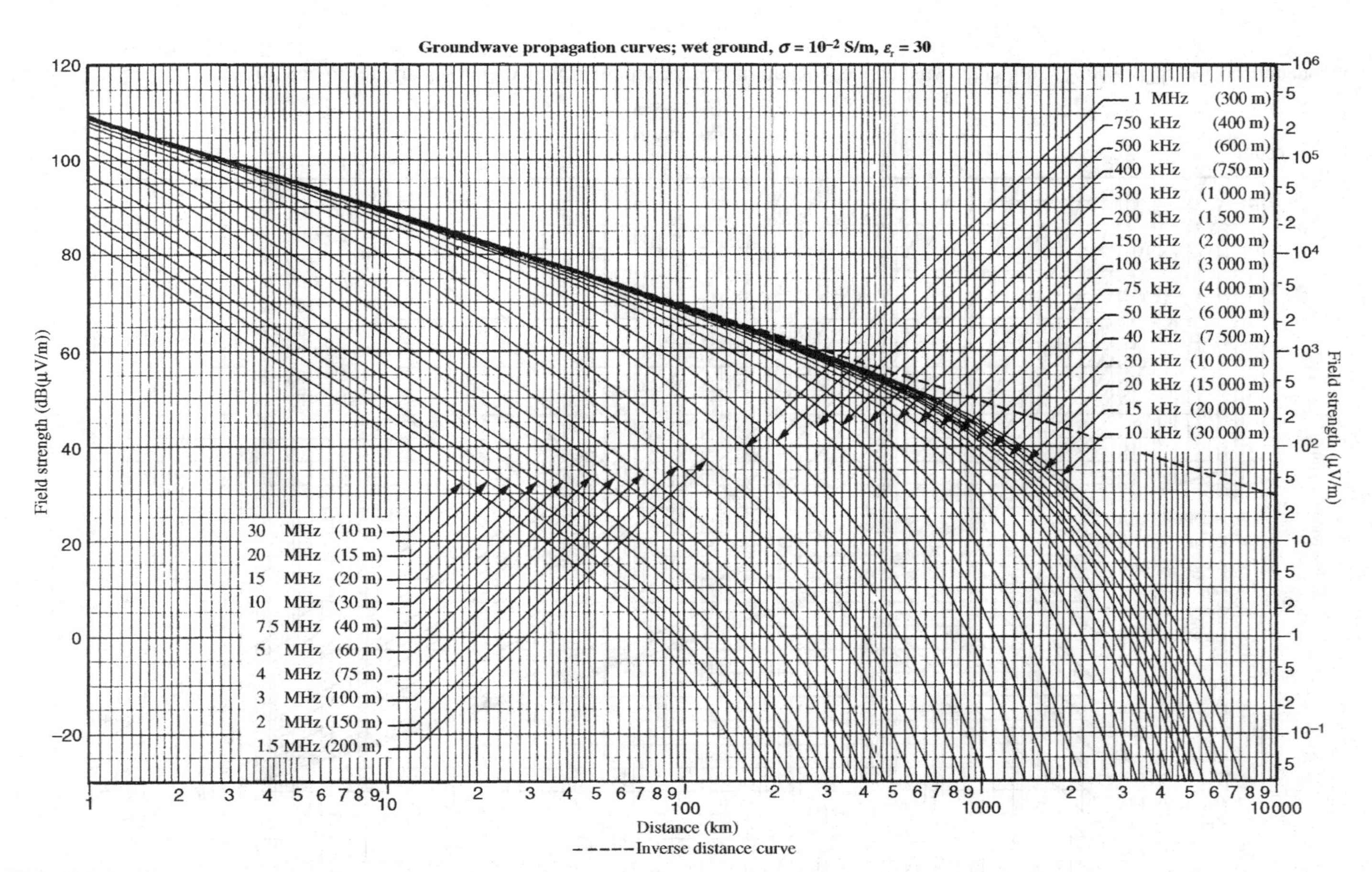

FIGURE 9.13 Groundwave curves for "wet ground" with relative dielectric constant 30, conductivity 0.01 S/m (*Source:* ITU-R Recommendation P. 368-9, used with permission.)

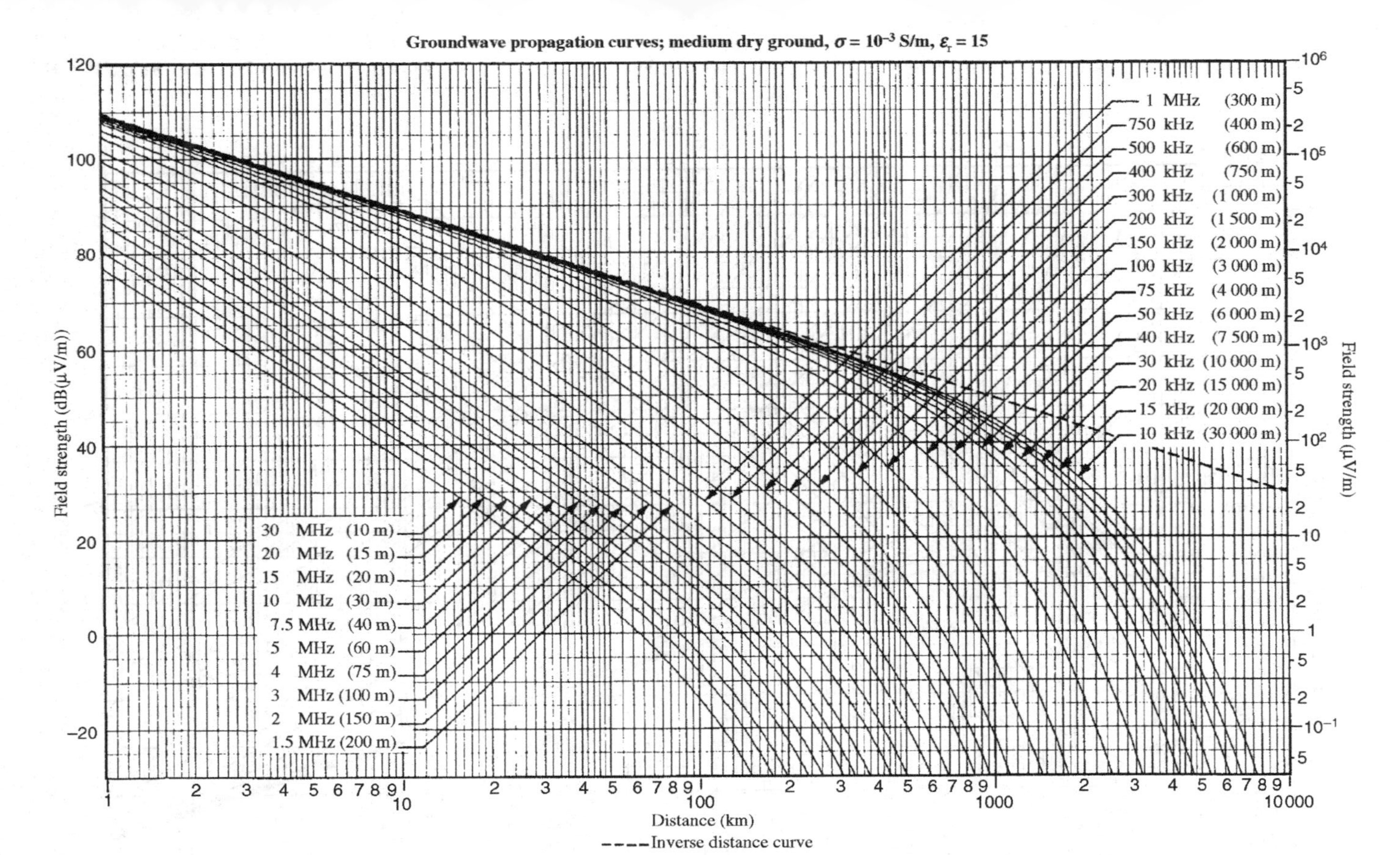

FIGURE 9.14 Groundwave curves for "medium dry ground" with relative dielectric constant 15, conductivity 0.001 S/m. (*Source:* ITU-R Recommendation P. 368-9, used with permission.)

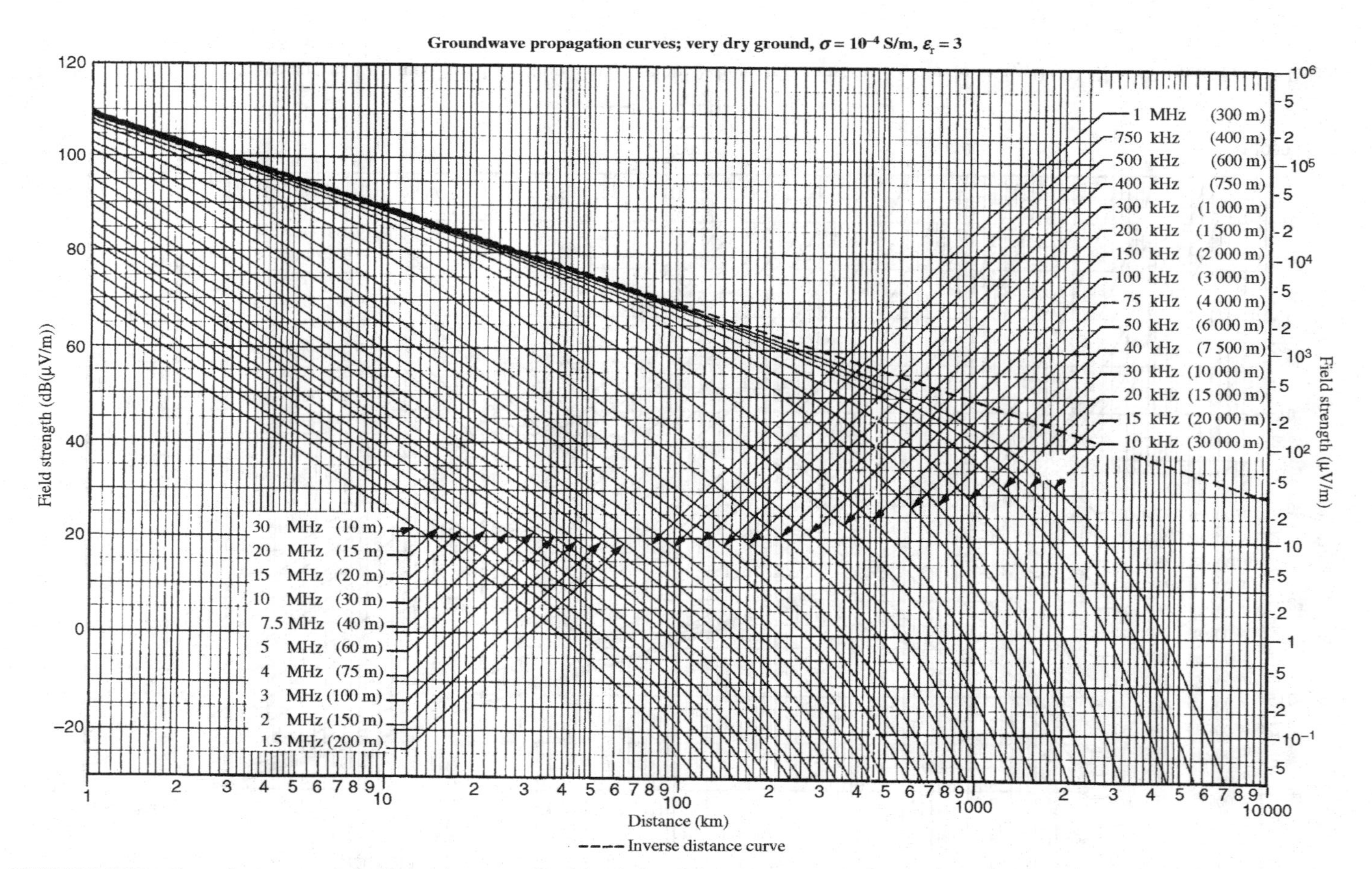

FIGURE 9.15 Groundwave curves for "very dry ground" with relative dielectric constant 3, conductivity 0.0001 S/m. (*Source:* ITU-R Recommendation P. 368-9, used with permission.)

wavelength. The resulting field transitions from $1/d$ to $1/d^2$ to exponential decay with distance at a rate that depends on the frequency and ground dielectric parameters. Use of the "numerical" distance clarifies the relative influence of these physical parameters. Groundwave contributions are usually not significant at VHF and higher frequencies. At lower frequencies, the distance to which they are important varies, generally increasing with decreasing frequency, and also depending on the ground constants and other system parameters. If ionospheric propagation is strong, ionospheric signals often dominate the groundwave mechanism.

APPENDIX 9.A SPHERICAL EARTH GROUNDWAVE COMPUTATIONS

The "roots" $\underline{t}_s$ appearing in the spherical Earth theory can be obtained from a series expansion [12]:

$$\underline{t}_s = \sum_{n=0}^{\infty} \underline{a}_n \, \underline{\tau}^n, \tag{9.23}$$

where $\underline{\tau}$ is the spherical Earth dielectric parameter, $\underline{a}_0 = t_{s,0} \exp(-j\pi/3)$ (values of $t_{s,0}$ for $s = 1, 2, \cdots 5$ are given in Table 9.1), $\underline{a}_1 = \frac{1}{\underline{a}_0}$, and for $n > 0$,

$$\underline{a}_{n+1} = -\frac{1}{(n+1)\underline{a}_0} \sum_{m=1}^{n} \underline{a}'_m (n-m+1)\underline{a}_{n-m+1} \tag{9.24}$$

with $\underline{a}'_n = \underline{a}_n$ for all $n \neq 2$, but $\underline{a}'_n = \underline{a}_n - 1$ for $n = 2$. For $s > 5$, $t_{s,0}$ can be obtained from

$$t_{s,0} \approx \left[(3\pi/2)(s-0.75)\right]^{2/3}. \tag{9.25}$$

The series in equation (9.23) converges well for small amplitudes of $\underline{\tau}$; for larger amplitudes, a different series is more useful:

$$\underline{t}_s = \sum_{n=0}^{\infty} \underline{b}_n \, \underline{\tau}^{-n}, \tag{9.26}$$

TABLE 9.1 Values of $t_{s,0}$ and $t_{s,\infty}$ for s from 1 to 5

s	$t_{s,0}$	$t_{s,\infty}$
1	1.018792971647471	2.338107410459767
2	3.248197582179837	4.087949444130971
3	4.820099211178736	5.520559828095551
4	6.163307355639487	6.786708090071759
5	7.372177255047770	7.944133587120853

where $\underline{b}_0 = t_{s,\infty} \exp(-j\pi/3)$ (values of $t_{s,\infty}$ for $s = 1, 2, \cdots 5$ are given in Table 9.1), and

$$\underline{b}_1 = 1, \tag{9.27}$$

$$\underline{b}_2 = 0, \tag{9.28}$$

$$\underline{b}_3 = \frac{\underline{b}_0}{3}, \tag{9.29}$$

$$\underline{b}_4 = \frac{1}{4}, \tag{9.30}$$

$$\underline{b}_5 = \frac{\underline{b}_0^2}{5}, \tag{9.31}$$

$$\underline{b}_6 = \frac{7\underline{b}_0}{18}, \tag{9.32}$$

$$\underline{b}_7 = \frac{\underline{b}_0^3 + 1.25}{7}, \tag{9.33}$$

$$\underline{b}_8 = \frac{29\underline{b}_0^2}{60}, \tag{9.34}$$

$$\underline{b}_9 = \frac{\underline{b}_6 + \underline{b}_0^4 + \underline{b}_1^2 + 2\underline{b}_0\underline{b}_4 + 2\underline{b}_3\underline{b}_1}{9}. \tag{9.35}$$

For $s > 5$, $t_{s,\infty}$ can be approximated as

$$t_{s,\infty} \approx \left[(3\pi/2)(s - 0.25)\right]^{2/3}. \tag{9.36}$$

The series expansion (9.26) is typically useful when $|\underline{\tau}| > |\sqrt[4]{\underline{t}_{s,0}\underline{t}_{s,\infty}}|$ and that of equation (9.23) otherwise.

The height-gain functions for the spherical Earth theory $\underline{G}_s(y)$ are defined as

$$\frac{\underline{w}\left(\underline{t}_s - y\right)}{\underline{w}\left(\underline{t}_s\right)}, \tag{9.37}$$

where $\underline{t}_s$ is the sth spherical Earth "root" defined above, and

$$\underline{w}(\underline{t}) = \sqrt{\pi}\left[-j\underline{\mathrm{Ai}}(\underline{t}) + \underline{\mathrm{Bi}}(\underline{t})\right]. \tag{9.38}$$

$\underline{\mathrm{Ai}}$ and $\underline{\mathrm{Bi}}$ above are the Airy functions of the first and second kinds, respectively. For small values of y, the spherical Earth height-gain functions can be approximated by

$$\underline{G}_s(y) \approx 1 - \underline{\tau} y + \frac{\underline{t}_s y^2}{2} + \cdots \tag{9.39}$$

while for values of y such that $y \gg \left|\underline{t}_s\right|$,

$$\underline{G}_s(y) \approx \frac{\exp\left[-j\left(\frac{2y^{3/2}}{3} - \sqrt{y}\underline{t}_s - \pi/4\right)\right]}{\sqrt[4]{y}\,\underline{w}(\underline{t}_s)}\left\{1 - j\frac{\underline{t}_s^2}{4\sqrt{y}} + \cdots\right\}. \tag{9.40}$$

Given the large number of calculations required for groundwave predictions, computer programs have been developed to avoid much of the tedium and are recommended for general use. An example is the GRWAVE package available from the ITU.

REFERENCES

1. Sommerfeld, A. N., "Propagation of waves in wireless telegraphy," *Ann. Phys.,* vol. 28, pp. 665–737, 1909.
2. Sommerfeld, A. N., "Propagation of waves in wireless telegraphy II," *Ann. Phys.,* vol. 81, pp. 1135–1153, 1926.
3. Van der Pol, B., and H. Bremmer, "The diffraction of EM waves from an electrical point source round a finitely conducting sphere with applications to radiotelegraphy and the theory of the rainbow I," *Phil. Mag.* vol. 24. pp. 141–176, 1937.
4. Van der Pol, B., and H. Bremmer, "The diffraction of EM waves from an electrical point source round a finitely conducting sphere with applications to radiotelegraphy and the theory of the rainbow II," *Phil. Mag.* vol. 24, no. 164, pp. 825–864, 1937.
5. Van der Pol, B., and H. Bremmer, "The propagation of radio waves over a finitely conducting spherical Earth," *Phil. Mag.* vol. 25, pp. 817–834, 1938.
6. Norton, K. A., "The calculation of groundwave field intensity over a finitely conducting spherical Earth," Proc. *IRE*, vol. 29, p. 623, 1941.
7. Wait, J., "The ancient and modern history of electromagnetic groundwave propagation," *IEEE Antennas Propag. Mag.*, pp. 7–24, 1998.
8. Fishback, William T. "Methods for calculating field strength with standard refraction," in Kerr, Donald E. (ed.), *Propagation of Short Radio Waves*, McGraw-Hill, New York, 1951, pp. 112–140.
9. Li, R., "The accuracy of Norton's empirical approximations for groundwave attenuation," *IEEE Trans. Antennas Propag.,* vol. 31, pp. 624–628, 1983.
10. ITU-R Recommendation, P. 368–9, "Ground-wave propagation curves for frequencies between 10 kHz and 30 MHz," International Telecommunication Union, 2007.
11. Lichun, L., "A new MF and HF ground wave model for urban areas," *IEEE Antennas Propag. Mag.*, pp. 21–33, 2000.
12. Logan, N. A., and K. S. Yee, "A mathematical model for diffraction by convex surfaces," in *Electromagnetic Waves,* (R. E. Langer, ed.), University of Wisconsin Press, Madison, WI, 1962.

10

CHARACTERISTICS OF THE IONOSPHERE

10.1 INTRODUCTION[1]

The region of the Earth's atmosphere from approximately 40 km altitude out to several Earth radii is known as the ionosphere. In this region, solar and cosmic radiation ionize neutral particles to produce regions of space containing large numbers of positively charged ions and free electrons, with peak electron densities of about $N_e = 10^6$ particles/cm^3 at an altitude of about 300 km. The presence of free electrons has a significant effect on wave propagation at HF and lower frequencies, allowing for propagation over long distances through ionospheric reflections, as we will discuss in Chapter 11.

The physics of the ionosphere, essentially involving the chemistry of the upper atmosphere and its interaction with solar radiation, has been studied extensively [1], but can be very complex. Factors that can influence the ionosphere include the Earth's magnetic field, the amount of ionizing radiation obtained from the Sun (which can increase dramatically in periods of high "solar activity"), and atmospheric chemistry and mixing. In this chapter, we will study one of the basic theories that describes the electron/ion production and recombination rates in an ionospheric "layer" [2,3]. This theory, developed by Chapman, will turn out to be applicable mainly to the so-called

[1] Portions of this chapter are adapted from Ref. [3].

Radiowave Propagation: Physics and Applications. By Curt A. Levis, Joel T. Johnson, and Fernando L. Teixeira

E and F_1 regions of the ionosphere, but provides useful insight into the behavior of other regions as well. We will also consider the basic structure of the ionosphere and briefly examine the influence that the Sun and its solar cycle have on ionospheric ionization. Note that the terms "layer" and "region" for describing portions of the ionosphere are used interchangeably in the literature because the ionosphere was thought to consist of distinct layers in the early days of radio; in fact, the electron density is a continuous function of altitude. Nevertheless, certain regions are found to reflect or absorb radio waves in a relatively uniform way, distinct from other regions. While it is more accurate to refer to them as regions, the earlier nomenclature has persisted, and the terms "region" and "layer" are also used here without distinction.

10.2 THE BAROMETRIC LAW

We shall consider the production of ionized layers by solar radiation, mostly X-ray and ultraviolet. This radiation interacts with neutral molecules of the atmosphere to produce electron/ion pairs. It is therefore of interest to examine first the distribution of the molecules as a function of height.

Consider a small cylindrical volume of arbitrary cross section A and length dh, as shown in Figure 10.1. Let its axis be vertical. Denote the pressure at the bottom by p, corresponding to the height h. At the top, the height is $h + dh$ and the corresponding pressure is $p + dp$. Since atmospheric pressure decreases with height, dp will be a negative quantity when dh is positive.

The external atmosphere exerts a vertical, buoyant force on this volume by virtue of the pressure difference between the top and bottom surfaces. A force of magnitude Ap pushes up on the bottom, a force $A(p + dp)$ pushes down on the top, and the net upward differential force on the volume is therefore the buoyant force

$$dF_b = Ap - A(p + dp) = -Adp. \tag{10.1}$$

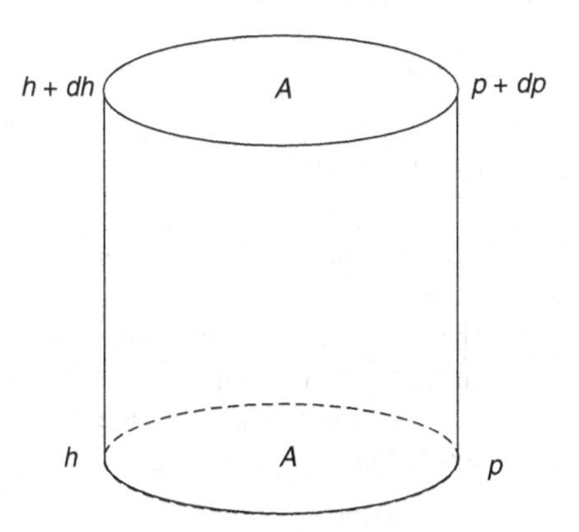

FIGURE 10.1 Differential atmospheric volume.

Assuming equilibrium to exist (i.e., no vertical wind), this force must be precisely balanced by the downward gravitational force. The differential weight of the gas in the volume is

$$dF_{\rm g} = \rho g dV = \rho g A dh, \tag{10.2}$$

where ρ is the mass density of the gas (mass per unit volume) and g is the gravitational constant. The density is given by

$$\rho = N m_s, \tag{10.3}$$

where N is the number density (number of molecules per unit volume) and m_s is the mean molecular mass. Thus, the differential gravitational force is

$$dF_{\rm g} = N m_s g \, A \, dh. \tag{10.4}$$

Equating this to the buoyant force gives

$$-dp = N m_s g \, dh. \tag{10.5}$$

From the ideal gas law of physics, $p = N k_{\rm B} T$, where $k_{\rm B}$ is Boltzmann's constant and T is the absolute temperature, so

$$N = \frac{p}{k_{\rm B} T}, \tag{10.6}$$

and use of this in equation (10.5) gives

$$\frac{dp}{p} = -\frac{m_s \, g}{k_{\rm B} \, T} dh, \tag{10.7}$$

which may be integrated to yield

$$\ln \, p = -\int \frac{m_s \, g}{k_{\rm B} \, T} dh, \tag{10.8}$$

or

$$\ln \, p(h) - \ln \, p(h_0) = -\int_{h_0}^{h} \frac{m_s \, g}{k_{\rm B} \, T} dh'. \tag{10.9}$$

The left-hand side is $\ln \, \left[p(h)/p(h_0) \right]$, so that we can write

$$p(h) = p(h_0) \exp \left(-\int_{h_0}^{h} \frac{m_s \, g}{k_{\rm B} \, T} dh' \right). \tag{10.10}$$

In the above integral, k_B is a true constant and g is very nearly constant since the scale of height variations considered is small compared to the distance from the Earth's center. On the other hand, m_s depends on h since heavier molecules are most abundant at low altitudes. Furthermore, the temperature T can be a quite complicated function of altitude. However, over limited distances the whole integrand may often be considered nearly constant. In that case, denoting

$$\frac{m_s \, g}{k_{\rm B} \, T} = \frac{1}{H}, \tag{10.11}$$

we get

$$p(h) = p(h_0)\, e^{-\frac{h-h_0}{H}}. \tag{10.12}$$

Finally, by again using the ideal gas law and assuming an isothermal condition, we obtain

$$N(h) = N(h_0)\, e^{-\frac{h-h_0}{H}} \tag{10.13}$$

and, by the use of equation (10.3),

$$\rho(h) = \rho(h_0)\, e^{-\frac{h-h_0}{H}}. \tag{10.14}$$

The dimensions of H are the same as those of h, that is, length. The pressure, mass density, and number density are seen to vary exponentially with height difference measured in H-units. H is therefore called "the local scale height": it scales the height difference and is local in the sense that the concept breaks down over height differences over which the mean molecular mass or the temperature varies significantly.

The reference height h_0 in this development is arbitrary. For some purposes, it may be assumed to be sea level. In the Chapman theory of the formation of ionized regions, it will turn out handy to choose the height of maximum ion production as the reference height. Figure 10.2a illustrates a typical plot of ρ and p versus altitude; note that a log scale is used so that the barometric curve should appear as a straight line with a slope determined by the scale height. The fact that scale height changes with altitude in the atmosphere is indicated by the curved nature of these plots versus height. Figure 10.2b displays the scale heights H associated with the pressure and density as a function of altitude. Figure 10.3 provides plots versus altitude of several atmospheric quantities. Information such as this is available from the U.S. National Oceanic and Atmospheric Administration (NOAA) [4].

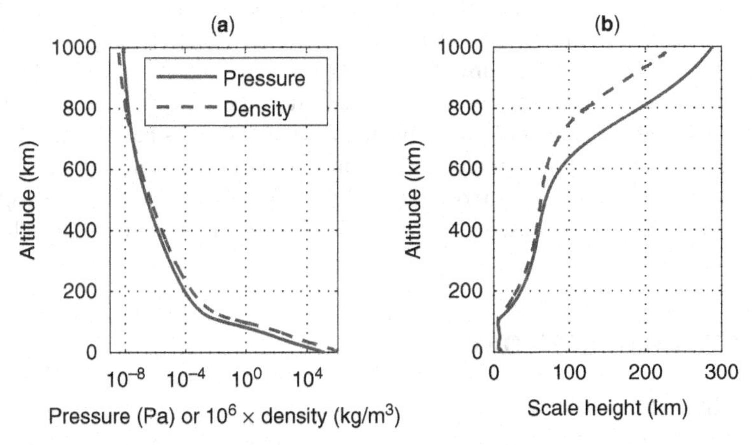

FIGURE 10.2 Typical plot of atmospheric density versus altitude. (*Source:* U.S. Standard Atmosphere, NOAA/NASA [4].)

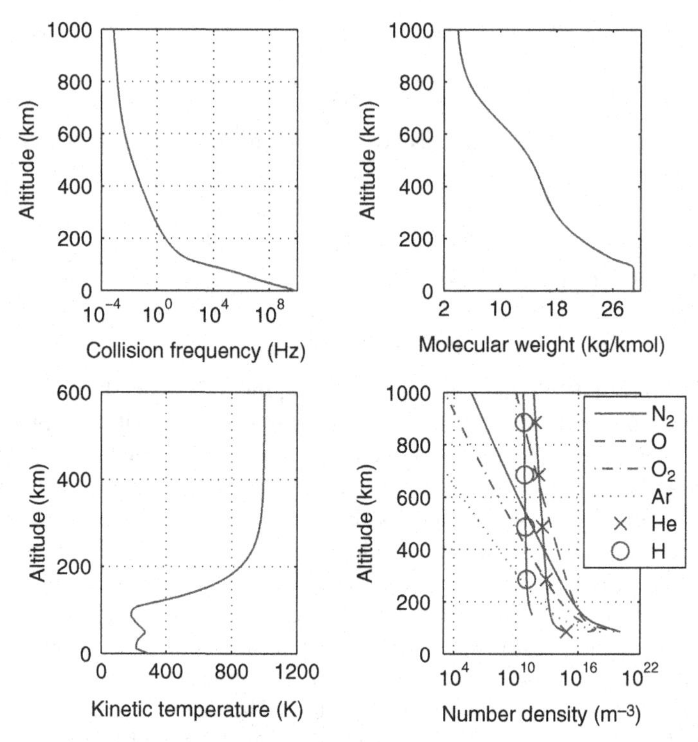

FIGURE 10.3 Sample atmospheric data. (*Source:* U.S. Standard Atmosphere, NOAA/NASA [4].)

These figures show the expected dependencies of decreasing pressure, density, and collision frequencies in the atmosphere as altitude increases. The decrease in mean molecular weight observed is due the change in chemical composition of the atmosphere for increasing altitudes, caused by decreasing fractions of molecular oxygen O_2 and nitrogen N_2. Atmospheric temperature also varies significantly with altitude; note the temperature is a measure of the average kinetic energy of molecules in the atmosphere, even for the very low atmospheric densities encountered at high altitudes. One broad division of the atmosphere into two regions is based on the amount of mixing of species. For altitudes below circa 90 km, turbulent mixing causes a more uniform atmospheric composition, and this region is labeled the *homosphere.* At higher altitudes, the atmosphere is less well mixed, and this region is thus labeled the *heterosphere.*

10.3 CHAPMAN'S THEORY

10.3.1 Introduction

Chapman was the first to formulate a quantitative theory of the formation of ionized regions in the atmosphere due to ionizing radiation from the Sun [2]. A qualitative picture of the process is relatively easy to obtain. At very high altitudes, there is plenty of solar flux but there are few molecules available to be ionized. Hence, ion production will be small at very high altitudes. At low altitudes, there are plenty of molecules to be ionized, but most of the flux has been used up at higher levels; again, ion production will be small. At some intermediate altitude, there will be sufficient molecules and sufficient flux to maximize ion production. Hence, the production of electron–ion pairs increases monotonically from low values at low altitudes to a maximum at some intermediate altitude and then decreases monotonically at higher altitudes.

The mathematical development of this relationship will proceed along the following steps:

(a) The amount of solar flux penetrating to a height h will be calculated. An important parameter will be the "optical depth", a measure of the opacity of the region above height h.
(b) By differentiation of the flux expression, the decrease in flux in an incremental height dh will be found.
(c) The flux decrease represents absorbed power, which is responsible for the ionization. From this principle, an expression for the ion production per unit volume at height h will be derived.
(d) The barometric law will be used to find the optical depth explicitly as a function of height. In order to simplify the equations, the reference height will be chosen as the height where the optical depth is unity.
(e) The height of maximum possible electron/ion production will be shown to be the reference height.
(f) From the law of electron/ion production and the assumption of equilibrium, laws of the variation of electron density will be derived.
(g) The electron/ion production and electron density depend on two variables: altitude and solar zenith angle. By developing suitable scaling laws, it will be found they can be represented as functions of a single variable.

10.3.2 Mathematical Derivation

(a) We first calculate the amount of solar flux S as a function of the height h. Consider a cylinder of arbitrary cross section A with its axis inclined with respect to the vertical by the solar zenith angle χ, as shown in Figure 10.4.

The flux density at the top (height $h + dh$) is $S + dS$, and that at the bottom is S. The loss of flux in the volume is therefore $A\,dS$. This flux is absorbed by the molecules in the cylinder. Let the average absorption cross section (i.e., the power absorbed per molecule per unit flux, averaged over the various molecular species) be denoted by σ (units of area). Then

$$A\,dS = N\,dV\,\sigma\,S, \tag{10.15}$$

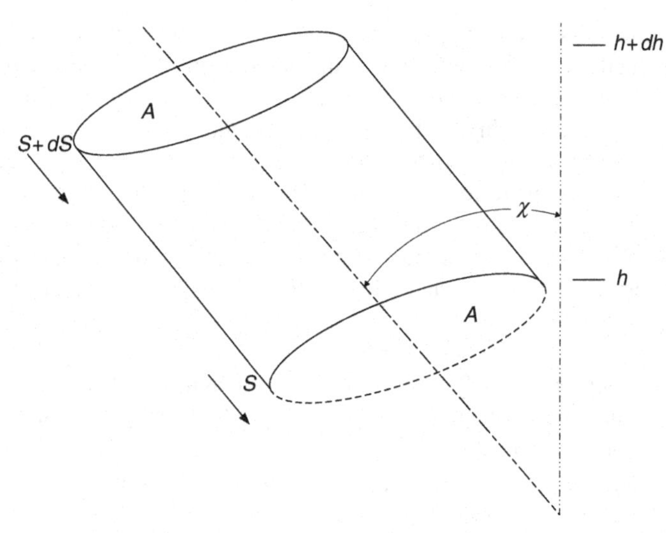

FIGURE 10.4 Tilted differential atmospheric volume.

where N is again the number density of molecules and dV is the differential volume given by

$$dV = A\,dh\,\sec\chi. \tag{10.16}$$

Use of this in equation (10.15) and division by AS gives

$$\frac{dS}{S} = N\sigma\,\sec\chi\,dh, \tag{10.17}$$

which can be integrated to give

$$\int_{\infty}^{h}\frac{dS}{S} = \ln\,S(h) - \ln\,S(\infty) = \sec\chi\int_{\infty}^{h} N(h')\,\sigma(h')\,dh'. \tag{10.18}$$

Define the optical depth $\tau(h)$ (a unitless quantity) by

$$\tau(h) \equiv -\int_{\infty}^{h} N(h')\,\sigma(h')\,dh', \tag{10.19}$$

then

$$\ln\frac{S(h)}{S(\infty)} = -\tau\sec\chi, \tag{10.20}$$

and therefore

$$S(h) = S_\infty\,e^{-\tau(h)\sec\chi}, \tag{10.21}$$

where $S_\infty = S(\infty)$ denotes the flux at a very large height, so that all absorption above it is negligible.

(b) The change in flux density per unit height increment can now be obtained by differentiation:

$$\frac{dS}{dh} = -S_\infty \frac{d\tau}{dh} \sec\chi \, e^{-\tau(h)\sec\chi}. \tag{10.22}$$

The value of $d\tau/dh$ follows from the definition of τ as an integral in equation (10.19):

$$\frac{d\tau}{dh} = -N(h)\,\sigma(h), \tag{10.23}$$

and thus we get

$$\frac{dS}{dh} = S_\infty \, N(h)\,\sigma(h) \sec\chi \, e^{-\tau(h)\sec\chi}. \tag{10.24}$$

(c) The power absorbed in the volume is given by

$$dP = A\,dS \;=\; A\,\frac{dS}{dh}\,dh. \tag{10.25}$$

If we denote the number of ions produced in the volume per unit time as $n_{\rm i}$, then

$$n_{\rm i} = \varsigma\,dP, \tag{10.26}$$

where ς denotes the ionization efficiency, that is, the number of ions produced per unit time per unit power. Thus,

$$n_{\rm i} = \varsigma\,A\,\frac{dS}{dh}\,dh \tag{10.27}$$

is the number of ions produced per unit time in the volume by the conversion of radiant flux energy to ionization energy.

Let q denote the rate of ion production per unit volume. Then

$$n_{\rm i} = q\,dV = q\,A\,dh\,\sec\chi. \tag{10.28}$$

By equating (10.27) and (10.28), we get

$$q(\chi, h)\sec\chi = \varsigma\,\frac{dS}{dh}, \tag{10.29}$$

and using equation (10.24) gives

$$q(\chi, h) = S_\infty \, N(h)\,\sigma(h)\,\varsigma(h)\,e^{-\tau(h)\,\sec\chi}. \tag{10.30}$$

So far the treatment has been quite general. Although some new notations were used, no restrictive assumptions were made. At this point, if an analytical (as opposed to numerical) approach is to be carried further, some simplifying assumptions are necessary.

(d) Assume that over the region of interest the absorption cross section σ and ionization efficiency ς do not vary appreciably. These quantities depend on the mixture of molecular species. Also, assume that $N(h)$ follows the barometric law of equation (10.13). Then from the definition of τ in equation (10.19), we get

$$\tau = -\sigma\, N(h_0) \int_{\infty}^{h} e^{-\frac{h'-h_0}{H}}\, dh',$$

$$\tau = \sigma\, N(h_0)\, H\, e^{-\frac{h-h_0}{H}}. \tag{10.31}$$

By putting $h = h_0$ in this equation, we find

$$\tau(h_0) = \sigma N(h_0)\, H, \tag{10.32}$$

so equation (10.31) can also be written as

$$\tau(h) = \tau(h_0)\, e^{-\frac{h-h_0}{H}}. \tag{10.33}$$

To eliminate the constants $\sigma N(h_0) H = \tau(h_0)$ from the remainder of this development, we now choose the reference height h_0, which is arbitrary, so that $\tau(h_0) = 1$. This choice also has physical significance. From equation (10.21), it can be seen that the reference height will then be the height for which, when the Sun is at zenith ($\chi = 0$), the flux $S(h_0)$ will have decreased to $1/e$ (or 37%) of its incident value S_∞. It will also be shown later that, for the Sun at zenith, the electron–ion production rate $q(\chi = 0, h)$ is maximum at $h = h_0$. Using the further simplifying notation

$$z = \frac{(h - h_0)}{H} \tag{10.34}$$

gives

$$\tau = e^{-z}, \tag{10.35}$$

$$N = N(h_0)\, e^{-z} \tag{10.36}$$

and from equation (10.32) with $\tau(h_0) = 1$,

$$N_0 \equiv N(h_0) = \frac{1}{\sigma H}. \tag{10.37}$$

Use of these in equation (10.30) gives

$$q(\chi, h) = \frac{\varsigma S_\infty}{H}\, e^{-z-e^{-z}\sec\chi}. \tag{10.38}$$

(e) We are now ready to prove the previous statement that q is maximized at $\chi = 0$, $h = h_0$. From the definition of z, equation (10.34), the condition $h = h_0$ is the same as $z = 0$. One small notational issue should be mentioned here. When the production rate $q(\chi, h)$ is expressed as a function of z, the *form* of the functional dependence changes since not h but $(h - h_0)/H$ is replaced by the new variable z. So we should write $\bar{q}(\chi, z) = q(\chi, h)$ to emphasize that they refer to two distinct mathematical expressions. However, since $\bar{q}(\chi, z)$ and $q(\chi, h)$ denote the same physical quantity, it is more convenient to adopt the "physicist practice" of using the same symbol for both and, if necessary, distinguishing them by their arguments

$$q(\chi, h) = q(\chi, z) = \frac{\varsigma S_\infty}{H} e^{-z-e^{-z}\sec\chi}. \tag{10.39}$$

To maximize $q(\chi, z)$, we note by inspection that $\sec\chi = 1$ maximizes the expression with respect to χ for all z, giving

$$q_{\max} = \frac{\varsigma S_\infty}{H} e^{-z-e^{-z}}. \tag{10.40}$$

To maximize this with respect to z, find

$$\frac{dq_{\max}}{dz} = \frac{\varsigma S_\infty}{H} \left[-1 - (-1)e^{-z}\right] e^{-z-e^{-z}}. \tag{10.41}$$

Requiring $\frac{dq_{\max}}{dz} = 0$ leads to $e^{-z} - 1 = 0$, and therefore $z = 0$ or $h = h_0$. The value of $q(\chi, z)$, maximized with respect to both solar zenith angle and height, is then

$$q_0 = q(\chi = 0, z = 0) = \frac{\varsigma S_\infty}{H} e^{-1}. \tag{10.42}$$

Hence,

$$\frac{\varsigma S_\infty}{H} = q_0 e, \tag{10.43}$$

and by the use of this in (10.39), we get

$$q(\chi, z) = q_0 e^{1-z-e^{-z}\sec\chi}. \tag{10.44}$$

(f) For ionospheric path calculations, we are not interested so much in the electron/ion production rate $q(\chi, z)$ as in the resulting electron density N_e, which will strongly affect electromagnetic wave propagation as discussed in Chapter 11. The function $N_e(\chi, h)$ can be derived by writing an equation for the net rate of change in electron density per unit volume as the difference between the rate of production q and the rate of electron loss r

$$\frac{d}{dt} N_e(\chi, h) = q(\chi, h) - r(\chi, h). \tag{10.45}$$

In the actual ionosphere, many electrons are produced and lost in the time required to produce an appreciable change in the resulting electron density: on the timescale appropriate to the terms on the right-hand side of (10.45), the ionosphere is in equilibrium. We can therefore set the left-hand side equal to zero and obtain

$$q(\chi, h) \approx r(\chi, h). \tag{10.46}$$

There are numerous reactions that can remove free electrons from the ionosphere, and indeed it is difficult to go very far into the properties of the ionosphere without considering its chemistry, but that is beyond the scope of this book. Two classes of electron–loss processes can be distinguished: in the first, known as recombination, the electron combines with a positive ion to form a neutral molecule; in the second, known as attachment, it attaches itself to a neutral molecule to form a negative ion.

The recombination process depends on electrons finding ions: for this kind of process,

$$r = \alpha N_e N_i, \tag{10.47}$$

where α is a recombination coefficient, N_e is the electron density, and N_i is the ion density. Since ionization produces electrons and positive ions in equal numbers, $N_i = N_e$ and

$$r = \alpha N_e^2. \tag{10.48}$$

From (10.46), we then have

$$N_e(\chi, h) = \sqrt{\frac{q(\chi, h)}{\alpha}}, \tag{10.49}$$

and using equation (10.44) and setting

$$N_0^2 = \frac{q_0}{\alpha} \tag{10.50}$$

gives

$$N_e(\chi, h) = N_0\, e^{\frac{1}{2}(1 - z - e^{-z} \sec\chi)}. \tag{10.51}$$

For the electron–molecule attachment process, the electrons have to find neutral molecules, and the number of these is independent of the number of electrons. Then the electron loss rate is

$$r = \bar{\beta}\bar{N}_e N \;=\; \beta\bar{N}_e, \tag{10.52}$$

where $\bar{N}_e$ is used to distinguish the attachment process electron density from N_e for the recombination process, $\bar{\beta}$ is an attachment coefficient, and

$$\beta \equiv \bar{\beta}N. \tag{10.53}$$

From the barometric equation, (10.13) or (10.36), it follows that

$$\beta = \beta(h_0)\, e^{-z}, \tag{10.54}$$

or

$$r = \beta(h_0)\, e^{-z}\, \bar{N}_e(\chi, z). \tag{10.55}$$

Setting the electron production and loss rates equal as demanded by (10.46), we have

$$\bar{N}_e(\chi, z) = \frac{q(\chi, z)}{\beta(h_0)}\, e^{z}, \tag{10.56}$$

and use of equation (10.44) gives

$$\bar{N}_e(\chi, z) = \bar{N}_e(h_0)\, e^{1 - e^{-z} \sec\chi}. \tag{10.57}$$

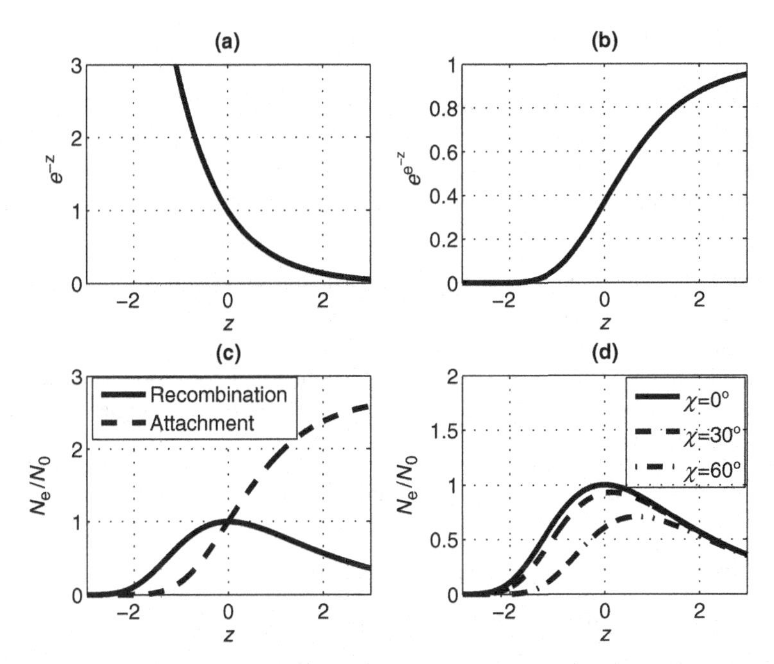

FIGURE 10.5 An illustration of functions involved in the Chapman theory: (a) a plot of e^{-z}, (b) a plot of exp $\left(e^{-z}\right)$, (c) N_e/N_0 for recombination and attachment layers, and (d) impact of the solar zenith angle χ on a recombination layer.

Equations (10.51) and (10.57) have achieved the goal of this rather lengthy derivation because it is the electron density that relates most strongly to radiowave propagation effects, as will be discussed in detail in the next chapter. The mass of an electron is such that it can respond to radio waves with substantial motion, especially at HF and lower frequencies. The mass of ions is much greater. Therefore, ionic motions are much smaller, and the influence of ions can be neglected in usual propagation problems.

The nature of the layers predicted by equations (10.51) and (10.57) is somewhat difficult to discern directly, due to the double exponential functions involved. Figure 10.5 provides an illustration to clarify the layer shapes for $\chi = 0$. Figure 10.5a is a plot of the function e^{-z}, one of the factors that determines the shape of the recombination layer of equation (10.51). This term is multiplied by the function exp (e^{-z}), plotted in Figure 10.5b, which ranges from 0 to unity as altitude increases. The product of these two clearly will produce a function that is maximum somewhere at middle altitudes; Figure 10.5c compares the electron density profile (divided by N_0) for recombination and attachment layers. The latter (equation (10.57)) is smaller for lower altitudes due to the increased atmospheric density and reattachment rate at lower altitudes. Figure 10.5d illustrates the influence of the solar zenith angle on a recombination layer; as the solar angle increases, the peak electron density decreases, and the height of the maximum increases in altitude.

(g) The electron production rate $q(\chi, z)$ and the resulting electron density distributions $N_e(\chi, z)$ and $\bar{N}_e(\chi, z)$ are functions of the two independent variables χ and z. It turns out, however, that values for arbitrary χ can be obtained rather simply from the values for $\chi = 0$ by use of scaling laws. The scaling law for $q(\chi, z)$ will now be developed.

Suppose we are interested in q for a particular set of values χ_1, z_1. Instead of calculating it directly from equation (10.44), let us calculate a related value: that of $q(0, z_1 - \ln \sec \chi_1)$, that is, the production rate that corresponds to a zenith angle of zero and a normalized height z that corresponds to the desired value z_1 reduced by the log secant of the desired zenith angle. By equation (10.44), we have

$$q(0, z_1 - \ln \sec \chi_1) = q_0 \, e^{1 - z_1 + \ln \sec \chi_1 - e^{-z_1 + \ln \sec \chi_1}}. \tag{10.58}$$

Since $e^{\ln u} = u$, we can transform this into

$$q(0, z_1 - \ln \sec \chi_1) = q_0 \sec \chi_1 \, e^{1 - z_1 - e^{-z_1} \sec \chi_1}. \tag{10.59}$$

Now evaluate $q(\chi_1, z_1)$ by (10.44); the result is

$$q(\chi_1, z_1) = q_0 \, e^{1 - z_1 - e^{-z_1} \sec \chi_1}. \tag{10.60}$$

Comparison of the last two relationships gives

$$q(0, z_1 - \ln \sec \chi_1) = \sec \chi_1 \, q(\chi_1, z_1). \tag{10.61}$$

Multiplying by $\cos \chi_1$, and then recalling that (z_1, χ_1) are arbitrary values of (z, χ), we can write

$$q(\chi, z) = \cos \chi \, q(0, z - \ln \sec \chi). \tag{10.62}$$

This is the desired scaling law. It says that a curve of q versus z for a zenith angle $\chi = 0$ is really all we need. The value of q for any value of z and χ can be obtained from the $\chi = 0$ curve at a z value equal to the true z reduced by $\ln \sec \chi$, multiplied by $\cos \chi$. Similar scaling laws can be obtained for electron densities under the previous assumptions of recombination (N_e) and attachment ($\bar{N}_e$). They are, respectively,

$$N_e(\chi, z) = \sqrt{\cos \chi} \, N_e(0, z - \ln \sec \chi) \tag{10.63}$$

and

$$\bar{N}_e(\chi, z) = \cos \chi \, \bar{N}_e(0, z - \ln \sec \chi). \tag{10.64}$$

10.4 STRUCTURE OF THE IONOSPHERE

The upper part of Figure 10.6 depicts the nomenclature used in describing the atmosphere, including the ionosphere. The electron density sketch is not meant

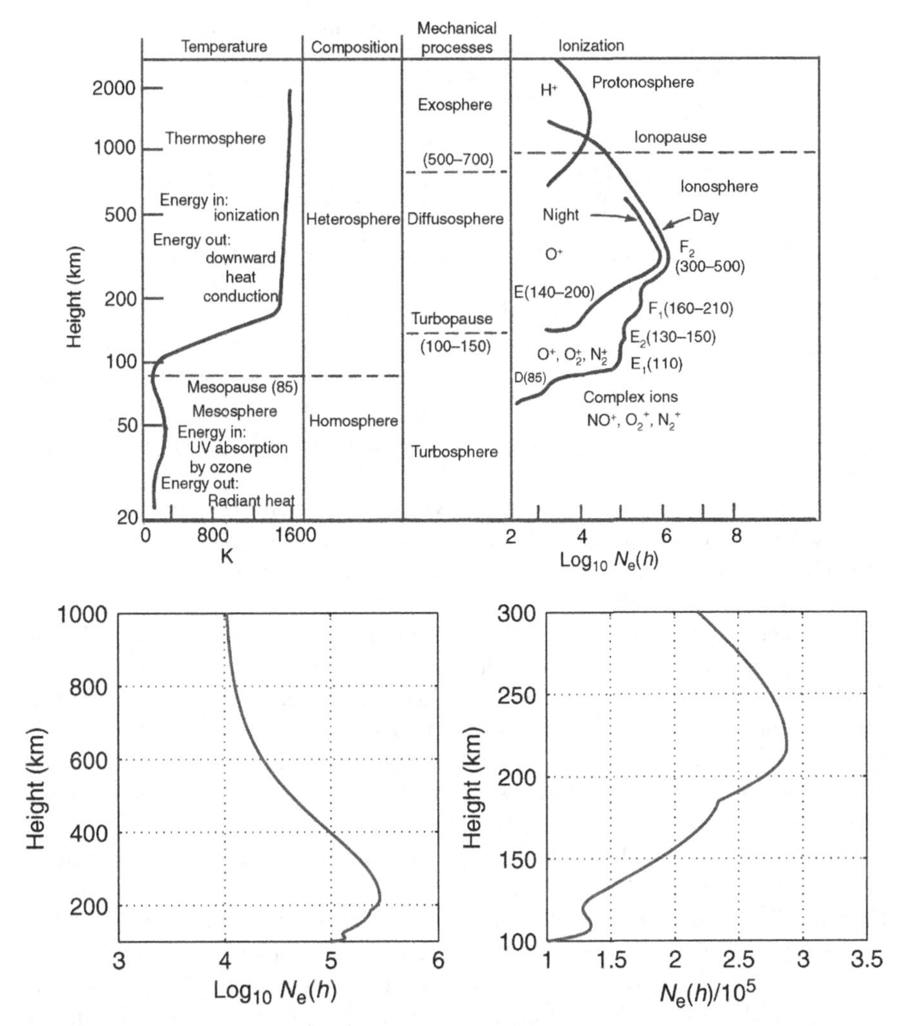

FIGURE 10.6 The upper plot depicts standard atmospheric nomenclature versus altitude. (*Source:* T. E. Van Zandt, "The Formation of the Ionosphere," Lecture 4 in *NBS Course in Radio Propagation: Ionospheric Propagation,* Central Radio Propagation Laboratory, National Bureau of Standards, U.S. Department of Commerce, Boulder, CO, 1961.) The lower plots are an example of the electron density distribution versus height generated by the International Reference Ionosphere for Columbus, OH, on 07/01/2007 (local noon). Left shows a larger range of altitudes while right shows more clearly the E region maximum near 110 km, the F_1 region from about 120 to 180 km, and the F_2 region above 180 km. The electron number densities (N_e) shown in these plots are in units of cm^{-3}.

to represent typical conditions but has been drawn to allow representation of all possible "layers". A more typical plot of electron density versus height is shown in the lower part of Figure 10.6, generated from the International Reference Ionosphere (IRI) for the location of Columbus, OH, on 07/01/2007 (12:00 PM).

TABLE 10.1 Typical Ionospheric Regions (or Layers)

Height Range (km)	Region Name	Possible Layers
50 to 90 km	D	D
90 to 120 or 150 km	E	E_1,E_2
Above 120 or 150 km	F	F_1, F_2

Note the two distinct curves for day- and nighttime electron densities: solar ionization during the day produces larger electron densities that gradually recombine at night through processes that vary with altitude. The particular molecular species that become ionized are indicated on the plot as well, with O_2 and N_2 molecules more prevalent at lower altitudes, and oxygen atoms at higher altitudes. Standard terminology usually defines three layers in the ionosphere, denoted by the letters D, E, and F with increasing altitude. The typical altitudes for these regions are indicated in the figure. The F and (rarely) E regions can also be subdivided into F_1 and F_2, and (rarely) E_1 and E_2 layers, with the lower subscript numbers referring to lower altitudes. Note the disappearance of the D layer at night. Table 10.1 summarizes information on ionospheric regions and layers. We will find in Chapter 11 that the D region primarily causes absorption and not reflection for MF and higher frequencies, so its absence at night allows MF frequencies to reach the E layer and be reflected. The F_1 and F_2 regions also combine into a single F region at night, and all layers have a reduced electron density since solar flux is no longer available to maintain the ionization process. Peak electron densities of about $N_e = 10^6$ particles/cm^3 occur in the F_2 layer. The N_e profile can have a secondary local maximum at about 100 km in the E region. In the D layer below 60 km, N_e drops precipitously to densities below $N_e = 10^3$ particles/cm^3.

In practice, it is found that the E and F_1 regions behave in a manner predicted rather well by equation (10.51). That equation, which specifies a recombination-type Chapman layer, is therefore of great importance. This is not true of equation (10.57), which gives the electron density of attachment-type Chapman layers: no extended ionospheric regions follow this law. Attachment may nevertheless be an important component of the ionospheric chemistry, especially in the D region. Ionization at higher altitudes than the F_1 layer is strongly influenced by the Earth's magnetic field and by mixing in the upper atmosphere, and is very complex. It is the F_2 region, however, that is of the greatest importance for HF communications, as we shall see in Chapter 11.

In addition to the day and night variations, ionospheric properties also vary with latitude, season, and solar cycles. Due to difficulties in the theoretical modeling of ionospheric properties, empirical methods are very important in predicting ionospheric propagation. A global network of ground-based ionospheric observation stations, which produce ionograms as described in Chapter 11, exists to allow for forecasting the expected behaviors (bottom-side sounders). Figure 10.7 illustrates the locations of many such ground stations.

This ground-based network can be augmented by a variety of space-based data from satellites (topside sounders). As the name indicates, bottom-side sounders refer to

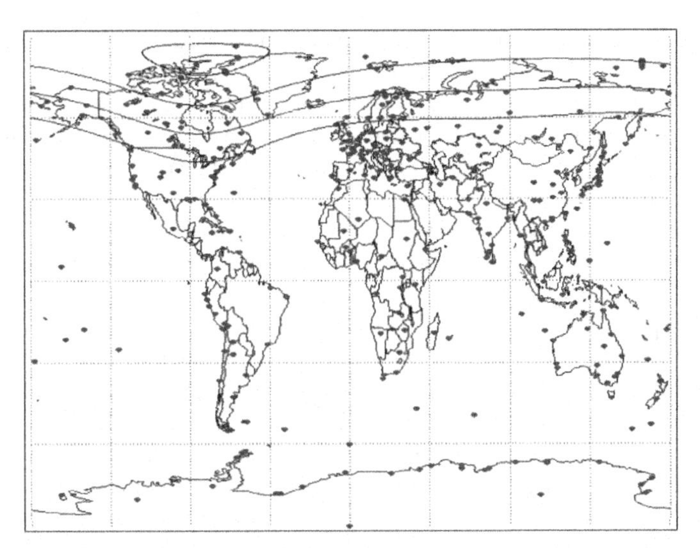

FIGURE 10.7 Sample locations of ionospheric observation stations. (*Source:* UK Solar System Data Centre, Science and Technology Facilities Council.)

those suited for gathering ionospheric data below the F_2 peak, while topside sounders are suited for gathering data above it. Ionospheric properties can be further monitored using dual-frequency signals received from networks of GPS (or other global navigation satellite systems) receivers or from low-Earth orbit satellites using radio occultation techniques, as discussed in Chapter 11.

One important source of ionospheric data is the International Reference Ionosphere (IRI) [5,6], an international project sponsored by the Committee on Space Research (COSPAR) and the International Union of Radio Science (URSI). This is a continuously updated empirical model of the ionosphere, based on a worldwide network of Earth-based and satellite-based ionospheric observation facilities. For a given geographical location, this model provides monthly averages of ionospheric parameters in the 50–2000 km height region. IRI information is currently distributed by NASA Goddard Space Flight Center and is freely available online.

Another useful ionospheric model is the NeQuick ionospheric electron density model [7], developed by the Abdus Salam International Centre for Theoretical Physics in Italy and by the University of Graz in Austria. Both models have been adopted by ITU-R Recommendation P.531, which deals with trans-ionospheric propagation [8].

10.5 VARIABILITY OF THE IONOSPHERE

From the Chapman theory, it is clear that geographic location, day of the year, and local time of day strongly affect the ionization of the E and F_1 regions through the Sun's zenith angle. It turns out that these parameters also strongly affect the F_2 region, although it is not a Chapman layer. The ionosphere is also subject to less predictable variations. These are due, in part, to atmospheric motions, and in part to variations in the emissions from the Sun. The atmosphere at ionospheric heights is not stationary. It exhibits currents and tides, somewhat analogous to those in the ocean though of course not stemming from the same causes. Though exhibiting a consistent pattern, these motions also vary. A propagation phenomenon that appears to be related to these motions is called "spread-F" [9]. When spread-F is active, F region reflections come from a great many altitudes instead of a single layer, and these altitudes depend rather randomly on the frequency of the signal. This makes broadband communications difficult.

Another propagation effect that appears to be at least partially related to atmospheric circulation is "sporadic E", denoted by E_S. This manifests itself as a thin layer at the altitude of the normal E region, that is, near 100 km height, with an enhanced electron density comparable to that in the F region. As a result, signals intended for long-range communication via the F layer are reflected before they reach that layer, thus interfering with long-range communications. On the other hand, communication using E_S is enabled at shorter ranges, but cannot be depended upon because E_S occurs only sporadically. Radio amateurs sometimes exploit this method of communication.

Since solar flux is the cause of ionization in the ionosphere, variations in solar flux can have strong effects on atmospheric ionization and ionospheric propagation. Only a very brief introduction to these effects can be given here. The reader is referred to the literature [10] for a more detailed discussion. Ionization is primarily produced by the ultraviolet and X-ray portions of the solar flux, but particles ejected from the Sun also cause ionization, especially during periods of high solar activity. Although the amount of visible radiation from the Sun changes little from day to day, ultraviolet and X-ray radiation may change appreciably, resulting in variability in ionospheric propagation. Periods of high solar activity have been found to correlate well with the appearance of "sunspots" on the Sun's surface. These are dark regions that are of lower temperature than the surrounding surface. The average number of sunspots has been found to follow an 11-year cycle. Figure 10.8 illustrates the averaged sunspot number observed from 1940 to 2008 and also shows how this number correlates with the total averaged observed solar radiowave flux measured at $f = 2.8$ GHz (corresponding to $\lambda = 10.7$ cm), a widely used index of solar activity. Note that daily sunspot numbers can vary significantly from the averaged values. Sometimes no sunspots or flares are visible over a period of several weeks or months. Typically, the average annual 10.7 cm solar flux varies between about 70 s.f.u. (when no sunspots are present) and 250 s.f.u. The sunspot number is commonly used in empirical formulas for ionospheric propagation.

Solar events such as flares and coronal mass ejections are also correlated with the sunspot number. Figure 10.9 is an image of the Sun's surface with such features. In these events, large numbers of particles are emitted from the Sun's surface and spread throughout space to reach the planets. During such solar bursts, the measured solar flux density can reach much higher values.

The Earth's magnetic field, with a magnitude between about 3×10^{-5} and 5×10^{-5} T, is essentially that of a dipole. The Earth's magnetic field interacts with

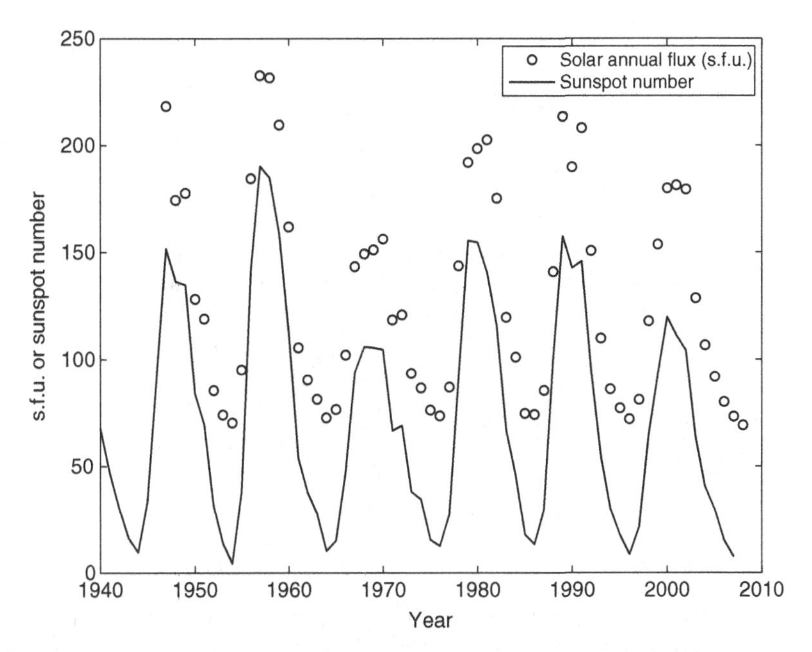

FIGURE 10.8 Time series of observed annual sunspot numbers and average annual 10.7 cm solar fluxes in solar flux units, where 1 s.f.u. = 10^{-22} W/(m^2 Hz).

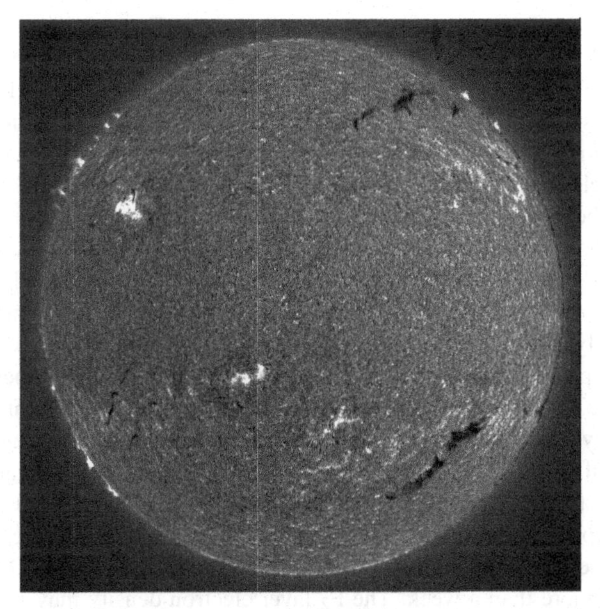

FIGURE 10.9 Solar emission at the hydrogen alpha line (656 nm). The dark areas are sunspots. (*Source:* Big Bear Solar Observatory/New Jersey Institute of Technology. Printed with permission.)

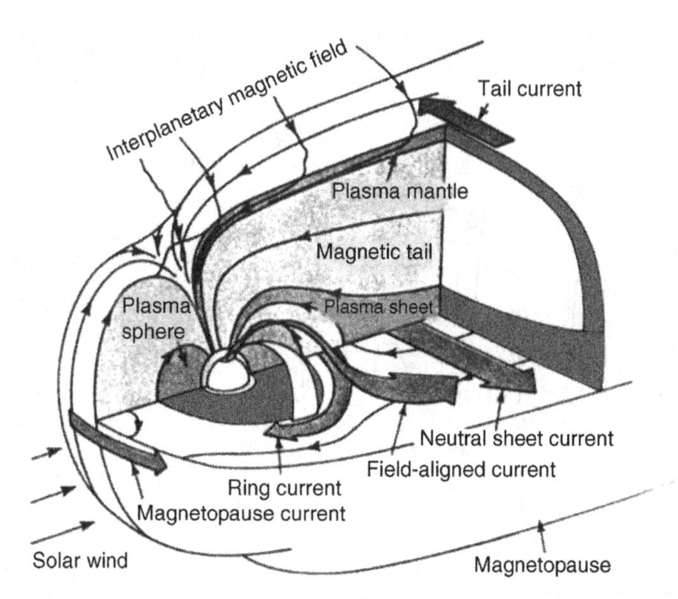

FIGURE 10.10 Illustration of the solar wind interaction with Earth's magnetic field. (Reprinted by permission from "Magnetospheric Currents," by Thomas A. Potemra, Johns Hopkins APL Technical Digest, vol. 4, no. 4, pp. 278 (1983). Copyright Johns Hopkins University Applied Physics Laboratory.)

these charged particles from the Sun ("solar wind") to redirect them eventually to higher latitude regions, as illustrated in three dimensions in Figure 10.10. Thus, solar activity most seriously impacts ionospheric propagation at higher latitudes, but can still cause serious outages for all latitudes during large "solar storms". Large quantities of solar particles can also cause damage to satellites and disrupt satellite communications, and the currents produced by the motion of these particles can produce large magnetic fields that can in some cases disrupt power and electronic systems as well. Figure 10.11 depicts the structure of the outer ionosphere (magnetosphere), which is strongly influenced by the Earth magnetic field. At high altitudes, the Earth's magnetic field is strongly distorted by the solar wind.

Streams of protons and electrons ejected from solar flares and prominences cause an increase in the solar wind. When these charged particles impact the Earth's magnetic field, they may cause a shock wave that propagates through the magnetic field, causing substantial changes and increases. Such events are called "geomagnetic storms" or "magnetic storms". They often last for several days. When they cause major ionospheric disturbances, these are called "ionospheric storms." Their effect on communication can be profound, but is very complicated. The mid-latitude D layer may be disturbed for more than a week. The F_2 layer electron density may increase during one phase of the storm and decrease during another. Detailed discussion of magnetic and ionospheric storms is beyond the scope of this book; the reader is referred to the literature [10].

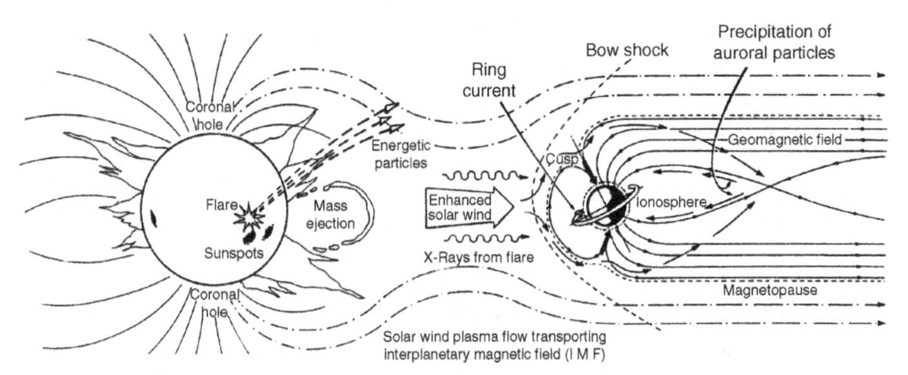

FIGURE 10.11 Diagram of the interaction of the solar wind with the Earth magnetic field. The Sun diameter and the solar-terrestrial distance are not to the same scale as the magnetic field. (Adapted from Hunsucker, R. D., and J. K. Hargreaves, *The High-Latitude Ionosphere and its Effects on Radio Propagation,* p. xix, Copyright Cambridge University Press, 2003, after "Synoptic Data for Solar-Terrestrial Environment," The Royal Society, September 1992. Reprinted with permission.)

Many of the charged particles emitted by the Sun may become trapped in the Earth's magnetic field and enter the ionosphere near the magnetic poles. This may cause an increased electron density in the D region that (as discussed in Chapter 11) can cause absorption of radio waves. When severe, this effect is called "polar cap absorption," or PCA. PCA events may last for several days and may, at times, extend as far south as 40° latitude. Trans-polar ionospheric radio propagation becomes impossible during such events.

In addition to particle emissions, solar flares also emit strong ultraviolet and X-ray radiation. The daytime regions then show an increased electron density in the D and E regions lasting from a few minutes to a few hours with greatest intensity near the equator, where the Sun is nearly overhead. These occurrences are called "sudden ionospheric disturbances" or SIDs. The result is enhanced communication at VLF frequencies and fading, sometimes to the point of complete blackout, at HF. The reasons for these propagation effects will become apparent in Chapter 11.

Due to the important effects that solar activity has on ionospheric and trans-ionospheric propagation, several sources are available for solar activity monitoring, analysis, and forecasting, including the Space Weather Prediction Center at NOAA and the Solar Data Analysis Center at NASA.

REFERENCES

1. Rishbeth, H., "Basic physics of the ionosphere," in *Propagation of Radiowaves*, second edition (L. Barclay, ed.), The Institution of Electrical Engineers, London, 2003.
2. Chapman, S., "The absorption and dissociative or ionizing effect of monochromatic radiation in an atmosphere on a rotating earth," *Proc. Phys. Soc.,* vol. 43, pp. 26–45, 1931.

3. Davies, K., *Ionospheric Radio Propagation,* National Bureau of Standards Monograph 80, U.S. Department of Commerce, 1965.
4. *U.S. Standard Atmosphere*, National Oceanic and Atmospheric Administration, Washington, DC 1976.
5. Reinisch B., and D. Bilitza, "Karl Rawer's life and the history of IRI," *Adv. Space Res.*, vol. 34, no. 9, pp. 1845–1950, 2004.
6. Bilitza D., and B. Reinisch, "International Reference Ionosphere 2007: improvements and new parameters," *Adv. Space Res.*, vol. 42, no. 4, pp. 599–609, 2007.
7. Hochegger, G., B. Nava, S. M. Radicella, and R. Leitinger, "A family of ionospheric models for different uses," *Phys. Chem. Earth*, vol. 25, no. 4, pp. 307–310, 2000.
8. ITU-R Recommendation P.531-9, "Ionospheric propagation data and prediction methods required for the design of satellite services," International Telecommunication Union, 2007.
9. Fritts D. C., et al., "Overview and summary of the Spread F experiment (SpreadFEx)," *Ann. Geophys.*, vol. 27, pp. 2141-2151, 2009.
10. Hargreaves, J. K., *The Solar-Terrestrial Environment: An Introduction to Geospace—The Science of the Terrestrial Upper Atmosphere, Ionosphere, and Magnetosphere*, Cambridge University Press, Cambridge, 1992.

11

IONOSPHERIC PROPAGATION

11.1 INTRODUCTION[1]

The nature of the interaction of electromagnetic waves with the ionosphere depends strongly on the frequency of operation. Therefore, the choice of the most convenient model of the resulting propagation phenomena also depends on the frequency.

In the VLF band, for example, the most convenient way to describe the propagation is by means of the so-called "spherical waveguide" mode. This formalism has the ionosphere (modeled as a magnetic conductor) acting as the upper boundary and the Earth surface (modeled as an electric conductor) acting as the bottom boundary of an effective curved "waveguide". This formalism is particularly convenient at VLF because the transverse dimensions of the effective waveguide are of the same order of magnitude as the wavelength of operation, which means that the propagation can be well described using a few low-order modes. Ionospheric VLF signals can be reasonably stable, but two major obstacles exist in this frequency band. First, such low frequencies require large antennas with high transmit powers. Second, the information (or bit) transmission rates achieved are relatively low because of the low carrier frequency. Despite these disadvantages, ionospheric VLF propagation has

[1] Portions of this chapter are adapted from K. Davies, *Ionospheric Radio Propagation*, National Bureau of Standards Monograph 80, 1965.

Radiowave Propagation: Physics and Applications. By Curt A. Levis, Joel T. Johnson, and Fernando L. Teixeira

been used in applications related to long-distance communications, navigation, and standard frequency dissemination. With the advent of communication satellites, long-distance fiber-optic systems, and global navigation satellite systems such as GPS, the relative importance of VLF/LF ionospheric links for communications and navigation has diminished considerably. An important exception is submarine communications, which still routinely rely on VLF transmitters.[2]

The waveguide mode formalism can also be used in the LF band, but the computations become more cumbersome due to the larger number of modes necessary for an adequate description of field behaviors. The LF band can be considered as a transition region, and above 100 kHz the skywave mode of operation discussed below for the MF and HF bands is more applicable [1,2].

Under suitable conditions, electromagnetic waves in the MF and HF bands launched from the Earth may be refracted by the ionosphere and returned to the Earth at long distances away from the transmitter. For this "skywave" mode, the small wavelengths compared to the physical scales involved make a ray description more convenient. In many ways, this mode is similar to tropospheric ducting, except that the "reflection" occurs in the stratosphere. As a result, transcontinental distances can be spanned by relatively low-powered transmissions using modest equipment. Such ionospheric links are more suited to MF and HF frequencies because—as we will see later in this chapter—at the VHF band and above the waves penetrate fully through the ionosphere and fail to return back to Earth. Before the advent of satellite communications, MF and HF ionospheric links were the principal means of transcontinental communications. Even though their relative importance has diminished, MF and HF ionospheric links are still an important tool for applications such as "shortwave" broadcasting (especially for less developed countries), amateur radio links (ham radio), diplomatic and military communications, and aid agencies. Because of the relatively modest equipment required, ionospheric skywave links also provide a robust backup mode of long-range communications if fiber-optic or space-based communications infrastructure is disrupted.

Some peculiar effects are observed in ionospheric skywave links. Broadband signals can suffer a great deal of distortion, and therefore these links are most frequently used for narrowband signals, although broadband transmission is possible if distortion is compensated. Even with voice transmissions, the received voice quality may vary considerably over a period of a few seconds as the condition of the ionosphere changes. Ionospheric disturbances may at times create problems with communications in these frequency bands, especially in polar regions, for periods ranging from hours to days.

In this chapter, we will first focus on understanding MF and HF ionospheric links. Later in the chapter, we will also consider the effect of the ionosphere as a source of *perturbations* on higher frequency Earth–space links traversing the ionosphere, that is, on trans-ionospheric links.

To understand how MF and HF ionospheric links behave, one must first examine how waves travel in an ionized medium and obtain expressions for the refractive index

[2] A number of other applications (unrelated to communications) of ionospheric VLF/LF propagation modes also exist, such as worldwide remote detection and location of lightning discharges.

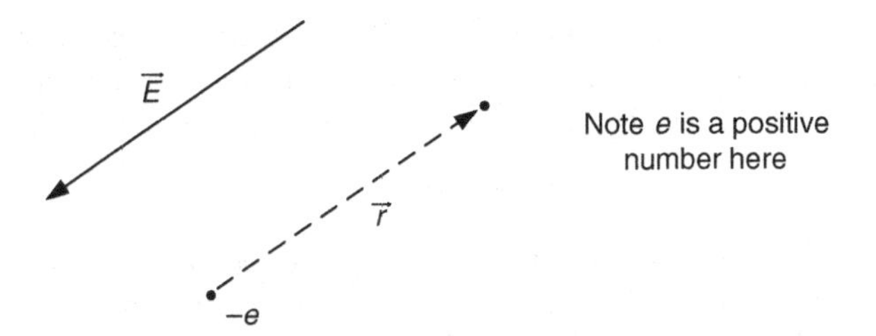

FIGURE 11.1 Electron motion in an applied field.

of such a medium. This can then be combined with knowledge of the ionosphere's properties to predict the behavior of MF and HF ionospheric links. Since the magnetic field of the Earth is relatively weak, it is useful to consider first the case of an ionized medium without an externally applied magnetic field before proceeding to the more general, but more complicated, case of a magnetoionic medium.

11.2 DIELECTRIC PROPERTIES OF AN IONIZED MEDIUM

When an electric field is applied in a medium with free charges, the charges (free electrons, positive ions, and/or negative ions) are accelerated. Because the ions have much greater mass than the electrons, ionic motions are relatively small and can be ignored for most purposes. We shall neglect them here.

Let the displacement of an electron due to the applied electric field be denoted as $\overline{r}$, as illustrated in Figure 11.1. The result is the same as if the electron had remained where it was and a dipole had been added as shown in Figure 11.2. The dipole moment induced by the electric field due to the motion of the electron is $-e\overline{r}$, where $e = 1.602 \times 10^{-19}$ C is the electron charge magnitude. If there are N free electrons per unit volume and the average electron displacement is $\overline{r}$, the volume polarization that results is

$$\overline{P} = -Ne\,\overline{r}. \tag{11.1}$$

Recall now that it is precisely such volume polarization due to induced dipole moments that determines the behavior of dielectrics. The equation of motion of an electron can be used to find the polarization. An electron in the presence of an electric field will

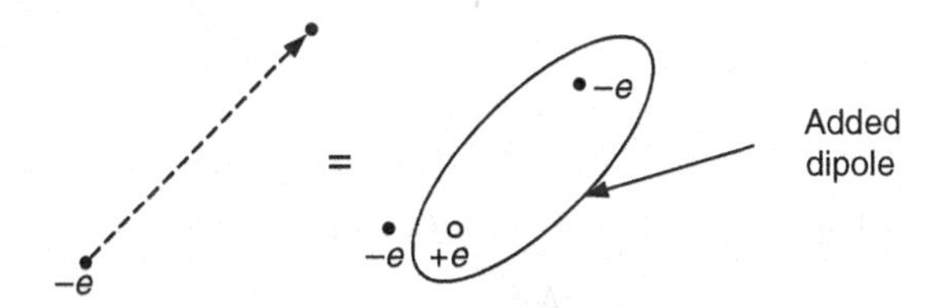

FIGURE 11.2 Representation as dipole moment.

experience a force $\overline{F}_{\rm e} = -e\overline{E}$, causing the electron to accelerate parallel to the electric field (note forces on the electron due to the magnetic field associated with $\overline{E}$ are much smaller and can be neglected). The electron position $\overline{r}$ is therefore written as

$$\overline{r} = \hat{x}\, x, \tag{11.2}$$

where the electric field direction is taken as $\hat{x}$. In addition to the accelerating force of the electric field, the electron can also experience a frictional force $\overline{F}_{\rm c}$ resulting from the effects of collisions with neutral molecules. Now from the equation of motion $\overline{F} = m_{\rm e}\overline{\mathcal{A}}$, with $\overline{\mathcal{A}}$ the acceleration, we have

$$\hat{x} m_{\rm e} \frac{\partial^2}{\partial t^2} x = \overline{F}_{\rm c} + \overline{F}_{\rm e} = \left[-\nu m_{\rm e} \frac{\partial}{\partial t} x - e\, E_x \right] \hat{x}, \tag{11.3}$$

where $m_{\rm e}$ is the electron mass (9.109×10^{-31} kg), and the parameter ν is called the electron collision frequency (this name should not be interpreted too literally).

Now consider a time-harmonic electric field

$$\overline{E} = \text{Re}\left\{ \hat{x} \underline{E}_x e^{j\omega t} \right\}. \tag{11.4}$$

The resulting electron displacement will also be time harmonic:

$$x = \text{Re}\left\{ \underline{x} e^{j\omega t} \right\}. \tag{11.5}$$

Recall for an arbitrary phasor $\underline{f}$ that

$$\frac{d}{dt} \text{Re}\left\{ \underline{f} e^{j\omega t} \right\} = \text{Re}\left\{ j\omega \underline{f} e^{j\omega t} \right\} \tag{11.6}$$

so that differentiation with respect to time is equivalent to multiplying by $j\omega$. The phasor equivalent of equation (11.3) is therefore

$$m_{\rm e}(-\underline{x}\omega^2) + \nu m_{\rm e}(j\omega \underline{x}) = -e\underline{E}_x\,, \tag{11.7}$$

or

$$\underline{x} = \frac{e\underline{E}_x}{m_{\rm e}\omega^2} \left(\frac{1}{1 - j\frac{\nu}{\omega}} \right). \tag{11.8}$$

Substituting this into the phasor equivalents of (11.1) and (11.2)

$$\underline{\overline{P}} = -Ne\, \underline{\overline{r}}, \tag{11.9}$$

$$\underline{\overline{r}} = \hat{x}\, \underline{x} \tag{11.10}$$

gives

$$\underline{\overline{P}} = -\frac{Ne^2 \underline{E}_x}{m_{\rm e}\omega^2} \left(\frac{1}{1 - j\frac{\nu}{\omega}} \right) \hat{x}. \tag{11.11}$$

This can be simplified by defining the *angular plasma frequency* ω_N (rad/s) as

$$\omega_N^2 = \frac{Ne^2}{\epsilon_0 m_{\rm e}} \approx 3183N \tag{11.12}$$

with N in electrons per cubic meter (so that $f_N = \omega_N/2\pi = 8.98\sqrt{N}$ Hz), and the dimensionless constants

$$X = \left(\frac{\omega_N}{\omega}\right)^2 \tag{11.13}$$

and

$$Z = \frac{\nu}{\omega} \tag{11.14}$$

to give

$$\underline{\overline{P}} = -\hat{x}\epsilon_0 \underline{E}_x \frac{X}{1 - jZ}. \tag{11.15}$$

For isotropic dielectrics, one can define a relative dielectric constant $\underline{\epsilon}_r$ (which as in Chapter 9 denotes $\underline{\epsilon}^e/\epsilon_0$) through

$$\underline{\overline{D}} = \epsilon_0 \underline{\overline{E}} + \underline{\overline{P}} \equiv \underline{\epsilon}_r \epsilon_0 \underline{\overline{E}}, \tag{11.16}$$

with $\underline{\epsilon}_r$ real for lossless media, but complex for lossy media. The relative dielectric constant determines the propagation constant $\underline{k}_1$ of a plane wave in the medium since

$$\underline{k}_1 = \underline{n} k_0, \tag{11.17}$$

where $\underline{n} \equiv \sqrt{\underline{\epsilon}_r}$ is the refractive index. In the case of complex $\underline{n}^2$, the root should be chosen so that exponentially decaying fields (due to losses) ensue.

From equations (11.15) and (11.16) follows

$$\underline{n}^2 = \underline{\epsilon}_r = 1 - \frac{X}{1 - jZ}. \tag{11.18}$$

In the absence of collisions, $Z = 0$, and we can simplify the formula for the relative permittivity to

$$\begin{aligned} \underline{\epsilon}_r &= 1 - \frac{\omega_N^2}{\omega^2} \\ &\approx 1 - \frac{3183N}{\omega^2} \\ &\approx 1 - \left(\frac{8.98}{f_{\text{MHz}}}\right)^2 \frac{N}{10^{12}} \end{aligned} \tag{11.19}$$

with N again in electrons per cubic meter in the final two forms.

Figure 11.3 plots equation (11.19) versus frequency with the electron number density as a parameter. From (11.19) and Figure 11.3, it is evident that the relative permittivity is negative when the frequency of the wave is less than the plasma frequency of the medium. According to (11.18), this results in an imaginary refractive index. This means that any wave transmitted into such a medium will be attenuated exponentially, without any phase change, and will become negligible in a relatively short distance. On the other hand, (11.19) and Figure 11.3 show that the relative permittivity approaches unity if the frequency of the wave is much greater than the

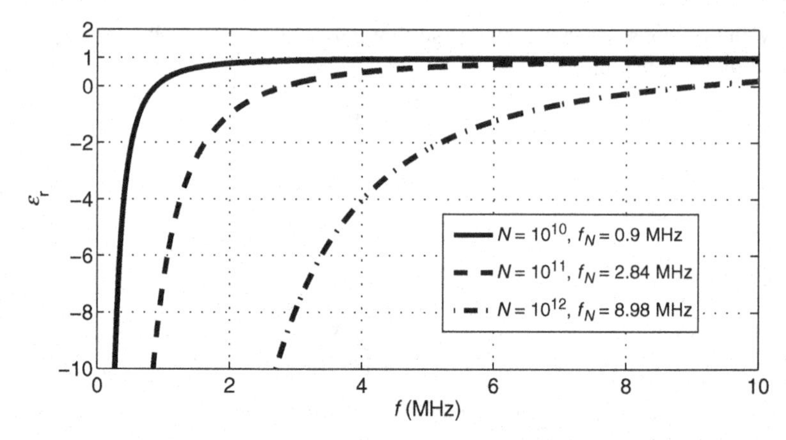

FIGURE 11.3 Relative permittivity versus frequency for a lossless, nonmagnetized ionosphere. Electron density N in inverse cubic meters is indicated in legend, along with the associated plasma frequency f_N.

plasma frequency. Then the refractive index approaches that of free space, and the wave is less strongly affected by the ionized medium. This is why VHF and higher frequency signals travel through the ionosphere instead of being reflected. Some effects remain even at VHF and higher frequencies, as discussed later in this chapter in connection with Earth–satellite communication and navigation systems. Also, the influence of the ionosphere increases (i.e., the relative permittivity is more different from unity) as the electron number density is increased.

11.3 PROPAGATION IN A MAGNETOIONIC MEDIUM

In the presence of a static magnetic field $\overline{H}_0$, electrons moving at velocity $\overline{v}$ experience a force proportional to $\overline{v} \times \mu\overline{H}_0$. The magnitude of this force is zero when the electron moves parallel to $\overline{H}_0$ and maximum when it moves transversely with respect to $\overline{H}_0$. Such a medium has a built-in directionality, which characterizes an anisotropic medium.

Are plane waves still possible solutions of Maxwell's equations in such a medium? If so, under what condition? What is the impedance they see, and what is their propagation constant (or, equivalently, the refractive index)? We shall answer these questions by writing Maxwell's equations and the equation of motion for free electrons in the medium. The divergence equations are automatically satisfied for a plane wave that has $\overline{\underline{D}}$ and $\overline{\underline{B}}$ perpendicular to the propagation direction. This leaves three vector equations: the two curl equations and one equation of motion. Taking one component at a time, this gives nine linear scalar equations in the nine scalar variables that are the components of $\overline{\underline{E}}$, $\overline{\underline{H}}$, $\overline{\underline{P}}$. From there on it is a matter of algebraic manipulation to eliminate variables, and this can be done in a variety of ways. Along the way one finds, first, that the characteristic impedance is given by $\underline{\eta}_1 = \omega\mu_0/\underline{k}_1$. Then one finds that the equations are consistent (i.e., have a solution) only for two specific values of the polarization ratio $\underline{R}$, and these depend both on the medium parameters and on the direction of propagation. This means that the medium will allow propagation of plane waves only for particular polarizations specified by the medium. These are known as characteristic polarizations. Finally, an explicit formula for the refractive index for each characteristically polarized wave is obtained. This formula is known by a variety of names, but most often it is called the Appleton or Appleton–Hartree equation.

11.3.1 Mathematical Derivation of the Appleton–Hartree Equation

Assume that the plane wave can exist and that it propagates in the z direction. Then, by the properties of a plane wave, the z dependence is given by $e^{-j\underline{k}_1 z}$ and there is no x or y dependence. From Maxwell's equations, we have

$$\nabla \times \underline{\overline{H}} = j\omega\underline{\overline{D}} = j\omega(\epsilon_0\underline{\overline{E}} + \underline{\overline{P}}).$$

Taking the x component of the above

$$\frac{\partial \underline{H}_z}{\partial y} - \frac{\partial \underline{H}_y}{\partial z} = j\omega(\epsilon_0\underline{E}_x + \underline{P}_x). \tag{11.20}$$

Since there is no y dependence, $\partial/\partial y \equiv 0$ for any field polarization component. Since the z dependence for all quantities is $e^{-j\underline{k}_1 z}$, we have

$$j\underline{k}_1\underline{H}_y = j\omega\underline{E}_x\left(\epsilon_0 + \frac{\underline{P}_x}{\underline{E}_x}\right). \tag{11.21}$$

Treating the other two components similarly gives

$$j\underline{k}_1\underline{H}_x = -j\omega\underline{E}_y\left(\epsilon_0 + \frac{\underline{P}_y}{\underline{E}_y}\right), \tag{11.22}$$

and

$$0 = j\omega\left(\epsilon_0\underline{E}_z + \underline{P}_z\right) = j\omega\underline{D}_z\,. \tag{11.23}$$

Similarly, from

$$\nabla \times \underline{\overline{E}} = -j\omega\mu_0\underline{\overline{H}}, \tag{11.24}$$

one gets

$$j\underline{k}_1\underline{E}_y = -j\omega\mu_0\underline{H}_x\,, \tag{11.25}$$

$$j\underline{k}_1\underline{E}_x = j\omega\mu_0\underline{H}_y\,, \tag{11.26}$$

$$0 = -j\omega\mu_0\underline{H}_z\,. \tag{11.27}$$

Equations (11.23) and (11.27) show that there is no longitudinal component of $\underline{\overline{D}}$, $\underline{\overline{H}}$, and $\underline{\overline{B}}$, but (11.23) does not guarantee $\underline{E}_z = 0$. The plane wave will be transverse only with respect to the $\underline{\overline{D}}$, $\underline{\overline{H}}$, and $\underline{\overline{B}}$ vectors, which differs from the isotropic case.

From (11.25) and (11.26) follows

$$\frac{\underline{E}_x}{\underline{H}_y} = \frac{\omega\mu_0}{\underline{k}_1} = -\frac{\underline{E}_y}{\underline{H}_x}. \tag{11.28}$$

Thus, the characteristic impedance is given by $\underline{\eta}_1 = \omega\mu_0/\underline{k}_1$, just as for an isotropic medium. If (11.26) is solved for $\underline{E}_x$ and the result is used in (11.21), one obtains

$$\underline{k}_1^2 = \omega^2\mu_0\epsilon_0\left(1 + \frac{\underline{P}_x}{\epsilon_0\underline{E}_x}\right).$$

Hence, the square of the refractive index is given by

$$\underline{n}^2 \equiv \frac{\underline{k}_1^2}{k_0^2} = 1 + \frac{\underline{P}_x}{\epsilon_0\underline{E}_x}. \tag{11.29}$$

Similarly, if (11.25) is solved for $\underline{E}_y$ and the result used in (11.22), one obtains

$$\underline{n}^2 = 1 + \frac{\underline{P}_y}{\epsilon_0\underline{E}_y}. \tag{11.30}$$

Equations (11.29) or (11.30) can be used to find $\underline{n}^2$ if the ratios $\underline{P}_x/\underline{E}_x$ and $\underline{P}_y/\underline{E}_y$ are known. For a meaningful solution they must be equal, so that (11.29) and (11.30) give the same result, requiring

$$\frac{\underline{P}_x}{\underline{E}_x} = \frac{\underline{P}_y}{\underline{E}_y}, \tag{11.31}$$

from which we can obtain

$$\frac{\underline{P}_x}{\underline{P}_y} = \frac{\underline{E}_x}{\underline{E}_y} \equiv \underline{R}, \tag{11.32}$$

where $\underline{R}$ is the polarization ratio of the wave.

Up to this point, we have considered only Maxwell's equations for a plane wave propagating in the z direction in a generalized medium. To determine the ratios $\underline{P}_x/\underline{E}_x$, $\underline{P}_y/\underline{E}_y$, which we need to solve (11.29) or (11.30) for $\underline{n}^2$, we write the equation of motion of an "average" electron, since it is the displacement of electrons due to $\overline{E}$ that causes the polarization $\overline{P}$. The equation of motion is

$$m_e\frac{\partial^2}{\partial t^2}\overline{r} + \nu m_e\frac{\partial}{\partial t}\overline{r} = -e\overline{E} + e\mu_0\overline{H}_0 \times \frac{\partial}{\partial t}\overline{r}, \tag{11.33}$$

where the second term on the left-hand side again represents the viscous or frictional effects of collisions. The electric and magnetic field forces on the electron are represented on the right-hand side, and $\overline{H}_0$ represents the Earth's static magnetic field intensity (amplitudes of $\mu_0\overline{H}_0$ typically range from 30 to 60 μT). In order to be complete, (11.33) should also include another term corresponding to the force produced by the (time-varying) magnetic field of the wave. However, the magnetic field magnitude of the wave turns out to be negligible compared to the magnitude of the Earth's magnetic field, and hence that term is ignored here.

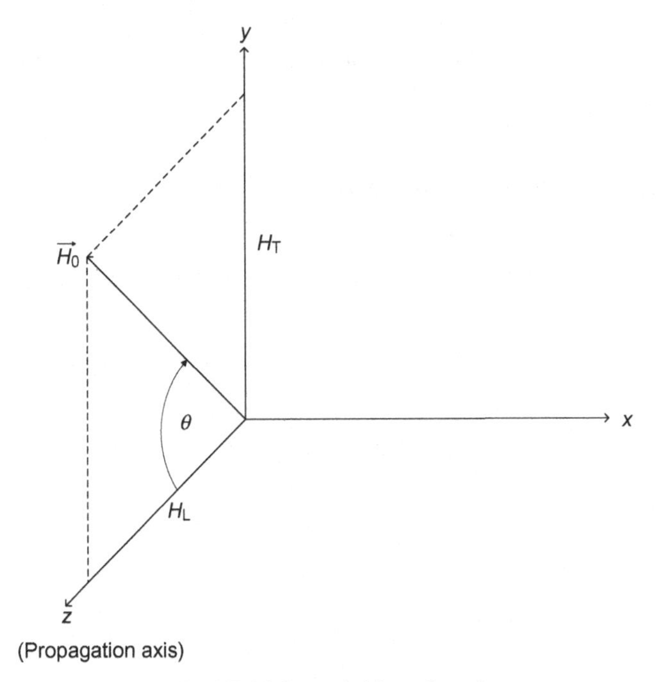

FIGURE 11.4 Definition of y axis.

It is now necessary to be more specific with respect to the coordinate system. So far we have required the z axis to be in the propagation direction but left the x and y axes arbitrary. Let us now fix the y axis by noting that the propagation direction (z axis) and the vector $\overline{H}_0$ together define a plane. Let the y axis lie in this plane as shown in Figure 11.4.

In the figure, we have drawn the y axis so that the angle θ, between the propagation direction and the Earth's magnetic field, is acute. This is not necessary, and y could have been directed downward as well, making θ obtuse. In either case, the x axis must be taken to satisfy a right-handed coordinate system. The longitudinal component of $\overline{H}_0$ is defined as $H_{0z} = H_0 \cos\theta \equiv H_\text{L}$ and the transverse component as $H_{0y} = H_0 \sin\theta \equiv H_\text{T}$.

Note that this choice of coordinates in no way restricts the generality of the derivation. This is true because no matter what the propagation direction may be, we can always take the z axis parallel to it, and no matter what the $\overline{H}_0$ direction may be (as long as it is not in the z direction), we can always take the y axis so that $\overline{H}_0$ lies in the yz plane. Then, we can choose the x axis to complete a right-handed orthogonal coordinate system. Clearly, this is a recipe for any combination of propagation direction and magnetic field direction as long as they are not parallel. If they are parallel, any choice of y and x axes that results in a right-handed coordinate system is acceptable.

With this coordinate system established, the equation of motion (11.33) can now be broken into three scalar equations. Starting with the x component, one gets

$$m_{\mathrm{e}}\frac{\partial^2}{\partial t^2}x + \nu m_{\mathrm{e}}\frac{\partial}{\partial t}x = -e\left(E_x + \mu_0 H_{\mathrm{L}}\frac{\partial}{\partial t}y - \mu_0 H_{\mathrm{T}}\frac{\partial}{\partial t}z\right). \qquad (11.34)$$

Since the system is linear, a sinusoidal field

$$E_x(\overline{r}, t) = \mathrm{Re}\left[\underline{E}_x(\overline{r})e^{j\omega t}\right]$$

will result in sinusoidal displacements and (11.34) can be replaced by its phasor counterpart

$$m_{\mathrm{e}}(-\omega^2\underline{x}) + \nu m_{\mathrm{e}}(j\omega\underline{x}) = -e\left(\underline{E}_x + j\omega\underline{y}\mu_0 H_{\mathrm{L}} - j\omega\underline{z}\mu_0 H_{\mathrm{T}}\right). \qquad (11.35)$$

The polarization $\underline{\overline{P}}$ is produced by the electron displacements via

$$\underline{\overline{P}} = -Ne\underline{\overline{r}}, \qquad (11.36)$$

and therefore

$$\underline{x} = -\frac{\underline{P}_x}{Ne}, \qquad (11.37)$$

$$\underline{y} = -\frac{\underline{P}_y}{Ne}, \qquad (11.38)$$

$$\underline{z} = -\frac{\underline{P}_z}{Ne}. \qquad (11.39)$$

We can substitute these relationships back into (11.35) to get

$$\frac{m_{\mathrm{e}}\omega^2\underline{P}_x}{Ne} - j\frac{\omega\nu m_{\mathrm{e}}\underline{P}_x}{Ne} = -e\left[\underline{E}_x - \frac{j\omega\mu_0\underline{P}_y H_{\mathrm{L}}}{Ne} + \frac{j\omega\mu_0\underline{P}_z H_{\mathrm{T}}}{Ne}\right]. \qquad (11.40)$$

Multiplication by ϵ_0/e gives

$$\left(\frac{\epsilon_0 m_{\mathrm{e}}}{Ne^2}\right)\omega^2\underline{P}_x\left(1 - j\frac{\nu}{\omega}\right) = -\epsilon_0\underline{E}_x + j\omega\left(\frac{\epsilon_0 m_{\mathrm{e}}}{Ne^2}\right)\left(\frac{\mu_0 e H_{\mathrm{L}}}{m_{\mathrm{e}}}\right)\underline{P}_y$$
$$-j\omega\left(\frac{\epsilon_0 m_{\mathrm{e}}}{Ne^2}\right)\left(\frac{\mu_0 e H_{\mathrm{T}}}{m_{\mathrm{e}}}\right)\underline{P}_z. \qquad (11.41)$$

To simplify this equation, note that $\epsilon_0 m_{\mathrm{e}}/Ne^2$ has dimension (sec)^2 and $\mu_0 e H_{\mathrm{L}}/m_e$ has dimension (sec)$^{-1}$. As in the isotropic plasma case, we define the plasma frequency ω_N by

$$\omega_N^2 \equiv \frac{Ne^2}{\epsilon_0 m_{\mathrm{e}}}, \qquad (11.42)$$

and additionally define two new quantities, the longitudinal and transverse gyrofrequencies, ω_{L} and ω_{T}, respectively, as

$$\omega_{\mathrm{L}} \equiv -\frac{\mu_0 e H_{\mathrm{L}}}{m_{\mathrm{e}}}, \qquad (11.43)$$

$$\omega_{\mathrm{T}} \equiv -\frac{\mu_0 e H_{\mathrm{T}}}{m_{\mathrm{e}}}. \tag{11.44}$$

Note that the magnitudes of these quantities are the angular frequencies at which an electron traveling in a direction perpendicular to the respective magnetic field component would complete a circular orbit, hence the name. Equation (11.41) now becomes

$$\left(\frac{\omega}{\omega_N}\right)^2 \underline{P}_x\left(1 - j\frac{\nu}{\omega}\right) = -\epsilon_0 \underline{E}_x - j\frac{\omega\omega_{\mathrm{L}}}{\omega_N^2}\underline{P}_y + j\frac{\omega\omega_{\mathrm{T}}}{\omega_N^2}\underline{P}_z\ . \tag{11.45}$$

Finally, substituting the dimensionless quantities

$$X = \left(\frac{\omega_N}{\omega}\right)^2, \tag{11.46}$$

$$Y_{\mathrm{L}} = \frac{\omega_{\mathrm{L}}}{\omega}, \tag{11.47}$$

$$Y_{\mathrm{T}} = \frac{\omega_{\mathrm{T}}}{\omega}, \tag{11.48}$$

$$Z = \frac{\nu}{\omega}, \tag{11.49}$$

and multiplying through by X gives

$$\epsilon_0 X \underline{E}_x = -\underline{P}_x(1 - jZ) - jY_{\mathrm{L}}\underline{P}_y + jY_{\mathrm{T}}\underline{P}_z\,. \tag{11.50}$$

The reason for the lengthy algebra discussed previously now becomes apparent: we have now an equation involving only the $\overline{\underline{E}}$ and $\overline{\underline{P}}$ components needed for finding $\underline{n}^2$ and dimensionless constants X, Y_{L}, Y_{T}, and Z that specify the medium properties and the Earth's magnetic field.

Going back to the y and z components of the equation of motion (11.33) and treating them in the same way gives

$$\epsilon_0 X \underline{E}_y = -\underline{P}_y(1 - jZ) + jY_{\mathrm{L}}\underline{P}_x\,, \tag{11.51}$$

$$\epsilon_0 X \underline{E}_z = -\underline{P}_z(1 - jZ) - jY_{\mathrm{T}}\underline{P}_x\,. \tag{11.52}$$

The objective is now to eliminate variables to calculate $\underline{n}^2$ by equations (11.29) or (11.30). Let us arbitrarily choose (11.30) so that we wish to retain $\underline{P}_y$, $\underline{E}_y$ and eliminate $\underline{P}_x$, $\underline{E}_x$, $\underline{P}_z$, and $\underline{E}_z$. Equation (11.23) has not been used so far and therefore is available. Equations (11.21) and (11.26) as well as (11.22) and (11.25) were used before to eliminate $\overline{\underline{H}}$ components, but they were redundant as shown by (11.29) and (11.30). The redundancy resulted in (11.32), which is therefore independent and available.

Solving (11.23) for $\underline{E}_z$ gives

$$\underline{E}_z = -\frac{\underline{P}_z}{\epsilon_0}, \tag{11.53}$$

which can be used to eliminate $\underline{E}_z$ from (11.52) to give

$$X\underline{P}_z = \underline{P}_z(1 - jZ) + jY_{\rm T}\underline{P}_x \,. \tag{11.54}$$

Solving for $\underline{P}_z$ and substituting the result in (11.50) eliminates $\underline{P}_z$ and gives, with the use of (11.32),

$$\epsilon_0 X\underline{R}\,\underline{E}_y = -\underline{R}\,\underline{P}_y(1 - jZ) - jY_{\rm L}\underline{P}_y + Y_{\rm T}^2\underline{R}\,\underline{P}_y/(1 - X - jZ)\,. \tag{11.55}$$

This equation has $\underline{P}_y$ in each term on the right and $\underline{E}_y$ on the left. Hence, it can be solved for the desired $\underline{P}_y/\underline{E}_y$ ratio, but it is not the only such equation! Using (11.32) to eliminate $\underline{P}_x$ from (11.51) and multiplying by $\underline{R}$ gives

$$\epsilon_0 X\underline{R}\,\underline{E}_y = -\underline{R}\,\underline{P}_y(1 - jZ) + jY_{\rm L}\underline{R}^2\underline{P}_y \,. \tag{11.56}$$

Clearly, the solution for $\underline{P}_y/\underline{E}_y$ will be acceptable only if it satisfies both (11.55) *and* (11.56). Noting the left-hand sides are the same for these two, one can equate the right-hand sides to obtain

$$-jY_{\rm L} + \frac{Y_{\rm T}^2\underline{R}}{1 - X - jZ} = jY_{\rm L}\underline{R}^2 \,. \tag{11.57}$$

Recall that X, $Y_{\rm L}$, $Y_{\rm T}$, and Z depend only on the properties of the plasma and the magnetic field $\overline{H}_0$, fixed by the physical scenario. Thus (11.57) must be considered a condition on $\underline{R}$. Solving the quadratic equation gives

$$\underline{R} = \frac{-j}{Y_{\rm L}}\left\{\frac{Y_{\rm T}^2}{2(1 - X - jZ)} \mp \left(\frac{Y_{\rm T}^4}{4(1 - X - jZ)^2} + Y_{\rm L}^2\right)^{\frac{1}{2}}\right\}. \tag{11.58}$$

The meaning of this equation is that only plane waves with these two *characteristic* polarization states are able to retain their polarization state as the wave propagates through the medium.

If (11.58) is satisfied, that is, when the polarizations are characteristic of the medium, (11.55) and (11.56) give the same value for $\underline{P}_y/\underline{E}_y$,

$$\frac{\underline{P}_y}{\underline{E}_y} = -\frac{\epsilon_0 X}{1 - jZ - jY_{\rm L}\underline{R}}, \tag{11.59}$$

which can then be used in (11.30) to give

$$\underline{n}^2 = 1 - \frac{X}{1 - jZ - jY_{\rm L}\underline{R}} \tag{11.60}$$

in which $\underline{R}$ must have the characteristic values given by (11.58). When these are substituted in (11.60), one finally obtains the Appleton–Hartree formula

$$\underline{n}_\pm^2 = 1 - \frac{X}{1 - jZ - \left[\frac{\frac{1}{2}Y_{\rm T}^2}{(1-X-jZ)}\right] \pm \left\{\left[\frac{1}{4}\frac{Y_{\rm T}^4}{(1-X-jZ)^2} + Y_{\rm L}^2\right]^{\frac{1}{2}}\right\}}. \tag{11.61}$$

Note that the refractive index calculated with the top sign of (11.61) goes with the polarization given by the top sign in (11.58).

11.3.2 Physical Interpretation

In general, with an anisotropic medium, such as the ionosphere in the Earth's magnetic field, there are two modes of plane wave propagation. Each of these has a particular polarization that depends entirely on the properties of the ionosphere and magnetic field and that can be calculated by (11.58) when these quantities are known. Each of these modes has a particular phase velocity and attenuation constant that can be calculated from the refractive index (the square root of (11.61)). If losses are present (ν and therefore Z are not zero), the square of the refractive index will be complex, otherwise it will be real. The two modes travel at different phase velocities. Hence, they are refracted differently and travel different paths, and when both are present, they do not stay in phase from place to place. In other words, they do not add constructively (in phase) at every point. A wave that does not have one of these characteristic polarizations does not retain its polarization state as it traverses the medium.

In describing propagation through the ionosphere, it is often possible to neglect the Earth's magnetic field and treat the ionosphere as isotropic. When this is not possible, one has to determine the characteristic modes (polarizations) by (11.58), express the incident field in terms of these modes, trace each one separately through the ionosphere, and then recombine them as they exit the ionosphere.

The Appleton–Hartree formula, (11.61), is quite useful for numerical calculations. However, it is too messy for most analytical developments. In the developments to follow, some simplifications will be made. For example, in calculating losses due to absorption the magnetic field may play a rather minor role and may often be neglected. Furthermore, absorption may be unimportant in determining the path of a ray. When such approximations are made, the results will not be precise, but nevertheless suggestive of what happens in the real ionosphere.

11.3.3 Ordinary and Extraordinary Waves

In many cases (i.e., various X, Y_L, Y_T, and Z values), including those most commonly encountered in the ionosphere, one of the solutions is not too different from that with no static magnetic field ($\overline{H}_0 = 0$). This solution is called the *ordinary (o-)wave*. The other is called the *extraordinary (x-)wave*. "Extraordinary" here does not mean unusual, and both waves are usually present.

When the characteristic polarization ratios or refractive indices are computed directly from (11.58) and (11.61), the solution using, say, the top sign in (11.58) and (11.61) may correspond to the ordinary wave for $0 < X < 1$ and abruptly jump to the extraordinary wave for $X > 1$. The converse is then true for the other sign. Together, they always give both waves. This odd behavior is simply due to the behavior of the complex square root, which is a two-valued function. Most computers choose the principal root for the complex square root function, regardless of the physics of the situation.

The behavior of the refractive index as a function of frequency is often presented using graphs of $\underline{n}^2 \equiv (n_R - jn_I)^2$, or of the pairs n_R and n_I or n_R^2 and n_I^2. Plots

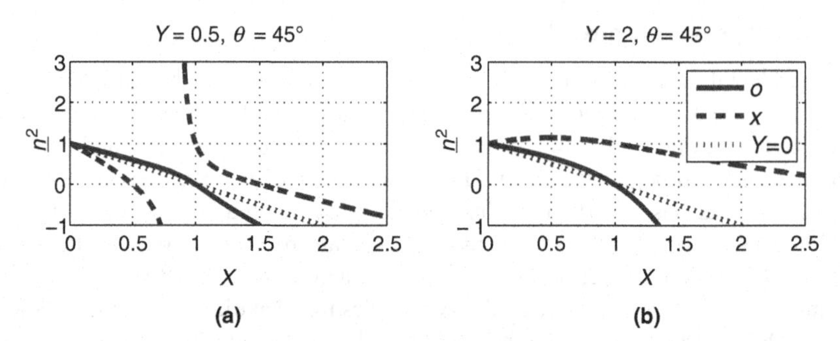

FIGURE 11.5 Sample dispersion curves for $\theta = 45^\circ$, $Z = 0$, and (a) $Y = 0.5$ and (b) $Y = 2$. The case with $Y = 0$ is also included for comparison; note $Y_L = Y\cos\theta$ and $Y_T = Y\sin\theta$.

of $\underline{n}^2$, n_R^2, or n_I^2 versus frequency are called *dispersion curves*.[3] It is also common in the ionospheric literature to plot these quantities as a function of $X = \left(\frac{\omega_N}{\omega}\right)^2$ instead of ω. When this is done, it is important to recognize that larger X values can correspond either to lower frequencies or to larger electron densities. Figure 11.5 provides example plots of $\underline{n}^2$ versus X for $Z = 0, \theta = 45^\circ$ (so that $Y_T = Y_L = Y/\sqrt{2}$), and for $Y = 1/2$ (Figure 11.5a) and 2 (Figure 11.5b). The nonmagnetic curve is also included and is seen to be most similar to the ordinary wave, while the extraordinary wave exhibits larger variations, including, in some cases, behaviors reminiscent of resonant systems.

For the lossless case, the ordinary wave can always be identified as the one that passes through the point $(X, \underline{n}^2) = (1, 0)$. When loss is large, the dispersion curves may have little resemblance to the lossless case, and direct identification of the two waves can be difficult. In principle, the ordinary wave can always be identified as that for which the dispersion curves deform continuously so as to pass through the point $(X, \underline{n}^2) = (1, 0)$ as Z is gradually reduced to zero.

11.3.4 The Q_L and Q_T Approximations

When propagation is parallel to the Earth's magnetic field ($Y_T = 0$), the characteristic polarizations are found from (11.58) to be circular, $\underline{R} = \pm j$. This condition is called *longitudinal propagation*. For propagation perpendicular to the Earth's magnetic field ($Y_L = 0$), the characteristic polarizations are linear, $R = 0$ and $R = \infty$. This condition is called *transverse propagation*. In both cases, the expression (11.61) for the refractive indices greatly simplifies. For these approximations to be useful, Y_T and Y_L need not vanish completely. It is only necessary that the terms in which they appear, respectively, become negligible. These conditions are known as *quasi-longitudinal*

[3] In the ionospheric literature, the real part n_R and the negative imaginary part n_I of the refractive index are often denoted by the symbols μ and χ, respectively. We refrain from this usage here since the latter symbols are used for the permeability and the susceptibility.

(Q_L) and *quasi-transverse* (Q_T) propagation. In particular, the Q_L approximation is often very useful, and it can be valid for a surprisingly large range of angles. A frequently used simplification of (11.61) for the lossless Q_L case is

$$n_{\pm}^2 = 1 - \frac{X}{1 \pm |Y_L|}, \tag{11.62}$$

which is valid for $Y_T^2 \ll 2\,|Y_L|\,|\,1 - X\,|$, and the $+$ applies to the ordinary wave. For a 10 MHz wave in a plasma having $\omega_N = 3 \times 10^7$ rad/s and gyrofrequency $\mu_0 e H_0 / m_e = 8 \times 10^6$ rad/s, for example, this condition is satisfied for angles as large as $\theta = 50°$. In this situation, the characteristic polarizations of the wave can still be described adequately as quasi-circular.

Under Q_L conditions, an incoming linearly polarized wave will be subject to a progressive change in the direction of polarization. This occurs because a linearly polarized wave can always be decomposed into left- and right-circular polarized waves propagating along the same direction, as seen in Chapter 3. These two circularly polarized waves, which are the characteristic waves of the medium, have different phase velocities. As a result, their relative phase angle at any given point along the direction of propagation gradually changes as the wave propagates. This causes the direction of the resulting linearly polarized wave, which is the superposition of the characteristic waves, to rotate as it progresses in space. This phenomenon is called *Faraday rotation*, and it is particularly relevant for satellite communication links, for example, because it can cause severe cross-polarization interference in dual-polarized systems that rely on linear polarized antennas. For this reason, circular polarized systems are more commonly used for satellite transmissions in regions where ionospheric Faraday rotation occurs. In such systems, left- and right-circular polarized waves can carry different signals with little mutual interference when the Q_L condition is satisfied. Additional discussion of Faraday rotation effects is provided later in this chapter.

11.4 IONOSPHERIC PROPAGATION CHARACTERISTICS

Ionospheric links over long distances have some unusual properties. One is the existence of a *maximum usable frequency* (MUF): for a given propagation path as the carrier frequency is increased, eventually a frequency is reached above which the signal disappears quite abruptly. Another is the *skip distance*: for a given frequency the signal may be received over a wide range of distances, but there exists a minimum distance for the signal to be received at that frequency. The signal seems to "skip" over the shorter distances and can be received only further away. A third characteristic is *selective fading*: even for narrowband modulation, some sidebands fade while others are enhanced in a pattern that changes continuously in time. Consequently, analog voice communications often have a strange tonal quality. To understand these qualities of ionospheric propagation, we need to lay some groundwork. The approach will be the superposition of two sets of curves. One is the "ionogram" that gives information about the pertinent ionospheric conditions. The other is a set of "transmission curves" that are based on Snell's law for the particular path. Together, they give information about the propagation possibilities for a particular path under given ionospheric conditions. The development of these concepts is rather long, but important because of the physical insights it provides. Your patience is requested!

11.5 IONOSPHERIC SOUNDING

Although Chapman's theory discussed in the previous chapter provides some idea of the behavior of ionized layers, especially the E and F_1 regions, it is not sufficiently accurate to describe these layers for propagation predictions, and it fails completely when the F_2 region is involved. Ionospheric sounding is an observational tool to provide information about local properties of the ionosphere. Historically, the conventional form of ionospheric sounding has been through the use of ground-based frequency-swept pulsed radars (Figure 11.6).

Most ground-based ionospheric sounders (ionosondes) direct their beams vertically upward and are denoted as vertical sounders. Only ground-based vertical sounders will be considered in any detail here, although oblique sounders can also be used. In the former case, the receiving system is collocated with the transmitting system (monostatic radar). Ground-based ionosondes are also called "bottom-side" sounders, as they collect data from the lower regions of the ionosphere, up to the F

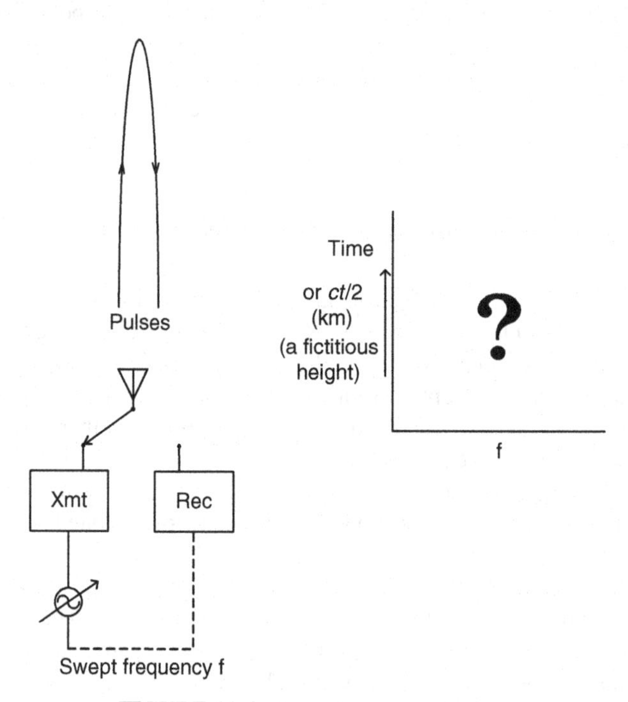

FIGURE 11.6 Ionospheric sounder.

region electron density maximum. "Topside" sounders, on the other hand, are space-based sounders that collect data from the top portions of the ionosphere.

A ground-based ionosonde transmits pulsed signals up to the ionosphere and records properties of the pulses that reflect and return to the ground. The center frequency of these pulses is changed in time, usually in small increments from pulse to pulse, and the time delay between the transmission of each pulse and its reception after reflection from the ionosphere is recorded as a function of frequency. The resulting time versus frequency record is called an *ionogram*. Ionospheric sounders are very important in ionospheric research in general, and they are also important for understanding the peculiarities of ionospheric propagation. While the vertical axis of an ionogram really represents time, it is customary to multiply this time by $c/2$, where c is the speed of light in vacuum, and label the vertical scale as a height. As will be seen in what follows, the waves do not actually travel at the velocity c, and the height of reflection in an ionogram is therefore called the *virtual height*, generally denoted h', to distinguish it from the true maximum height h of an actual ray path.

11.5.1 Ionograms

To determine how an ionogram might look, the conditions for "reflection" in the ionosphere and the time required to get to the reflection point and back need to be examined.

Snell's law requires, for an oblique path in a horizontally stratified medium, that

$$n_{\mathrm{R}}(h)\sin\theta(h) = \sin\theta_{\mathrm{I}}, \tag{11.63}$$

where $\theta(h)$ is the angle of the ray with respect to the vertical and θ_{I} is the corresponding incidence angle at the entrance of the ray into the ionosphere. For simplicity, it is assumed that $n = n_{\mathrm{R}} = 1$ just below the ionosphere. At the maximum height of penetration, $\theta = 90°$, so

$$n_{\mathrm{R}}(h_{\max}) = \sin\theta_{\mathrm{I}}. \tag{11.64}$$

For vertical incidence sounding, we have

$$\theta_{\mathrm{I}} = 0, \qquad n_{\mathrm{R}}(h_{\max}) = 0. \tag{11.65}$$

We therefore need to look at the conditions that force the real part of the refractive index, n_{R}, to vanish. Consider the lossless case, where $Z = 0$. For the *o*-wave, (11.61) becomes

$$n_{+}^{2} = 1 - \frac{X}{1 - \frac{Y_{\mathrm{T}}^{2}}{2(1-X)} + \sqrt{\frac{Y_{\mathrm{T}}^{4}}{4(1-X)^{2}} + Y_{\mathrm{L}}^{2}}}. \tag{11.66}$$

While it is not at all obvious, for $n = 0$ the solution is $X = 1$, as can be shown by expanding the square root with the binomial theorem and taking the limit as $X \to 1$. The *o*-wave will therefore be reflected from the height at which $X = 1$ (where $\omega = \omega_N$). For the *x*-wave, there are two solutions for $n_{-} = 0$: $X = 1 - Y$ and $X = 1 + Y$, where $Y^2 \equiv Y_{\mathrm{L}}^2 + Y_{\mathrm{T}}^2$ (also not obvious). The extraordinary rays are therefore reflected at heights where the electron density is such that $X = 1 - Y$ or $X = 1 + Y$. For a wave coming from the ground, the electron density, ω_N, and therefore X will initially be an increasing function of height, and the height for which $X = 1 - Y$ will be reached first. This is therefore the usual reflection height for the extraordinary wave. Under certain conditions, by a complex process called mode coupling, part of the x-wave may leak through the $X = 1 - Y$ region and propagate further upward, to be reflected at the height where the electron density satisfies the $X = 1 + Y$ condition. The wave then returns to the ground by the inverse process. Such a wave is sometimes called a z-wave although it is really just an x-wave arriving back at the ground by a rather complicated process.

To return to the question about what is the actual shape of an ionogram, it is now necessary to look at the velocity of propagation of the waves, since the ionogram actually represents the time for the wave to travel up to the reflection region and back down. Since the travel time of pulses is being considered, an understanding of the group velocity v_g will be needed. The following discussion will be restricted to the lossless, isotropic case, but still captures the dominant physical properties for more general cases.

It is convenient to define first a group velocity refractive index, n_g, for a lossless, isotropic ionosphere. From $n = k_1/k_0 = k_1 c/\omega$ and $v_p = \omega/k_1$ follows

$$n = c/v_p\,, \tag{11.67}$$

which shows how n is related to the phase velocity. A group refractive index can be defined analogously:

$$n_g = c/v_g\,. \tag{11.68}$$

A lemma is now needed:

$$n_g = 1/n \tag{11.69}$$

that can be proven as follows. From Chapter 3, recall $v_g = (dk_1/d\omega)^{-1}$. Use of this in (11.68) gives

$$n_g = c\frac{dk_1}{d\omega} = c\frac{d}{d\omega}(nk_0). \tag{11.70}$$

Use of

$$k_0 = \omega/c \tag{11.71}$$

and the expression for the refractive index for the lossless, isotropic case

$$n = \sqrt{1 - X}, \tag{11.72}$$

where

$$X = \frac{\omega_N^2}{\omega^2} \tag{11.73}$$

gives an expression explicit in ω. After some differentiation and algebra, the desired result (11.69) is obtained.

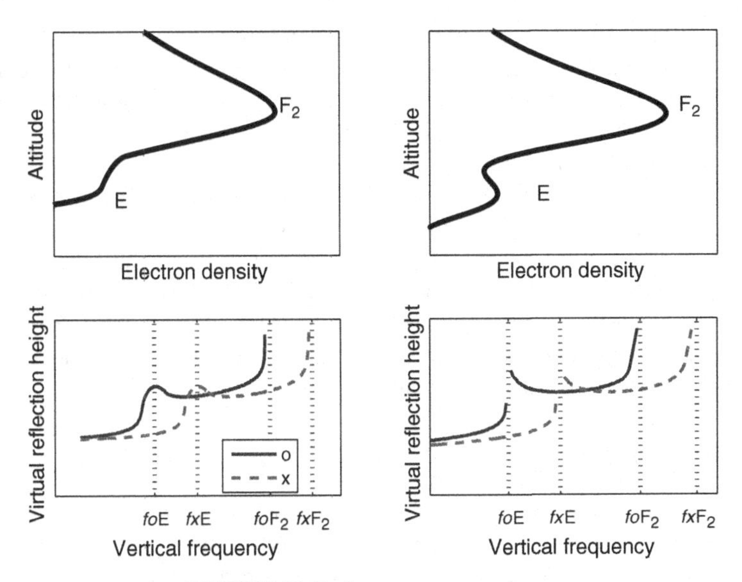

FIGURE 11.7 Ionogram curve shape.

Recall now that, for vertical incidence, reflection occurs at a height for which $n = 0$, that is, for which $n_g = \infty$ by (11.69) and therefore $v_g = 0$ by (11.68). This makes perfect physical sense: as the (pulse) wave approaches the reflection height, it gradually slows down, stops, and comes back down. While the lemma was proved only for the isotropic, lossless case, this principle should be quite general and is, indeed, observed in actual ionograms. With this understanding that the waves slow down as they approach their maximum height, the shape of an ionogram can now be predicted.

Sketched in Figure 11.7 are ionograms for two assumed ionospheric profiles, one (at right) with a prominent E region "nose" and the other with a much smaller nose. The shape of each ionogram trace arises as follows. At a given frequency, the ionosonde pulse propagates up from the ground where $N = 0$ into regions of increasing N, ω_N, and therefore X until a value of X causing reflection is encountered at height h. For frequencies that have reflection heights h in a region where the h versus N curve is slowly varying (i.e., nearly vertical in a plot of h versus N as in Figure 11.7), the wave will travel a relatively large distance in the region where it is slowed down most, producing a large time delay. This causes a hump in the ionogram, which is a plot of virtual height (h'), proportional to travel time, against frequency. The "steeper" the h versus N curve, the longer the delay, and the higher the hump. Where the h versus N curve becomes vertical, it can be shown that the delay becomes infinite, and a cusp in the ionogram results.

Generally, two traces are observed: one for the o-wave and one for the x-wave. The frequency that produces the greatest delay for each ionospheric region is called the critical frequency; thus, f_oE in the figure denotes the critical frequency for the o-wave and the E region; f_xF$_2$ denotes the critical frequency of the x-wave for the F$_2$ region. (When the magnetic field is neglected, so that there is only one wave, the notations f_cE, f_cF$_1$, and so on are used for the critical frequencies.)

How does one determine which trace corresponds to the o-wave and which to the x-wave? Recall that the critical frequency for both waves corresponds to the reflection height for which the h versus N curve is steepest, and therefore the same height and electron density N. Neglecting collision loss, this occurs for the o-wave when $X = 1$ (see above), and for the x-wave when $X = 1 - Y$. Thus, the value of X at the critical frequency is smaller for the x-wave than for the o-wave. But since X is proportional to N, which is the same for both waves, and inversely proportional to frequency, the critical frequency of the x-wave is greater than that of the o-wave for a given ionospheric region. This allows identification of the traces in an ionogram: at the critical frequencies, the x-wave appears to the right of the o-wave. By the same principle, if a z-wave exists, since it is reflected at $X = 1 + Y$, it will appear to the left of the o-wave at the critical frequencies.

11.5.2 Examples of Actual Ionograms

We digress here to show and interpret some actual ionograms. The simplest ionogram is the one shown in Figure 11.8. The higher sets of curves are due to multiple bounces; for example, the middle set appears for each frequency at precisely twice the virtual height and corresponds to reflection at the ionosphere, then by the ground, and again at the ionosphere, for two complete round trips. These multiple-bounce traces give no further information and may be disregarded. This ionogram was obtained during nighttime in Puerto Rico, which lies at moderate geomagnetic latitude. The nighttime ionogram shows only a single F layer. Since E- and F_1-region ionization is caused primarily by solar ultraviolet and X-ray radiation, this is to be expected.

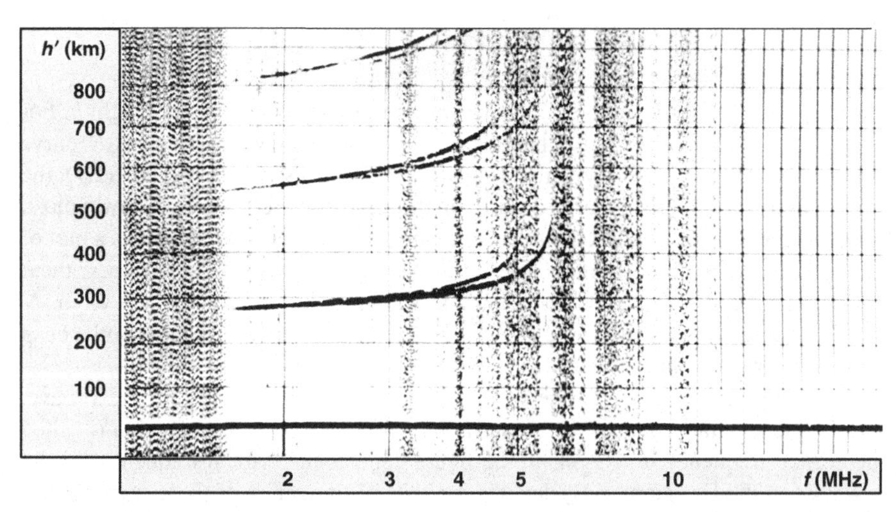

FIGURE 11.8 Ionogram for Puerto Rico, nighttime. (Courtesy of the National Geophysical Data Center, provided by Mr. Ray Conkright.)

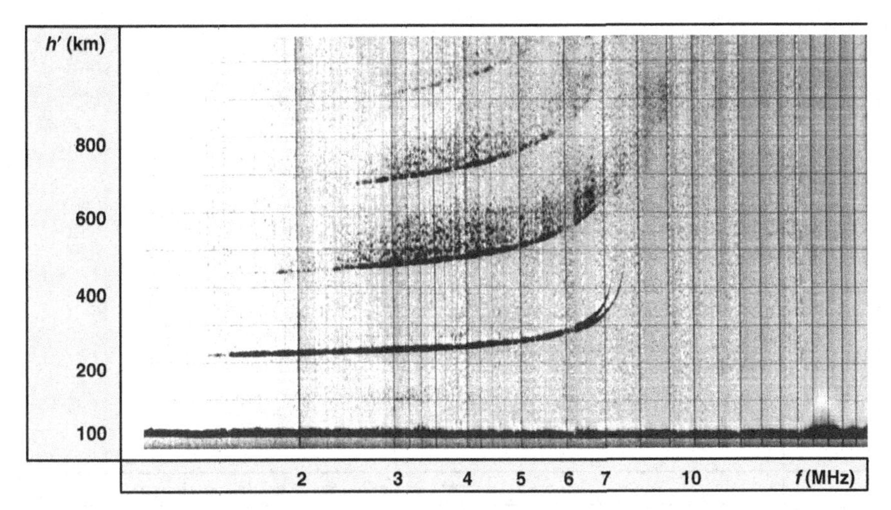

FIGURE 11.9 Ionogram for Huancayo, Peru, at nighttime with multiple-bounce "spreading" echoes. (Courtesy of the National Geophysical Data Center, provided by Mr. Ray Conkright.)

The ionogram shown in Figure 11.9 is similar except that the multiple-bounce traces have many echoes with randomly greater delays than expected. This ionogram is from a station in Huancayo, Peru, located in a valley between mountains. The reflections from the "rough" surface represented by this mixed ground are probably responsible for this "spreading" of the echoes. The same effect can occur at some locations even on the single-bounce trace because at the frequencies considered (a few megahertz), it is difficult and expensive to construct high-gain antennas, and it may therefore be possible for energy to reach the ionosphere via ground reflection, and return the same way, even on the "direct" bounce.

The situation for Figure 11.10 is very different: the echoes are "spread" widely even for the direct bounce. This ionogram was also taken in Huancayo, where this effect is not common. It must be attributed to an ionospheric effect; the phenomenon is called "spread-F" and is due to a patchiness in the F layer that is only partially understood. It occurs often, but not exclusively, near the geomagnetic equator. Because Huancayo is located near the equator, this is a case of equatorial spread-F.

Another abnormality that may be observed is the presence of "sporadic-E" reflections that occur in the ionogram shown in Figure 11.11. The sporadic-E echoes are those that occur in this ionogram at a virtual height of about 100 km for almost all frequencies. The symbol E_s denotes sporadic-E. The normal E region reflections below 3 MHz are missing; therefore, the E layer is said to be "blanketed" by the E_s. The F_1 and F_2 reflections above about 3 MHz are not blanketed by E_s on this particular ionogram. This ionogram was also taken in Huancayo during the daytime.

An ionogram obtained at high geomagnetic latitude (Churchill, Canada) is shown in Figure 11.12. This ionogram shows both sporadic-E and spread-F layers. Ionograms from high latitudes are often complex and more difficult to interpret.

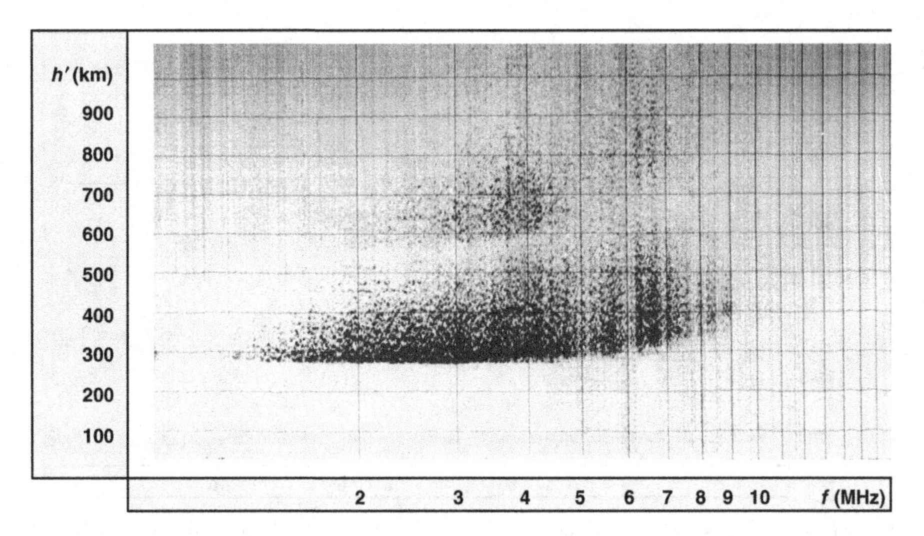

FIGURE 11.10 Ionogram for Huancayo at nighttime, with equatorial "spread-F". (Courtesy of the National Geophysical Data Center, provided by Mr. Ray Conkright.)

Ionospheric sounding is a valuable research tool for probing the ionosphere. Many ionospheric sounding data are accessible through the Internet [3]. However, the main motivation for introducing ionograms here is to explain the puzzling peculiarities of ionospheric propagation, such as the skip and MUF phenomena. For this purpose, some additional theorems governing ionospheric propagation are needed. They are (1) the Secant law, (2) Breit and Tuve's theorem, and (3) one of Martyn's theorems, all

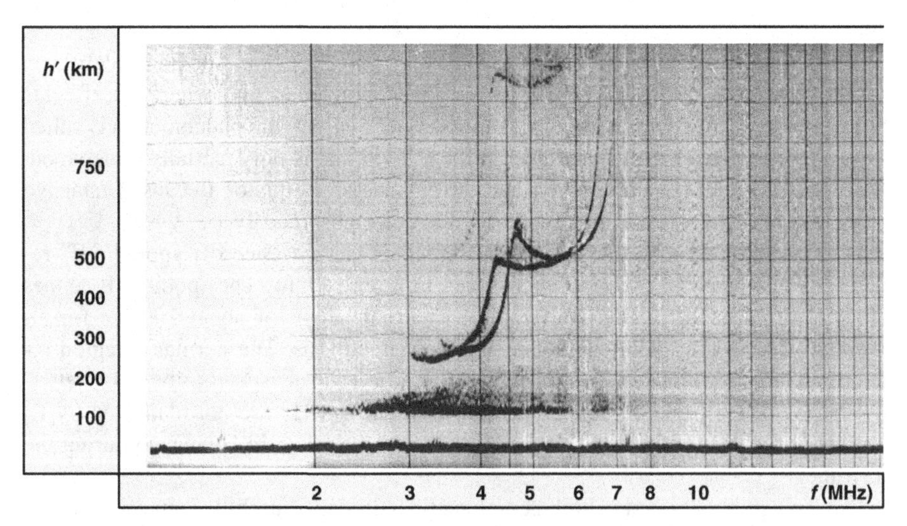

FIGURE 11.11 Ionogram for Huancayo at daytime, with "sporadic-E". (Courtesy of the National Geophysical Data Center, provided by Mr. Ray Conkright.)

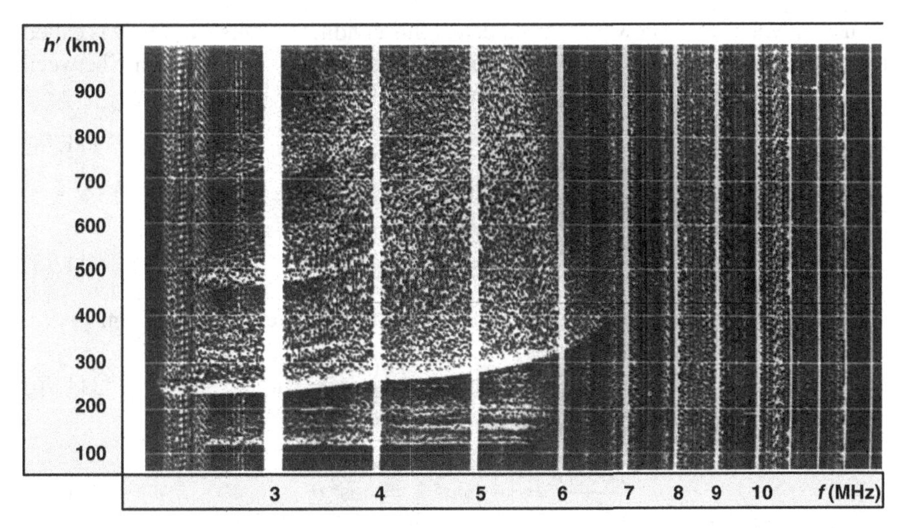

FIGURE 11.12 Ionogram for Churchill, Canada, at nighttime, with both sporadic-E and spread-F present. (Courtesy of the National Geophysical Data Center, provided by Mr. Ray Conkright.)

to be considered next. It is important to note that they will be developed on the basis of plane stratification of the ionosphere, no Earth magnetic field effects, and lossless conditions. These are idealized assumptions that yield only an approximation of the actual situation, but a very useful one.

11.6 THE SECANT LAW

Consider two simultaneous propagation experiments, one oblique and the other vertical, with reflections occurring above the same point on the ground, as in Figure 11.13. The frequency of the oblique experiment f_{ob} is arbitrary. Let the height of highest penetration be $h = h_1$. Now let the frequency f_{v} of the vertical experiment be adjusted so that its height of penetration is also h_1 (this is a thought experiment, so

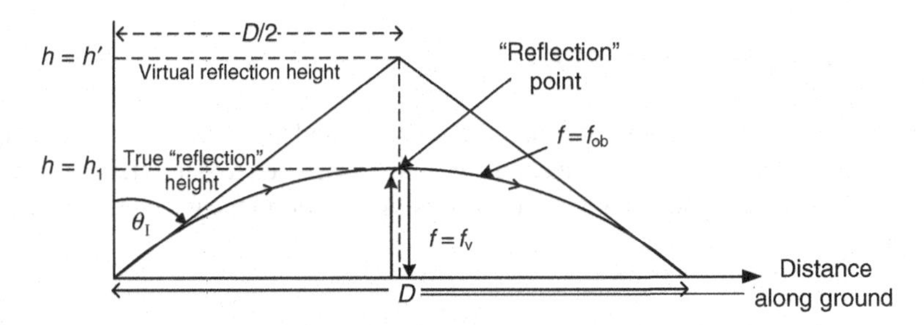

FIGURE 11.13 Oblique and vertical waves.

let us not worry about how one would detect this condition). This frequency is called the equivalent vertical frequency. The Secant law states that the relationship between the oblique frequency and the equivalent vertical frequency is

$$f_{\rm ob} = f_{\rm v} \sec\theta_{\rm I} \,. \tag{11.74}$$

Proof: For the oblique case, Snell's law gives

$$n(h_1)\sin\frac{\pi}{2} = n(h_1) = \sin\theta_{\rm I} \,. \tag{11.75}$$

Neglecting loss and magnetic effects, and using (11.72) gives, in succession,

$$n^2(h_1) = 1 - X(h_1) = 1 - \frac{f_N^2(h_1)}{f_{\rm ob}^2} = \sin^2\theta_{\rm I} \,, \tag{11.76}$$

$$\frac{f_N^2(h_1)}{f_{\rm ob}^2} = 1 - \sin^2\theta_{\rm I} = \cos^2\theta_{\rm I} \,, \tag{11.77}$$

$$f_{\rm ob} = f_N(h_1)\sec\theta_{\rm I} \,. \tag{11.78}$$

The vertical case is the limit of the oblique case as $\theta_{\rm I} \to 0$, giving from equation (11.78)

$$f_{\rm v} = f_N(h_1) \tag{11.79}$$

so that f_N can be replaced with $f_{\rm v}$ to yield the Secant law, (11.74), which completes the proof.

Note that $\sec\theta_{\rm I}$ can also be defined from the geometry as

$$\sec\theta_{\rm I} = \sqrt{\left(\frac{D}{2h'}\right)^2 + 1} \,, \tag{11.80}$$

where D is the horizontal distance traversed by the curved oblique ray and h' is a "virtual" reflection height defined by drawing a straight ray from the transmitter at angle $\theta_{\rm I}$ up to the middle of the path. Later in the chapter, we will consider relationships between propagation along the straight ray/virtual reflection height path and along the curved ray path beyond the geometrical properties used here.

11.7 TRANSMISSION CURVES

Transmission curves are obtained from the Secant law and the geometry of a particular path. In what follows, a plane Earth geometry will be assumed for simplicity, although extension to the true spherical geometry is possible [1]. From the geometry and using the Secant law, it follows that

$$\frac{f_{\rm ob}}{f_{\rm v}} = \sec\theta_{\rm I} = \sqrt{\left(\frac{D}{2h'}\right)^2 + 1} \,, \tag{11.81}$$

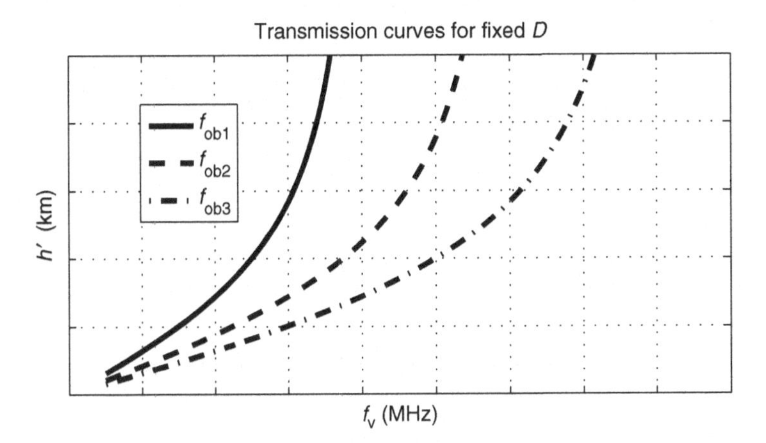

FIGURE 11.14 Transmission curves.

which can be solved for h' to yield

$$h' = \frac{D}{2\sqrt{(f_{\text{ob}}/f_{\text{v}})^2 - 1}}. \tag{11.82}$$

Transmission curves are plots created using equation (11.82). They show the virtual height h' versus the equivalent vertical frequency f_{v} for a fixed value of distance D, and with the oblique frequency f_{ob}, that is, the candidate frequency to be used over this path, as a parameter. A typical plot is shown in Figure 11.14.

Transmission curves are useful because ionospheric propagation over path length D using frequency f_{ob} implies a reflection from the ionosphere at the virtual height h' and equivalent vertical frequency f_{v} read from the transmission curve. This curve is determined entirely by the path geometry; still needed is information about the required state of the ionosphere to achieve propagation on this frequency for the specified path.

Note that the coordinates of transmission curves are vertical frequency and virtual height, just as they are for an ionogram. It would be tempting to superpose the transmission curves with an ionogram from the path midpoint. The frequencies of the ionogram can then be equated with the equivalent vertical frequencies of the oblique path. But such a superposition would be premature at this point. To be sure, the definition of equivalent vertical frequency guarantees that the true heights of reflection of the oblique and vertical waves are the same; however, it has not yet been shown that the virtual heights of the two are the same: we defined the virtual height using time delays on the vertical path, while the oblique virtual height has been defined using a straight ray geometry. In both ionograms and transmission curves, we have used the symbol h' for virtual height, but until it is shown that these are indeed the same, this is merely sloppy notation!

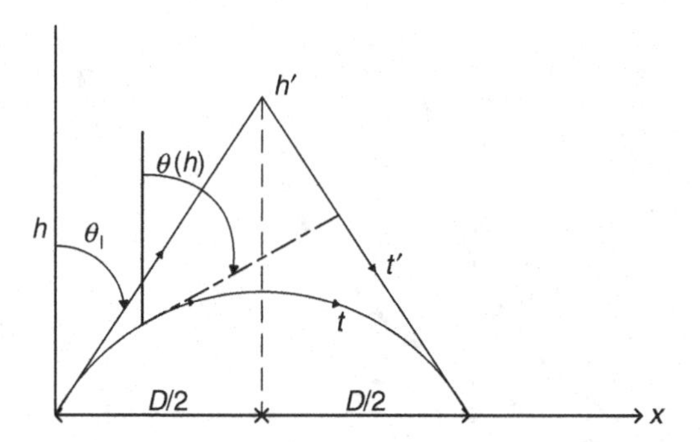

FIGURE 11.15 Breit and Tuve's geometry.

11.8 BREIT AND TUVE'S THEOREM

The theorem of Breit and Tuve states that the time t required for a wave to travel over an actual ray path with maximum height h in the lossless, isotropic ionosphere is the same as the time t' required for a ray, starting at the same incidence angle, to travel at the speed of light in vacuum over the corresponding virtual straight-line path with maximum height h' at the center.[4]

Proof: From the geometry of Figure 11.15, the distance traversed by the virtual ray from the ground to height h' is given by $(D/2)\csc\theta_{\mathrm{I}}$. Since the virtual ray is assumed to travel at the speed of light in free space, the total travel time will be

$$t' = 2\frac{(D/2)\csc\theta_{\mathrm{I}}}{c} = (D/c)\csc\theta_{\mathrm{I}}\,. \tag{11.83}$$

For the actual ray, the element of time corresponding to the traversed distance ds is

$$dt = \frac{ds}{v_{\mathrm{g}}(h)} = \frac{dx}{v_{\mathrm{g}}(h)\sin\theta(h)}. \tag{11.84}$$

By the use of (11.68), (11.69), and (11.63), this can be transformed to

$$dt = \frac{dx}{cn\frac{1}{n}\sin\theta_{\mathrm{I}}} = \frac{dx}{c\sin\theta_{\mathrm{I}}} \tag{11.85}$$

and integration yields

$$t = \frac{\csc\theta_{\mathrm{I}}}{c}\int_0^D dx = \frac{D}{c}\csc\theta_{\mathrm{I}} = t', \tag{11.86}$$

where (11.83) has been used in the last step. This completes the proof of Breit and Tuve's theorem. We have found that the straight ray virtual path that was previously

[4]In ionospheric parlance, the word "path" is sometimes used to denote the time taken to travel a geometric path. In this theorem, the term means the geometric path of a ray.

defined geometrically incurs the same time delay (assuming propagation in a vacuum) as the true curved ray path. The virtual path is therefore meaningful in a time delay sense. It still remains to determine relationships between the virtual reflection heights on the oblique paths and the equivalent frequency vertical path.

11.9 MARTYN'S THEOREM ON EQUIVALENT VIRTUAL HEIGHTS

Martyn's theorem on equivalent virtual heights states that, given an oblique transmission reflected from a lossless, isotropic ionosphere at some height h and the equivalent vertical transmission (i.e., at the equivalent vertical frequency and at the center of the path), the virtual reflection height $h'_{\rm ob}$ of the oblique wave will be the same as the virtual reflection height $h'_{\rm v}$ of the equivalent vertical wave. This is equivalent to deriving a relationship between the time delays on the vertical and oblique true ray paths; the relationship obtained will enable the superposition of ionograms and transmission curves.

To prove the theorem, a preliminary lemma is required:

$$n_{\rm ob}(h)\cos\theta(h) = n_{\rm v}(h)\cos\theta_{\rm I} \tag{11.87}$$

in a lossless, isotropic ionosphere.

Proof of Lemma: It follows from (11.72) that

$$n_{\rm ob}^2(h) = 1 - \frac{f_N^2(h)}{f_{\rm ob}^2}. \tag{11.88}$$

Similarly, for the vertical wave the relation is

$$n_{\rm v}^2(h) = 1 - \frac{f_N^2(h)}{f_{\rm v}^2}. \tag{11.89}$$

With the Secant law, $f_{\rm v}$ can be eliminated from (11.89) to give

$$n_{\rm v}^2(h) = 1 - \frac{f_N^2(h)}{f_{\rm ob}^2\cos^2\theta_{\rm I}}, \tag{11.90}$$

which can be solved for f_N^2,

$$f_N^2 = f_{\rm ob}^2\cos^2\theta_{\rm I}[1 - n_{\rm v}^2(h)]. \tag{11.91}$$

By substituting this into (11.88), using Snell's law (11.63) and some algebra, (11.87) is obtained.

Proof of Theorem: Let $t_{\rm ob}$ be the time required for the oblique wave to complete the travel from A to C in Figure 11.16 and let $t_{\rm v}$ be the time for the equivalent wave to travel up to the same true reflection point and back down. Then $t_{\rm ob}$ is given by

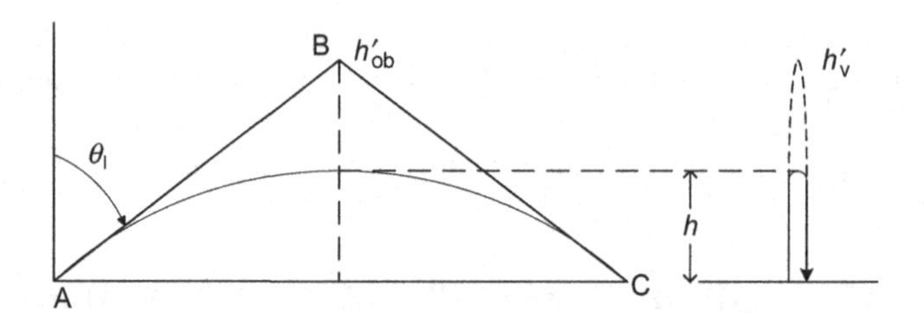

FIGURE 11.16 Martyn's theorem.

$$\frac{t_{ob}}{2} = \int_{t(h_1=0)}^{t(h_1=h)} dt = \int_{S(h_1=0)}^{S(h_1=h)} \frac{dS_1}{v_{g,ob}} = \int_{S(h_1=0)}^{S(h_1=h)} \frac{dS_1}{cn_{ob}}$$
$$= \frac{1}{c}\int_0^h \frac{dh}{\cos\theta(h)\, n_{ob}(h)}. \tag{11.92}$$

By the lemma, this becomes

$$\frac{t_{ob}}{2} = \int_0^h \frac{dh}{\cos\theta_I c n_v(h)} = \sec\theta_I \int_0^h \frac{dh}{c/n_{g,v}(h)} = \sec\theta_I \int_0^h \frac{dh}{v_{g,v}(h)}. \tag{11.93}$$

The integrand in the last integral is just the differential of time, and the integral therefore gives $t_v/2$, yielding

$$c\frac{t_{ob}}{2}\cos\theta_I = \frac{ct_v}{2}. \tag{11.94}$$

By the theorem of Breit and Tuve, t_{ob} can be replaced with t'_{ob}, the time taken by the virtual wave to travel the route ABC. Moreover, since the virtual ray is assumed to travel at velocity c, the quantity $(ct_{ob}/2)$ is the distance AB. Therefore, the left-hand side of (11.94) becomes

$$c\frac{t_{ob}}{2}\cos\theta_I = h'_{ob}, \tag{11.95}$$

and the result

$$h'_{ob} = h'_v \tag{11.96}$$

follows, showing that the virtual height for an oblique propagation geometry is the same as that for the equivalent vertical frequency. Finally, here is the justification for superposing the ionogram and transmission curves!

11.10 MUF, "SKIP" DISTANCE, AND IONOSPHERIC SIGNAL DISPERSION

Figure 11.17 is an ionogram, neglecting magnetic field effects, superimposed on a set of transmission curves. The transmission curves specify the horizontal propagation distance and, via the Secant law, the physics of refraction. They contain no

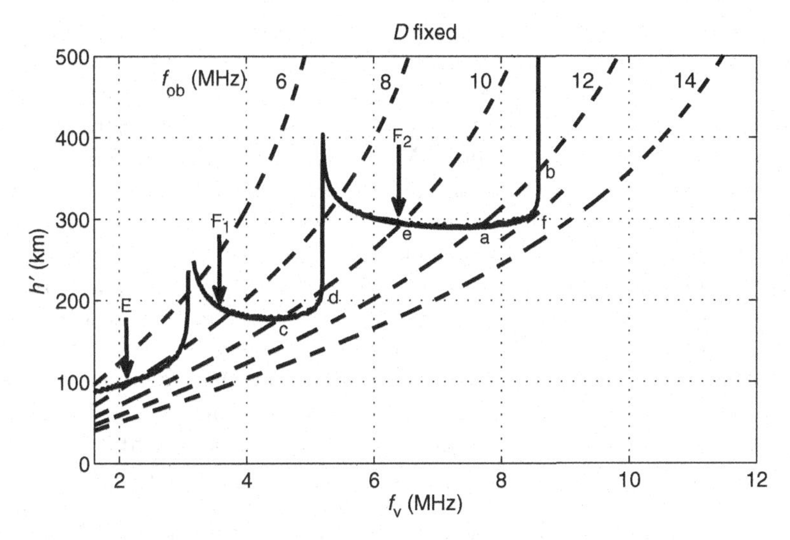

FIGURE 11.17 Superposition of transmission curves and ionogram.

information about the particular ionospheric conditions. The ionogram, on the other hand, furnishes precisely that information. Together, that is, by the points they have in common, they determine the possibility of transmitting signals over the distance D at frequency f_{ob}.

At frequencies f_{ob} above about 13 MHz, there are no points in common to the ionogram and transmission curves. Thus, if a frequency above 13 MHz is chosen as the operating frequency for these conditions, no signal will be received at distance D. No reception over this distance would be possible until the frequency is lowered to about 13 MHz. This frequency is called the basic maximum usable frequency (BMUF) for these ionospheric conditions and this propagation distance.

Referring again to Figure 11.17, at an operating frequency f_{ob} of 12 MHz, two propagation paths are possible, as shown by the intersections a and b. The ray corresponding to point b reaches a much higher virtual (and also actual) height than the one corresponding to a; it is called the high-angle or Pedersen ray. At 10 MHz, propagation is possible via a low-angle ray reflected from the F_1 region (intersection point c), a Pedersen F_1- region ray (intersection point d), and an F_2- region ray (intersection point e). However, the F_1 region reflections are unlikely to allow much energy to penetrate to the F_2 region, and the F_2 ray is therefore not likely to be effective; the F_2 region is said to be "screened" by the F_1 region for these conditions. At 6 MHz in Figure 11.17, E region transmission is possible, the F_1 layer is screened by the E layer, and transmission by F_2 reflection is impossible. Each ionospheric region and both the *o*- and *x*-waves (if magnetic effects are important) have their own maximum usable frequency. From Figure 11.17, the maximum usable frequency for F_1 transmission would be around 11 MHz. The notation F1(D)MUF is sometimes used to refer to these values, where D is the distance in kilometers. The BMUF is the highest of all these maximum usable frequencies.

In addition to the BMUF, it is a common practice to define the *operational* MUF (or simply MUF) as the highest frequency for which operation would be possible between two locations *under a specific set of radio link conditions* such as antenna gains, transmit power, required signal-to-noise ratio, and so on. The relative difference between the BMUF and the operational MUF is typically about 10–35%.

Because attenuation due to absorption decreases with frequency (as will be discussed later), it is tempting to operate just below the MUF. However, ionospheric conditions and hence the MUF depend on the season, hour of the day, geographical location, and sunspot activity. As a result, for propagation prediction purposes, the *predicted* MUF is taken as the *monthly median* of the observed MUF (also called the maximum observed frequency or MOF), that is, the highest frequency for which a skywave mode is available on 50% of the days in a month. This is equivalent to saying that, on any given day, communications may succeed at the predicted MUF only half of the time. To increase the probability of success, a suggested best operating frequency is chosen below the MUF. This latter frequency is known by various names such as frequency of optimum transmission (FOT), optimum working frequency (OWF), or optimum transmission frequency (OTF). For E and F_1 links, the FOT is taken as 95% of the MUF, while for F_2 links it is determined from a table in ITU-R Recommendation P.1239-1 [18]. Note that, while likely to be near optimum in a statistical sense, there is no guarantee that the FOT will be optimal, or even satisfactory, at a particular time.

An illustration of ray path changes, for a fixed distance, as frequency is varied is shown in Figure 11.18 for a simplified two-region ionosphere. At the highest frequency, f_a, the electron density in both layers is insufficient to cause the real part of the refractive index to vanish. Frequency f_a is above the critical frequencies of both layers, and the ray is not "reflected" and penetrates into space, with some refraction by both layers. As the operating frequency is lowered, the MUF of the top layer is

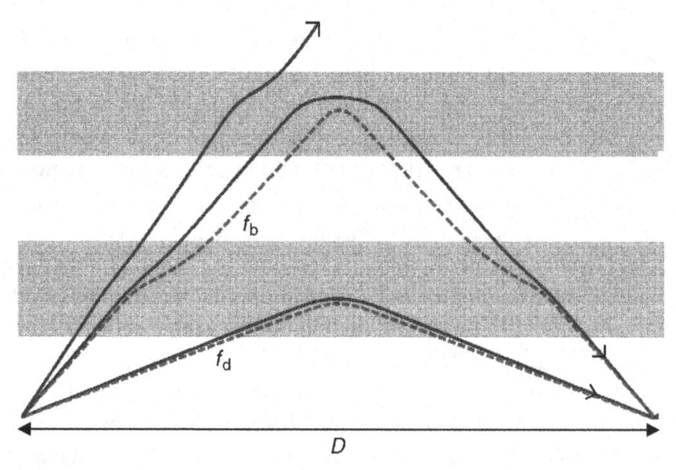

FIGURE 11.18 Ray paths for fixed distance D and varying frequency.

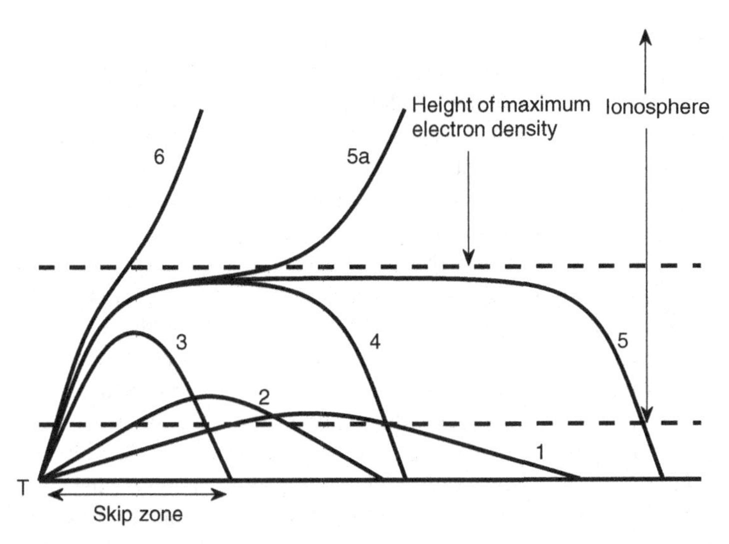

FIGURE 11.19 Ray paths for various distances at fixed frequency. (Adapted from Figure 4.6 of Davies, K., *Ionospheric Radio Propagation,* National Bureau of Standards Monograph 80, 1965.)

reached, and below this value ray paths are shown for the *o*- and *x*-waves as sketched for frequency f_b. When the frequency is lowered below the MUF of the bottom layer, the ray paths are as shown for frequency f_d, and the upper layer is screened by the lower.

In Figure 11.18, the frequency was varied and paths for a specific distance *D* were considered. Actual antennas at MF and HF usually radiate over a wide range of vertical angles. Consider now the ray paths at a fixed frequency as a function of incidence angle on the ionosphere as in Figure 11.19. Assuming the operating frequency is greater than the critical frequency of the layer, a steeply incident ray (small incidence angle) will pass through the layer with little deviation, as shown by path 6. An increase in the incidence angle increases the deviation as shown by path 5a. A further increase in the incidence angle causes the ray to return at long distances as a high-angle ray, as in path 5. Such long distances can also be reached by a low-angle ray, path 1. Note that at large distances the high-angle ray path is very high and the low-angle path is very low. As the incidence angle is increased for the high-angle ray (path 4) or decreased for the low-angle ray (path 2), both rays return at increasingly shorter distances. Finally, there exists an incidence angle at which the high-angle and low-angle rays coalesce (path 3). This condition corresponds to point *f* in Figure 11.17.

This description of Figure 11.19 has used up all the incidence angles without any ray returning at a shorter distance than that of path 3. The transmission cannot be received at shorter distances! This minimum distance is called the "skip distance" for the particular frequency and ionospheric conditions. Points within that distance are said to be within the "skip zone". From point *f* in Figure 11.17, it can be inferred that the frequency for which Figure 11.19 is drawn is the MUF for the skip distance. For points within the skip zone, the operating frequency is above the MUF, and communication can only be effected by lowering the operating frequency sufficiently. For a given oblique frequency, it is possible to predict the skip distance directly by overlaying the ionogram with transmission curves computed as D is varied for the specified f_{ob}. For this purpose, it is convenient to create transmission curves for a fixed oblique frequency with the path distance D as a parameter.

The high dispersion of ionospheric transmission, as evidenced by the strange and variable tone quality of analog voice transmissions, for example, can now also be explained. In general, there are four possible ionospheric ray paths from a transmitter to a receiver: high- and low-angle ordinary rays and high- and low-angle extraordinary rays. Wavelengths are on the order of tens of meters, while distances are hundreds or thousands of kilometers, that is, typically many thousand wavelengths. Therefore, it takes only a small change in the difference of refractive indices to greatly change the relative phases at the receiving site. At some sideband frequencies, the rays will add constructively, at others destructively, and this pattern changes as the ionosphere changes. Thus, wideband signals are distorted: the link is highly dispersive.

11.11 EARTH CURVATURE EFFECTS AND RAY-TRACING TECHNIQUES

The preceding discussion made three assumptions that served well to simplify the analysis and that work well in many (but not all) cases. First, we have often assumed that the ionosphere was lossless ($Z = 0$) and isotropic (i.e., Earth magnetic field effects were neglected). The results we have obtained remain approximately applicable in most cases even in the presence of these effects and require correction primarily for attenuation effects (as discussed later in the chapter) and to recognize the presence of the o- and x-waves.

Second was the assumption of a flat Earth. A misleading conclusion from this assumption is that paths are always available even for very large propagation distances. This is not true in reality because, for very large incidence angles θ_I, the Earth's curvature significantly changes the incidence angle at the ionosphere from that of a flat Earth. Typically, distances no larger than about 4000 km can be covered in "one-hop" ionospheric links, assuming ionospheric reflection at a height of about 400 km. For larger distances, *multihop ionospheric propagation* can be exploited; such links include several successive reflections at ionospheric layers and the Earth's surface along the path from the transmitter to the receiver [1,4].

The final assumption was that the ionosphere was horizontally uniform and a simple ray-tracing approach—as expressed by the theorems seen in this chapter—was applicable. More generally, there are situations where significant horizontal gradients may develop along the ionospheric propagation path. This is especially true, for example, during day/night transition times and away from mid-latitudes. In this case, more sophisticated ionospheric ray-tracing techniques can be employed

[1,5,6].[5] Ray-tracing techniques can be classified into analytical or numerical, depending whether the ray path integration is done numerically or analytically. Numerical ray tracing is in general more accurate but demands more computational resources than analytical ray tracing. A drawback for ray-tracing techniques is that they require very accurate ionospheric electron density models, beyond what is usually available in practice. Analytical ray-tracing techniques are still prevalently used in prediction tools for ionospheric propagation, as discussed in the next section.

11.12 IONOSPHERIC PROPAGATION PREDICTION TOOLS

As we have seen above, the behavior of a particular ionospheric link depends not only on the geographical locations of transmitter and receiver, but also on seasonal and diurnal variations and solar activity. Because of such spatial and temporal variability, it is difficult to apply the theory developed in this chapter for propagation *prediction* purposes without knowledge of the ionospheric conditions affecting the particular ionospheric link under consideration, at least in a statistical sense. As a result, various computer models for propagation prediction have been developed over the years based on detailed ionospheric description models [7, 8]. Perhaps the most widely used tools are those from the "IONCAP family," that is, those derived from the *Ionospheric Communications Analysis and Prediction Program* (IONCAP) [9] developed by the Institute of Telecommunication Sciences of the National Telecommunications and Information Administration (NTIA) and its predecessors. This program also serves as the basis of ITU-R Recommendation P.533 for predicting the expected performance of HF skywave propagation systems [10]. A program, REC533, implementing this prediction model is available from the ITU-R, and ITU-R Recommendation P.533 gives a detailed account of the algorithms on which the program is based. Other members of the "IONCAP family" include the *Voice of America Coverage Analysis Program* (VOACAP)—an improved version of IONCAP first developed for the use by "Voice of America" radio broadcast stations—and the *Ionospheric Communications Enhanced Profile Analysis and Circuit Prediction Program* (ICEPAC)—an updated version of IONCAP with a more accurate model for the electron density profile [11], especially at higher latitudes, that includes geomagnetic effects such as those associated with aurora. VOACAP and ICEPAC are available from the U.S. Department of Commerce and provide detailed point-to-point graphs and area coverage maps for many parameters related to ionospheric propagation quality such as signal-to-noise ratio (SNR), signal power, and MUF. All the output parameters are based on *monthly medians* for given location(s), time of day, and solar activity (input parameters). Since they are all based on approximate models, these programs do not necessarily produce the same results. VOACAP and ICEPAC are similar for low latitudes, but differ at high latitudes. In the latter case, ICEPAC results are expected to be more accurate, and hence are recommended.

[5] Careful ionospheric ray tracing is of particular importance for HF over-the-horizon radar systems.

Figure 11.20 shows typical output data from VOACAP for illustrative purposes. REC533 and ICEPAC provide outputs in an identical format. Figure 11.20 displays geographical distributions of the MUF for a transmitter located in Columbus ,OH, at 1:00 PM (18:00 UT) for the month of June, assuming two levels of solar activities with annual sunspot numbers of 10 (low solar activity) and 100 (high solar activity), respectively. The scenario with a higher number of sunspots allows a larger MUF to be used because higher solar activity implies larger ion production—in other words, a more refractive ionosphere—as seen in Chapter 10.

11.13 IONOSPHERIC ABSORPTION

Ionospheric rays are attenuated by spreading as they propagate and may also be attenuated by absorption due to collisions. In terms of the refractive index, the specific attenuation α, in nepers per unit length, is given by

$$\alpha = -\mathrm{Im}(\underline{k}_1) = -\mathrm{Im}(k_0\underline{n}) = k_0 n_\mathrm{I}\,. \tag{11.97}$$

In the general case, n_I must be found from the Appleton–Hartree formula. Things become much simpler when the magnetic field is neglected, so that

$$\underline{n}^2 = 1 - \frac{X}{1 - jZ}, \tag{11.98}$$

which can be separated into real and imaginary parts as

$$\underline{n}^2 = 1 - \frac{X}{1 + Z^2} - j\frac{X\,Z}{1 + Z^2}. \tag{11.99}$$

Noting that

$$\underline{n}^2 = (n_\mathrm{R} - jn_\mathrm{I})^2 = n_\mathrm{R}^2 - n_\mathrm{I}^2 - j2\,n_\mathrm{R}\,n_\mathrm{I}\,, \tag{11.100}$$

we can equate the imaginary parts of (11.99) and (11.100) to find

$$n_\mathrm{I} = \frac{X\,Z}{2n_\mathrm{R}(1 + Z^2)}. \tag{11.101}$$

When this is used in (11.97) and X and Z are replaced with their definitions in (11.13) and (11.14), the result is

$$\alpha = \left(\frac{e^2}{2\epsilon_0 m_e c}\right)\frac{\nu N}{n_\mathrm{R}\,(\omega^2 + \nu^2)} \approx \left(5.3 \times 10^{-6}\right)\frac{\nu N}{n_\mathrm{R}\,(\omega^2 + \nu^2)} \tag{11.102}$$

with N in electrons per cubic meter and ν and ω in rad/s.

No such simple formula can be given for the general case when the magnetic field is considered. However, for the Q_L case, it is found that a better approximation results if ω in (11.102) is replaced with $\omega \pm \omega_\mathrm{L}$, where the + sign refers to the ordinary ray.

Note that the specific attenuation decreases with frequency, so that operating at as high a frequency as is practically possible is desirable as discussed previously. From (11.102) it can be seen that there are two conditions for which attenuation will be

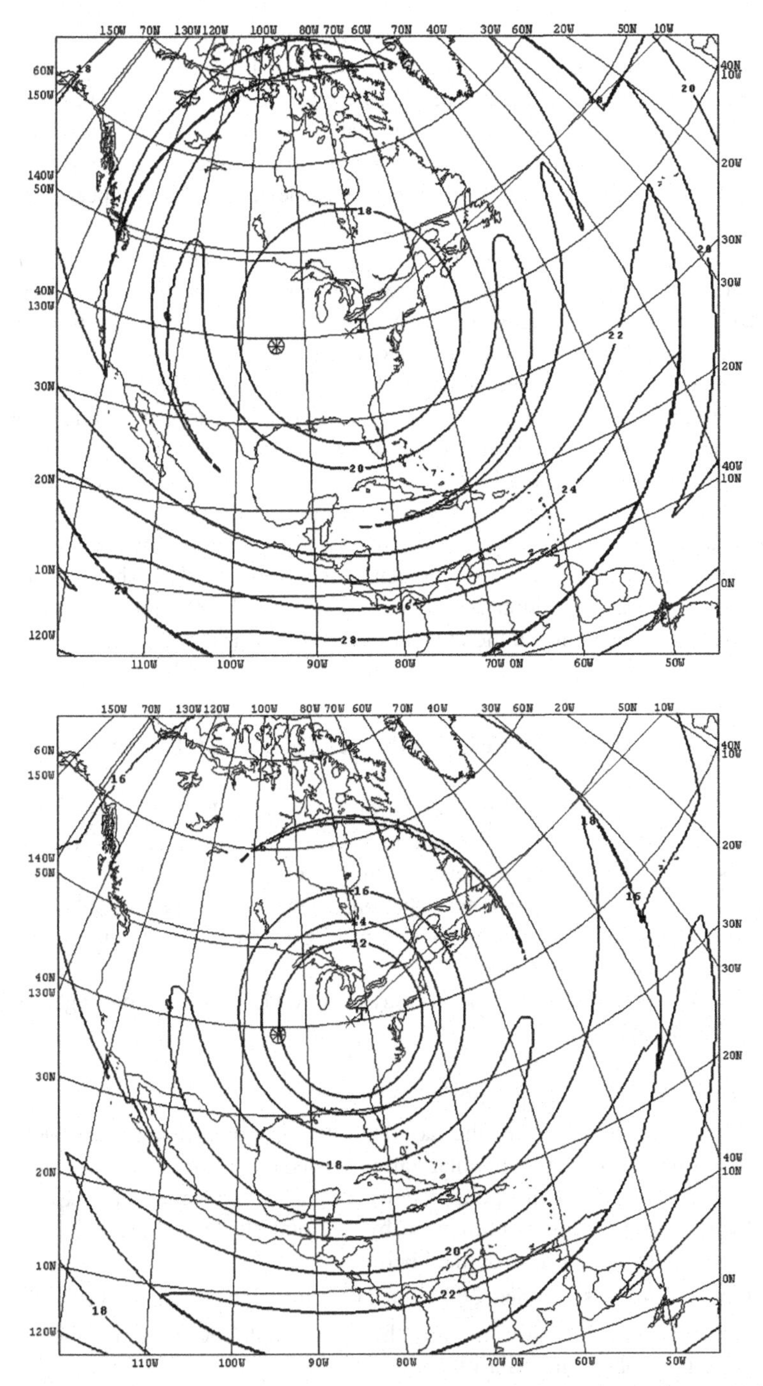

FIGURE 11.20 VOACAP prediction of the MUF in the month of June for transmissions from Columbus, OH, as a function of receiver location. The top and bottom plots use sunspot numbers of 100 and 10, respectively.

relatively high. The first occurs when n_R is small. From Snell's law and the conditions for reflection, (11.63)–(11.65), it can be shown that this occurs near the point of reflection for near-vertical paths. This can be observed in the ionograms. Near the critical frequencies (cusps in the ionograms and reflection locations where N versus height is slowly varying), the wave traverses a longer effective distance for which $n_R \simeq 0$, resulting in high absorption. This is evidenced by the traces becoming faint near the cusps. Since the wave is "turning around" in this region, and hence deviating from a free-space path, this kind of absorption is called *deviative absorption*. The other condition for high absorption corresponds to large values of the product νN in (11.102). This product is largest in the upper D and lower E regions between 80 and 100 km because at lower altitudes there is not enough ionization and at higher altitudes the reduced density does not allow many collisions. It tends to be maximized at local noon because E region ionization correlates well with the solar zenith angle. This kind of absorption is called *nondeviative*. The reader may have noticed that transmissions in the 535–1606 kHz U.S. AM radio broadcast band can often be received at extremely large distances at night, but not so during daytime. The reason is that absorption prevents such reception in the daytime, but at night E region ionization disappears ($N \simeq 0$ in (11.102)), and the signal can then be received. Finally, it should be pointed out that ionospheric absorption can also increase sporadically, for example, due to increased electron densities in the D layer caused by solar flare events.

11.14 IONOSPHERIC EFFECTS ON EARTH–SPACE LINKS

Signals at VHF and above penetrate fully through the ionosphere, and a skywave mode of communication cannot be established. In this case, a "repeater station" (i.e., a satellite) can be employed for long-distance communication links, and the ionosphere acts primarily as a source of undesired perturbations on the Earth–Space signal. These perturbations are significant for frequencies up to 12 GHz and are particularly important for satellite links operating below 3 GHz.

The following types of perturbations may occur on Earth–Space propagation links crossing the ionosphere:

(a) *Attenuation* due to absorption.
(b) *Faraday rotation*, that is, the fact that a linearly polarized wave undergoes a rotation of the direction of polarization caused by the different phase velocities of right- and left-hand polarized waves in the ionosphere, as explained in Section 11.3.4. This is influenced by the total electron content (TEC) along the propagation path, as defined below.
(c) *Group delay*, also influenced by the TEC along the propagation path.
(d) *Scintillations*, that is, variations of the signal amplitude and phase due to small-scale ionospheric irregularities.
(e) *Refraction*, which causes a change in the apparent angle of signal arrival.

All of these effects except (b) are also present in the troposphere, but to a different degree and with a different dependence on frequency. In the following, we will discuss these five ionospheric perturbation mechanisms individually, focusing on their impact on trans-ionospheric propagation in the UHF and SHF bands.

Other ionospheric effects such as RF carrier advance, Doppler shift, and phase dispersion can also occur, but typically they are of a lesser importance and will not be considered here.

11.14.1 Faraday Rotation

The physical mechanism behind Faraday rotation was introduced in Section 11.3.4. Ignoring refraction and for frequencies where the lossless Q_{L} approximation holds, the differential phase difference due to the different phase velocities of the ordinary and extraordinary waves that is established along an infinitesimal path ds produces an infinitesimal rotation of the polarization angle.

Consider an $\hat{x}$ polarized wave of unit amplitude propagating in the z direction as it enters the ionosphere:

$$\underline{\overline{E}} = \hat{x}e^{-jk_0 z}. \tag{11.103}$$

Under the Q_{L} approximation, the characteristic polarizations are right- and left-hand circular; resolving this field into the characteristic polarizations yields

$$\underline{\overline{E}} = e^{-jk_0 z}\left[\frac{1}{\sqrt{2}}\left(\frac{\hat{x}+j\hat{y}}{\sqrt{2}}\right) + \frac{1}{\sqrt{2}}\left(\frac{\hat{x}-j\hat{y}}{\sqrt{2}}\right)\right] \tag{11.104}$$

before entering the ionosphere. Each of the characteristic waves propagates at its own velocity in the ionosphere:

$$\underline{\overline{E}} = \frac{1}{\sqrt{2}}\left(\frac{\hat{x}+j\hat{y}}{\sqrt{2}}\right)e^{-jk_0 n_+ z} + \frac{1}{\sqrt{2}}\left(\frac{\hat{x}-j\hat{y}}{\sqrt{2}}\right)e^{-jk_0 n_- z}. \tag{11.105}$$

After propagating a small distance dz in the ionosphere (and defining $z = 0$ as the entry point), the field is

$$\underline{\overline{E}} = \frac{1}{\sqrt{2}}\left(\frac{\hat{x}+j\hat{y}}{\sqrt{2}}\right)e^{-jk_0 n_+ dz} + \frac{1}{\sqrt{2}}\left(\frac{\hat{x}-j\hat{y}}{\sqrt{2}}\right)e^{-jk_0 n_- dz}. \tag{11.106}$$

Now defining

$$d\Omega = \frac{k_0 dz}{2}\left(n_+ - n_-\right), \tag{11.107}$$

we have

$$\begin{aligned}\underline{\overline{E}} &= \frac{1}{2}e^{-jk_0 dz(n_+ + n_-)/2}\left[\left(\hat{x}+j\hat{y}\right)e^{-jd\Omega} + \left(\hat{x}-j\hat{y}\right)e^{jd\Omega}\right]\\ &= e^{-jk_0 dz(n_+ + n_-)/2}\left[\hat{x}\cos(d\Omega) + \hat{y}\sin(d\Omega)\right]\end{aligned} \tag{11.108}$$

so that $d\Omega$ turns out to be, indeed, the incremental rotation angle. Using the Q_L approximation for n_+ and n_- yields

$$d\Omega = \frac{\omega}{2c}\left(\sqrt{1 - \frac{X}{1+|Y_L|}} - \sqrt{1 - \frac{X}{1-|Y_L|}}\right) dz. \quad (11.109)$$

Assuming frequencies sufficiently high so that the conditions $Y_L \ll 1$ and $X \ll 1$ hold, and using $(1+x)^n \simeq 1 + nx$ for $x \ll 1$, the above can be approximated as

$$d\Omega \simeq \frac{\omega}{2c}\left[\left(1 - \frac{X}{2(1+|Y_L|)}\right) - \left(1 - \frac{X}{2(1-|Y_L|)}\right)\right] dz = \frac{\omega}{2c} X Y_L \, dz, \quad (11.110)$$

and upon using (11.42), (11.43), (11.46), and (11.47), $d\Omega$ can be expressed as

$$|d\Omega| \simeq \frac{\omega}{2c}\left(\frac{\omega_N}{\omega}\right)^2 \left|\frac{\omega_L}{\omega}\right| dz = \frac{1}{8\pi^2}\left(\frac{e^3}{\epsilon_0 m_e^2 c}\right) \frac{\mu_0 |H_L|}{f^2} N(z)\, dz, \quad (11.111)$$

where $N(z)$ is the electron density per unit volume. Finally, integrating the above along the propagation path, we obtain the following expression for the Faraday rotation (radians)

$$|\Omega| \simeq \frac{1}{8\pi^2}\left(\frac{e^3}{\epsilon_0 m_e^2 c}\right) \frac{|B_L|}{f^2} T_{ec}, \quad (11.112)$$

where $B_L = \mu_0 H_L$ is the longitudinal component of the magnetic flux density, and T_{ec} is the TEC in the ionosphere along the propagation path defined by

$$T_{ec} \equiv \int_s N(s)\, ds. \quad (11.113)$$

The propagation path has been modified in the above to allow waves propagating in an arbitrary direction ds. Substituting the values of the physical constants in (11.112), we have

$$|\Omega| \simeq 2.36 \times 10^4 \frac{|B_L|}{f^2} T_{ec}, \quad (11.114)$$

with Ω expressed in radians, f in hertz, B_L in tesla, and T_{ec} in electrons/m^2.

A plot of the Faraday rotation as a function of TEC and frequency is shown in Figure 11.21, assuming a high-latitude value of $B_L = 54$ μT. In the northern hemisphere, the Faraday rotation is counterclockwise as seen from an observer looking up. The average value of Ω has a cyclical (diurnal, seasonal, and solar cyclical) component that can be predicted and hence compensated by a proper adjustment of the relative tilt angle between the antennas. However, deviations from the average behavior occur as a result of geomagnetic storms that are difficult to predict. It should be noted that Faraday rotation occurs on the downlink as well as the uplink of an Earth–space system. To avoid Faraday rotation problems, circularly polarized antennas are often used since the characteristic polarizations undergo no changes during

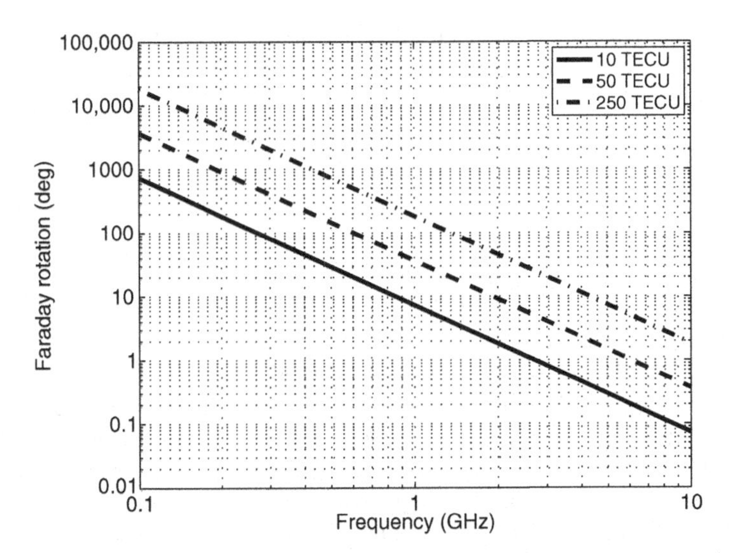

FIGURE 11.21 Faraday rotation as a function of frequency (in GHz) and TEC (in TEC units, where 1 TECU = 10^{16} electrons/m^2) $B_L = 54\mu$T.

propagation and the characteristic polarizations are approximately circular when the Q_L approximation holds.

11.14.2 Group Delay and Dispersion

Due to the presence of charged particles, electromagnetic signals propagating through the ionosphere are slowed down compared to signals in a vacuum. The infinitesimal delay time dt experienced by a signal propagating along an infinitesimal path ds in the ionosphere is influenced by the local index of refraction and can be written as

$$dt_{\pm} = \frac{ds}{v_g} = \frac{ds}{cn_{R\pm}}, \qquad (11.115)$$

where v_g is the group velocity. In the VHF and SHF frequency bands, the quantities Y_L, Y_T, and Z are very small and can be neglected in the refractive index expression (11.61). X also is small, but neglecting it would remove the entire effect of the plasma and is therefore not permissible. As a result, the infinitesimal excess delay time with respect to vacuum, denoted as $dt_{\pm}^{e}$, becomes

$$dt_{\pm}^{e} = \frac{ds}{cn_{R\pm}} - \frac{ds}{c} = \frac{ds}{c}\left(\frac{1-n_{R\pm}}{n_{R\pm}}\right) \simeq \left(1-\sqrt{1-X}\right)\frac{ds}{c} \simeq \frac{X}{2c}\,ds, \qquad (11.116)$$

where the fact that X is small has been used in expanding the square root by the binomial theorem. Substituting (11.42) and (11.46) in (11.116), we have

$$dt_{\pm}^{e} = dt^{e} \simeq \frac{1}{8\pi^2}\left(\frac{e^2}{\epsilon_0 m_e c}\right)\frac{N(s)}{f^2}\,ds. \qquad (11.117)$$

The total excess delay time is given by the integration of (11.117) along the propagation path, resulting with the use of (11.113) in

$$\Delta t^{\mathrm{e}} \simeq \frac{1}{8\pi^2}\left(\frac{e^2}{\epsilon_0 m_{\mathrm{e}} c}\right)\frac{T_{\mathrm{ec}}}{f^2}. \tag{11.118}$$

Substituting the values of the physical constants in (11.118), we have

$$\Delta t^{\mathrm{e}} \simeq 1.34 \times 10^{-7}\,\frac{T_{\mathrm{ec}}}{f^2}, \tag{11.119}$$

with Δt^{e} expressed in seconds, f in hertz, and T_{ec} in electrons/m^2.

Figure 11.22 illustrates the percentage of yearly average daytime hours when $\Delta t^{\mathrm{e}} > 20$ ns for a 1.6 GHz signal vertically incident on the ionosphere during a period of relatively high solar activity. The global distribution of Δt^{e} is influenced by the global T_{ec} distribution in the ionosphere. Since the excess delay time depends on frequency, different frequency components of a trans-ionospheric wideband signal will be subject to different delay times, resulting in dispersion. Figure 11.23 shows the difference in the time delay at the lower and upper frequencies of a pulse with temporal width τ centered at a specified frequency, for $T_{\mathrm{ec}} = 5 \times 10^{17}$ electrons/m^2.

Group delays are a source of error in global navigation satellite systems such as the Global Positioning System (GPS) because they introduce an external bias source to the pseudorange, that is, the calculated "distance" between the GPS receiver and the GPS satellite. This distance is obtained from the satellite-to-receiver travel time: the range error $\Delta\rho$ is given by $\Delta\rho = c\,\Delta t^{\mathrm{e}}$.

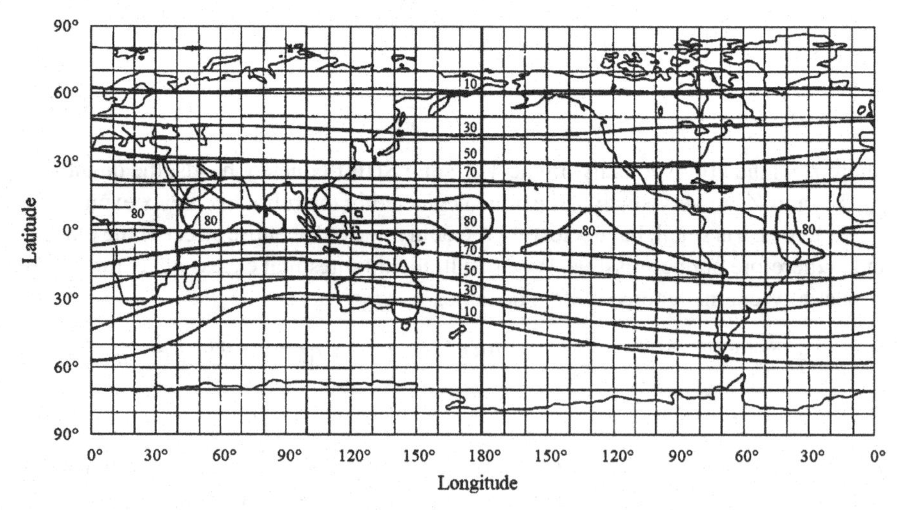

FIGURE 11.22 Contours of the percentage of yearly daytime hours when $\Delta t^{\mathrm{e}} > 20$ ns for a vertically incident 1.6 GHz signal, with sunspot number = 140. (*Source*: ITU-R Recommendation P.531-9 [12], used with permission.)

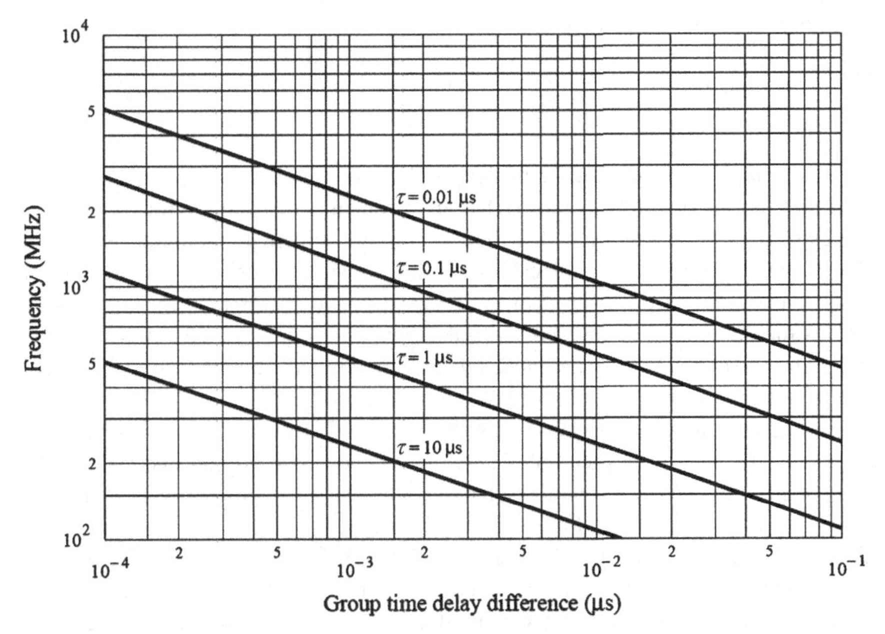

FIGURE 11.23 Difference in the one-way time delay between the lower and upper frequencies of the spectrum of a pulse of width τ transmitted through the ionosphere; pulse center frequency specified on vertical axis. (*Source*: ITU-R Recommendation P.531-9 [12], used with permission.)

11.14.3 Ionospheric Scintillations

Time-varying small-scale changes in the ionization density can cause fluctuations in the angle of arrival, amplitude, and phase of trans-ionospheric signals; these effects are called ionospheric scintillations and are analogous to similar effects in the troposphere discussed in Chapter 5. Here, we focus on the influence of ionospheric scintillations on signal amplitudes. Scintillation amplitudes greater than 15 dB are sometimes observed at 1 GHz. They decrease markedly at higher frequencies, and they vary greatly by geographical location. A detailed treatment is given in ITU-R Recommendation P.531-9 [12].

Statistics of the *instantaneous* variation of the signal intensity caused by ionospheric scintillations can be approximately described by a Nakagami-m probability density function [12], written as

$$f_\zeta(\zeta) = \frac{m^m\, \zeta^{m-1}\, e^{-m\zeta}}{\Gamma(m)}, \tag{11.120}$$

where $f_\zeta(\zeta)$ is the probability density function of the signal intensity (i.e., squared amplitude) ζ, $\Gamma(m)$ is the gamma function, and the mean value of the intensity ζ is normalized to one, that is, $\langle \zeta \rangle = 1$. (The $\langle\rangle$ symbol indicates a mean value.) The m-coefficient of the Nakagami-m distribution is given by $m = 1/\sigma_4^2$, where σ_4 is the "scintillation index" defined as

$$\sigma_4 \equiv \sqrt{\frac{\langle \zeta^2 \rangle - \langle \zeta \rangle^2}{\langle \zeta \rangle^2}}. \tag{11.121}$$

The scintillation index is related to the level of intensity fluctuations. As σ_4 approaches 1, the distribution of (11.121) approaches an exponential distribution for the intensity (corresponding to a Rayleigh distribution for the voltage magnitude, as seen in Chapter 8). Sometimes, $\sigma_4 > 1$ can occur due to wave focusing by the irregularities. For engineering purposes, a more descriptive parameter than the scintillation index is the fluctuation in the peak-to-peak signal intensity expressed in dB, denoted as I_4. The relationship between these two parameters can be approximated through $I_4 = 27.5\,\sigma_4^{1.26}$. Table 11.1 compares a few values of σ_4 and I_4.

The fading rate of ionospheric scintillations tends to vary from about 0.01 Hz to about 1 Hz. A "power spectral density" plot decomposes signal fluctuations into component frequencies, so that the relative contributions of fluctuations on different timescales can be discerned. Figure 11.24 illustrates a typical power spectral density of the scintillation in a geostationary satellite link at 4 GHz; higher values of the power spectral density indicate that signal fluctuations at the associated frequencies are more predominant. The power spectral density of the scintillations in Figure 11.24 exhibits a large range of frequency dependencies, with the fluctuation strength decreasing rapidly as their frequencies become higher (timescales become shorter.) Observations with dependencies ranging from $1/f$ to $1/f^6$ have been reported. In the absence of direct measurement data, ITU-R Recommendation P.513-9 suggests using $1/f^3$ for system applications [12]. Strong scintillations usually do not occur continuously but in distinct scintillation events. Ionospheric scintillations are more pronounced around a 20°-wide latitude ring centered on the geomagnetic equator, where they tend to occur predominantly in the late evening and early morning, and in regions above latitude 55°, where there is no preferential time and the behavior is more irregular.

It should be pointed out that the combined presence of rain attenuation, studied in Chapter 5, and ionospheric scintillation can cause effects—particularly in equatorial regions where both are stronger—that neither alone can produce. For example, the

TABLE 11.1 Conversion Table Between Some Values of the Scintillation Index σ_4 and the Intensity Fluctuation Expressed in dB

σ_4	I_4 (dB)
0.1	1.5
0.2	3.5
0.5	11
0.8	20
1.0	27.5

Source: ITU-R Recommendation P.531-9 [12].

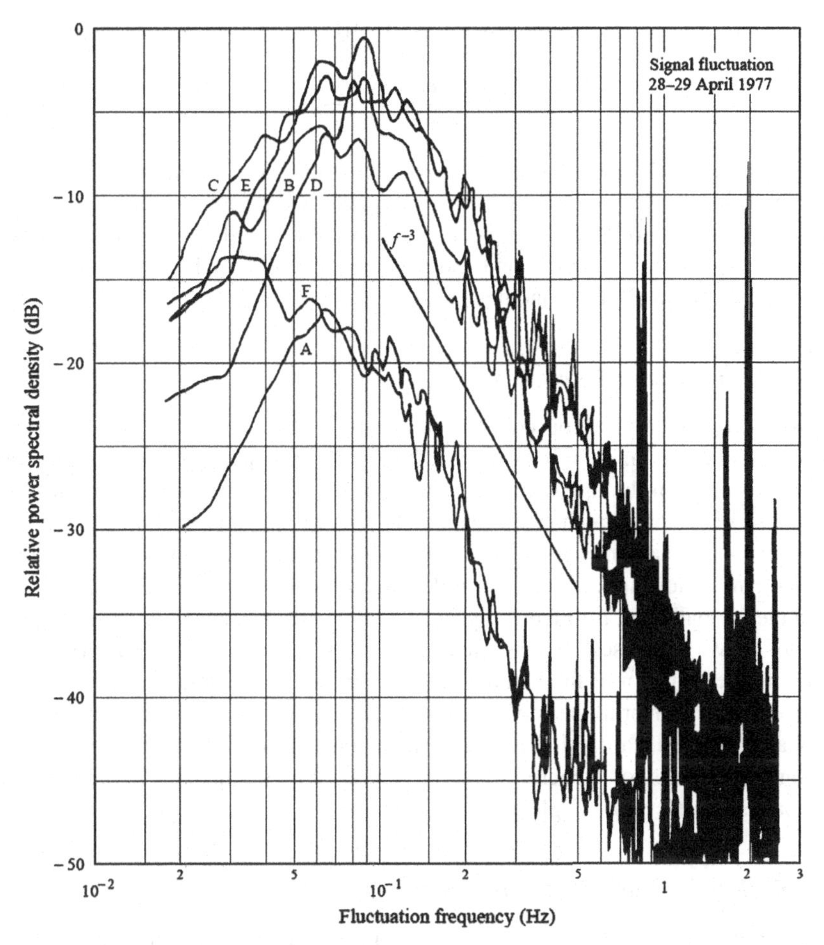

FIGURE 11.24 Power spectral density of the ionospheric scintillation for a 4 GHz geostationary satellite link. Note the $1/f^3$ behavior visible in the 0.1–1 Hz frequency range. The six different curves labeled A–F refer to a sequence of measurements taken 30 min or 1 h apart. (*Source*: ITU-R Recommendation P.531-9 [12], used with permission.)

combination can produce significant fluctuation in the cross-polarization signal, even though rain attenuation does not cause significant signal fluctuations and ionospheric scintillation does not cause significant depolarization.

11.14.4 Attenuation

We have previously considered ionospheric absorption in Section 11.9 when we discussed the conditions for deviative and nondeviative absorption. For trans-ionospheric signals at VHF band and above—where ω is much greater than the collision frequency ν—the attenuation constant α in (11.102) can be approximated as

$$\alpha \simeq \left(\frac{e^2}{2\epsilon_0 m_e c}\right)\frac{\nu N}{\omega^2}, \tag{11.122}$$

where $n_r \simeq 1$ (no refraction) was also used. Under these conditions, the total attenuation in dB for a one-way traverse of the ionosphere will be proportional to $1/f^2$. For vertical links at 30 MHz, for example, the typical (mid-latitude) ionospheric one-way attenuation due to absorption is on the order of 0.2–0.5 dB. At 1 GHz and above, the absorption is below 0.01 dB and hence can be neglected.

11.14.5 Ionospheric Refraction

The skywave mode is a consequence of strong ionospheric refraction that bends upgoing rays at MF and HF back toward the Earth. At higher frequencies, refraction causes a change in the apparent angle of arrival ϕ_a from the satellite signal at the ground terminal, but the change is generally small compared to the beamwidth of the ground station antenna and therefore usually has little effect on system operation. The amount of ionospheric refraction depends on the electron density distribution along the ray path and, for identical propagation conditions, the error in the apparent angle of arrival decreases with the square of the frequency in the VHF to SHF bands, similar to the behavior of the Faraday rotation and the group delay.

At 1 GHz, the error in the angle of arrival of a signal with elevation angle of 30° is less than 0.06 mrad. Table 11.2 lists some approximate peak ionospheric perturbations for (one-way) trans-ionospheric links in a few selected frequencies in the UHF and SHF bands. An elevation angle of 30° is assumed.

11.14.6 Monitoring TEC Distribution

We have seen that the TEC is a very important parameter in determining the effect of ionospheric perturbations on trans-ionospheric links. The TEC in (11.113) refers to total electron content integrated *along the ray path* from the satellite to the ground terminal. This is also referred to as the "slant TEC", as opposed to the "vertical TEC", which is integrated along a zenith path. The vertical TEC typically varies between 10^{16} and 200×10^{16} electrons/m^2 and is often expressed in TEC units (TECU), where 1 TECU $= 10^{16}$ electrons/m^2, as in Figure 11.21. For prediction purposes, estimates

TABLE 11.2 Approximate Peak Ionospheric Perturbation Effects

Perturbation	500 MHz	1 GHz	10 GHz
Faraday rotation	7.6 rad	1.9 rad	0.02 rad
Group delay	1 μs	250 ns	2.5 ns
Attenuation	0.2 dB	0.05 dB	$<10^{-3}$ dB
Refraction ($\Delta\phi_a$)	250 μrad	60 μrad	<1 μrad

cf. ITU-R Recommendation P.618-9 [13].

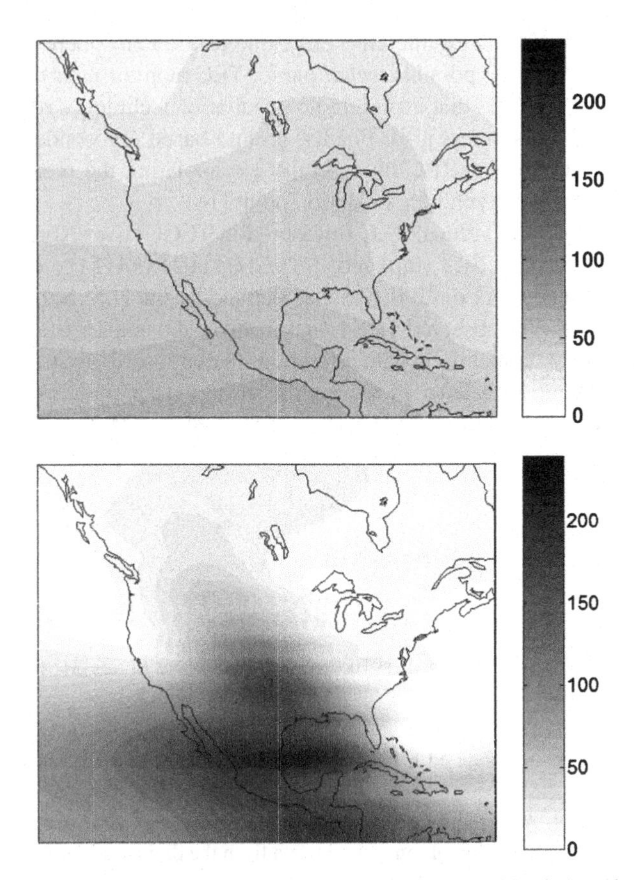

FIGURE 11.25 Map of the distribution of the vertical TEC over North America, under quiet (top) and stormy (bottom) ionospheric conditions. (*Source*: Ref. [17].)

for the vertical TEC can be obtained from computer models, such as the International Reference Ionosphere (IRI) model or the NeQuick ionospheric model mentioned in Chapter 10.

Neglecting refraction effects and for a sufficiently small off-zenith satellite look angle γ (so that the Earth curvature can be neglected and the ionosphere can be assumed horizontally stratified) the vertical TEC, $T_{\mathrm{ec}}^{\mathrm{V}}$, can be related to the slant TEC, T_{ec}, via

$$T_{\mathrm{ec}}^{\mathrm{V}} = T_{\mathrm{ec}} \cos \gamma. \tag{11.123}$$

One possible technique for monitoring the temporal (due to diurnal, seasonal, and solar cycle effects) and spatial (geographic) variability of the vertical TEC is by relying on a network of dual-frequency ground receivers of global navigation satellite systems such as GPS [14]. The dispersive nature of the ionosphere allows TEC estimation by GPS observations at two frequencies in the same location through equation (11.119)—here using the GPS error due to the ionosphere as a source of TEC information! Another possible technique for TEC monitoring is through the use of low-Earth orbit satellites that exploit radio occultation techniques relying on radio signals from GPS transmitters [15]. Finally, ground-based ionosonde data can also be used to estimate vertical TEC by carefully extrapolating the measured vertical electron density profile beyond the inflection point [16].

Figure 11.25 plots the vertical TEC (measured in TECU) over North America on 2 days: 10/28/2003 at 21:15UT (top) and 10/29/2003 at 21:00UT (bottom) from data obtained through a network of GPS stations operated by the U.S. National Geodetic Survey, NOAA [17]. The observed TEC values from the top map (in the range of about 50 TECU) are fairly typical in the late afternoon. In contrast, the bottom map shows the ionosphere about 24 h later, in the middle of an extremely strong ionospheric storm, with TEC values above 200 TECU in parts of central Mexico.

REFERENCES

1. Davies, K., *Ionospheric Radio*, IEE Electromagnetic Wave Series 31, Peter Peregrinus Ltd., London, 1989.
2. Milsom, J., "Surface waves, and sky waves below 2 MHz," in *Propagation of Radiowaves*, second edition (L. Barclay, ed.), pp. 357–383, The Institution of Electrical Engineers, London, 2003.
3. Reinisch, B. W., I. A. Galkin, G. Khmyrov, A. Kozlov, and D. F. Kitrosser, "Automated collection and dissemination of ionospheric data from the digisonde network," *Adv. Radio Sci.*, vol. 2, pp. 241–247, 2004.
4. Boithias, L., *Radiowave Propagation*, McGraw-Hill, New York, 1987.
5. Bennet, J. A., "Some reflections on the ray tracing of radio rays in the ionosphere," *Radio Sci. Bull.*, no. 327, pp. 10–12, 2008.
6. Cannon, P. S., and P. A. Bradley, "Ionospheric propagation," in *Propagation of Radiowaves*, second edition (L. Barclay, ed.), pp. 313–334, The Institution of Electrical Engineers, London, 2003.
7. Rush, C. M., "Ionospheric radio propagation models and predictions–a mini-review," *IEEE Trans. Antennas Propagat.*, vol. 34, no. 9, pp. 1163–1170, 1986.
8. Cannon, P. S., "Ionospheric prediction methods and models," in *Propagation of Radiowaves*, second edition (L. Barclay, ed.), pp. 335–355, The Institution of Electrical Engineers, London, 2003.
9. Teters, L. R., J. L. Lloyd, G. W. Haydon, and D. L. Lucas, "Estimating the Performance of Telecommunication Systems Using the Ionospheric Transmission Channel—Ionospheric Communications Analysis and Prediction Program User's Manual," National Telecommunication and Information Administration, Report NTIA 83–127, Boulder, CO, 1983.
10. ITU-R Recommendation P.533-9, "Method for the prediction of the performance of HF circuits," International Telecommunication Union, 2007.

11. Tascione, T. F., K. W. Kroehl, and B. A. Hausman, "A technical description of the ionospheric conductivity and electron density profile model (ICED version 196-II)," Syst. Doc vol. 7, Air Force Weather Agency, Scott Air Force Base, Illinois, 1987.
12. ITU-R Recommendation P.531-9, "Ionospheric propagation data and prediction methods required for the design of satellite services and systems," International Telecommunication Union, 2007.
13. ITU-R Recommendation P.618-9, "Propagation data and prediction methods required for the design of Earth–Space telecommunication systems," International Telecommunication Union, 2007.
14. Liu, Z., S. Skone, Y. Gao, and A. Komjathy, "Ionospheric modeling using GPS data," *GPS Solutions*, vol. 9, no. 1, pp. 63–66, 2005.
15. Angling, M. J., and P. S. Cannon, "Assimilation of radio occultation measurements into background ionospheric models," *Radio Sci.*, vol. 39, no. 1, RS1S08, 2004.
16. Reinisch, B. W., X. Huang, A. Belehaki, and R. Ilma, "Using scale heights derived from bottomside ionograms for modelling the IRI topside profile," *Adv. Radio Sci.*, vol. 2, pp. 293–297, 2004.
17. Spencer P. S. J., D. S. Robertson, and G. L. Mader, "Ionospheric data assimilation methods for geodetic applications," *Position, Location, and Navigation Symposium 2004*, Proceedings, pp. 510–517, Monterey, CA, April 26–29, 2004.
18. ITU-R Recommendation P.1239-1, "Reference ionospheric characteristics," International Telecommunication Union, 2007.

12

OTHER PROPAGATION MECHANISMS AND APPLICATIONS

12.1 INTRODUCTION

This chapter presents overviews of two of the more "unusual" propagation mechanisms (tropospheric and meteor scatter) mentioned briefly in Chapter 1, along with a discussion of propagation effects for two applications: tropospheric delays in satellite navigation systems and propagation effects on radar systems. While the material is presented only at an introductory level, it is intended to enable the reader to understand the basic aspects of these topics; sources of further information are also provided.

12.2 TROPOSPHERIC SCATTER

12.2.1 Introduction

At VHF and higher frequencies, the ionosphere is not a viable communications medium. In these frequency bands, the mode of signal transmission is characterized by three propagation mechanisms, depending on the distance, as sketched in Figure 12.1. For LOS conditions, the direct transmission mode of Chapter 5 is applicable, possibly with the addition of reflections from the ground, as discussed in Chapter 6. Beyond the radio horizon, diffraction (Chapter 7) takes over. The attenuation rate of the diffracted signal depends on the nature of the diffracting obstacle and is higher for relatively smooth ground than for a knife-edge-like mountain range. Even in the

Radiowave Propagation: Physics and Applications. By Curt A. Levis, Joel T. Johnson, and Fernando L. Teixeira

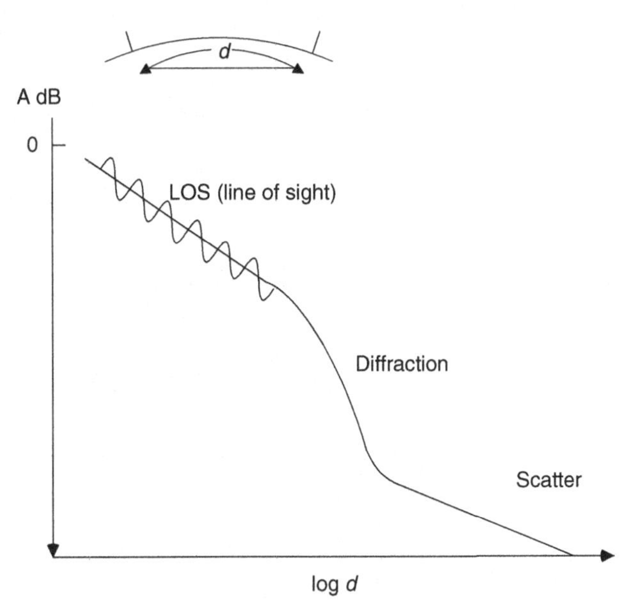

FIGURE 12.1 Typical plot of path attenuation versus distance d.

diffraction range of distances, another signal is present due to scatter from turbulent irregularities in the lower troposphere. At relatively short ranges, diffraction is dominant. Beyond some distance, which depends on both the diffracting obstacle and the tropospheric turbulence, the scattered signal becomes dominant and can be used for communication. Tropospheric scatter links can operate over long distances: as far as 1000 km for narrowband signals and up to 200–300 km for wider band signals [1]. However, signal fading on tropospheric scatter links is usually severe, and high-power transmitters, high-gain antennas, sensitive receivers, and often some form of diversity are needed to overcome fading problems. Limits on antenna sizes put the minimum practical frequency for "troposcatter" systems at around 200 MHz, while increasing losses with frequency and rain attenuation put an upper limit at approximately 5 GHz.

Since tropospheric scatter propagation depends on scattering from variations in the atmospheric refractivity, theoretical prediction of expected losses requires a very detailed knowledge of the local atmospheric structure. Obtaining such information is exceedingly difficult, given the wide variations in atmospheric properties that occur in time, location, and altitude. As a result, theoretical methods are of limited use in tropospheric scatter predictions. Measurements form the basis of the only available prediction models, which are empirical. Even their application to situations for which the measurements were not performed is of limited accuracy. To reduce uncertainties in empirical predictions, models are usually provided for the annual median loss along a tropospheric path, with additional information available on the fading rates to be expected.

The typical geometry of a tropospheric scatter link is illustrated in Figure 12.2. Stations beyond the line of sight direct their antennas near the horizon. The region

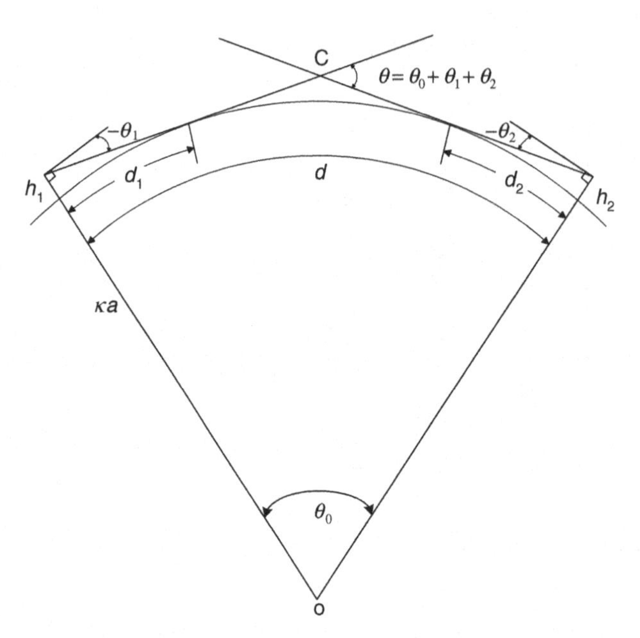

FIGURE 12.2 Geometry of a tropospheric scatter link.

that contains both the transmitting and receiving antenna beams, centered around the point C in the figure, is called the common volume. Atmospheric scatter that gives rise to the received signal results from this common volume. Measurements have shown that antennas should be directed 0.2–0.6 beamwidths above the horizon for optimal results due to the fact that the atmospheric irregularities that cause the scattering decrease with altitude. For simplicity, these antenna pointing angles are not shown in Figure 12.2, and they are not considered in the empirical model of Section 12.2.2 below.

It is clear that fading should be expected on a tropospheric scatter link since the scattered signals originate from a large number of sources within the common volume, and the position of these scatterers changes in time as the atmosphere varies. It is also clear that the tropospheric scatter mechanism is dispersive since the phase shift obtained in scattering from individual atmospheric irregularities will vary with frequency. Thus tropospheric links are always narrowband, and transmissible bandwidths decrease from around 10 MHz at short ranges to less than 1 MHz at long ranges, due to the increased number of scatterers inside the common volume. High transmitted powers, sensitive receivers, large antennas, and diversity techniques allow these fades to be tolerated, however, and reliable links can be created. The antenna-to-medium coupling loss term L_{coup}, discussed in Chapter 5, also becomes significant because the received fields are not plane waves since they are due to contributions from many scatterers spread over a substantial field of view, not due to a distant point source. This loss places a limit on the increase in link performance that can be obtained by increasing antenna gains.

As with ionospheric communication systems, the relative importance of troposcatter for long-distance communications has diminished since the advent of satellite systems. Nevertheless, troposcatter systems are still used for some long-distance terrestrial links where it is not feasible or cost-effective to employ many repeater stations between the transmitter and receiver, such as in links over the ocean to distant offshore oil platforms or islands. Fixed and, more recently, mobile troposcatter systems are also of importance in military applications, either to augment existing satellite systems or to provide a back up mode of operation.

12.2.2 Empirical Model for the Median Path Loss

ITU-R Recommendation P.617-1 [2] provides an empirical estimate of the annual median path loss for troposcatter links. This estimate is based on data from 200 MHz to 4 GHz, and it is written in decibels as

$$L_{\mathrm{p}} = 30\log_{10} f + 10\log_{10} d + 30\log_{10}\theta + N(H,h) + L_{\mathrm{coup}} + M, \quad (12.1)$$

where f is the frequency in MHz, d is the path length in kilometers, and θ is the scatter angle indicated in Figure 12.2, expressed in milliradians (mrad). The antenna-to-medium coupling loss L_{coup} is given by

$$L_{\mathrm{coup}} = 0.07\, e^{0.055(G_{\mathrm{T}}+G_{\mathrm{R}})}, \quad (12.2)$$

where G_{T} and G_{R} are the transmitter and receiver antenna gains in dBi. The term $N(H,h)$ also depends on the scatter angle and is given by

$$N(H,h) = 20\log_{10}(5+\gamma H) + 4.34\,\gamma h, \quad (12.3)$$

where $H = 2.5\times 10^{-4}\,\theta d$ and $h = 1.25\times 10^{-7}\,\theta^2 R_{\mathrm{e}}$, with θ and d expressed in mrad and km, respectively, and where $R_{\mathrm{e}} = \kappa a$ is the effective Earth radius for median refractive conditions, as discussed in Chapter 6, expressed in km. The meteorological structure parameter M depends on the local climate and varies between about 20 and 40 dB. The atmospheric structure parameter γ also depends on the climate and assumes values in the range 0.3 ± 0.03 km^{-1}. A table with values for M and γ in different climates can be found in Ref. [2]. Because the median path loss increases as θ increases, the antennas in a troposcatter link are generally positioned to minimize the scatter angle and maximize the common volume. Extensions of the annual median loss predictions of equation (12.1) to losses at other probabilities of occurrence are also described in Ref. [2].

12.2.3 Fading in Troposcatter Links

The above empirical model provides an estimate for the annual median path loss. In practice, the path loss in troposcatter links shows both daily and seasonal variations. Daily variations are caused by changes in tropospheric conditions such as pressure, moisture, and temperature. Similarly, seasonal variations occur as these conditions vary according to annual cycles. These seasonal variations depend on the local climate. In temperate climates, the path loss is larger during the winter; in desert climates, the path loss is larger during the summer. Path losses in equatorial climates have little seasonal variation. Regarding daily variations, the path loss is larger in the afternoon and decreases at night and in the morning [3].

Daily and seasonal variations correspond to a slow fading of the signal in decibels that can be described statistically in terms of a Gaussian distribution, as seen in Chapter 8. The variance of the slow fading tends to decrease as the path length increases because the common volume is then attained at higher altitudes, which show less daily variability in scatter properties than at lower altitudes. For a link with $d = 80$ km, the slow fading standard deviation is typically about 30 dB, whereas for $d = 500$ km, it decreases to around 10 dB [3]. Superimposed on the slow fading are short-term variations in the signal that can reach up to 20 fades per second. This fast fading is caused by fluctuations in the number, orientation, and position of the tropospheric fluctuations in the common volume, and is described statistically by a Rayleigh distribution (Chapter 8). Because fading in troposcatter links can be severe, diversity schemes become necessary. Spatial, frequency, and angle diversity can be used individually or in combination to combat troposcatter fading. For angle diversity, multiple antenna feeds can be spaced in the vertical direction to create multiple vertically spaced common volumes.

Tropospheric scattering can be a source of interference in other communication systems. In this case, the common volume can arise even from the intersection of the sidelobes of two antennas, or the sidelobe of one antenna and the mainlobe of a second antenna.

12.3 METEOR SCATTER

Meteors enter the Earth's upper atmosphere continually at speeds ranging from 10 to 75 km/s. At a height of approximately 120 km, meteors encounter an atmospheric density large enough that heat builds up between the meteor and the atmosphere to produce an ionization trail. These trails have an average length of 15 km but can reach up to 50 km.

Most of the meteors are quite small and disintegrate very quickly in the upper atmosphere. Nevertheless, it is estimated that about 10^{12} meteors enter the atmosphere every day that can produce an ionization trail sufficient to be of use in reflecting radio signals. In general, meteors suitable for this purposes have a mass greater than 1×10^{-7} g [4].

There are two mechanisms of reflection of radio signals by meteor trails. In *overdense trails*, that is with ionization density above 2×10^{14} electrons/m, the trail core behaves as a plasma and the incident wave is reflected from the region of the trail where the plasma frequency equals the radiowave frequency. *Underdense trails*, that is, those with ionization density below the above threshold, do not reach a plasma state. In this case, the incident wave excites the individual electrons that then reradiate the signal as a collection of small dipoles. The peak ionization density of a meteor trail depends on the meteor mass. Typically, only meteors with mass 1×10^{-3} g or greater can produce overdense trails. Because of the diffusion of the electrons into the atmosphere, overdense trails grow in volume while the density decreases, eventually becoming underdense. The relaxation time associated with this diffusion process depends on the local atmospheric density.

Because the number of underdense trails vastly exceeds the number of overdense trails, the reflected signal in meteor scatter systems is primarily caused by the former. Communication systems relying on meteor scatter are designed as if all reflections were caused by underdense trails. Because the duration of usable individual trails is typically very short, ranging from tens of milliseconds to a few seconds, communication links that rely on meteor scatter are often called *meteor burst communication* (MBC) systems. MBC links typically can have ranges of up to 1800 km and their useful frequency range is typically between about 20 and 120 MHz. This frequency range is limited from below by ionospheric attenuation and the need to minimize interference from ionospheric links, and from above by the fact that the received power for underdense trail scattering decreases as f^{-3}. The optimal performance occurs in the 40–55 MHz frequency range [5,6].

There is a seasonal and diurnal variability in the rate of incidence of meteors that produces a variability in MBC link performance. The highest incidence of meteors occurs at dawn and the lowest at sunset.[1] The reason is that the morning side of the Earth moves forward relative to the orbital movement around the Sun, whereas the evening side moves backward; therefore, more meteors are "swept" by the Earth in the morning. Seasonal variations of similar magnitude occur with a maximum flux rate in August and a minimum in February. This seasonal variation occurs because the Earth's orbit passes through a region of denser solar orbit material in August [5,6].

As already stated, MBC systems can only transmit bursts of data while a usable meteor trail is present. To determine when this is the case, each station continually sends a "probing" signal to the other station; when this signal is received, a suitable trail is present and data transmission can proceed. A usable trail must create at least approximately a specular reflection condition for the transmitter and receiver locations. When no usable trail is present, the transmission becomes idle ("wait time"). The wait time is much greater than the burst time: for burst durations of 0.1 s, the wait time is about 17 s on average; for burst durations of 0.4 s, it is about 140 s [7]. As a result, the average data rates are very small, in the range 10–300 bps, and MBC systems are incompatible with certain applications such as voice communications. On the other hand, they are relatively inexpensive for low data rate applications and can be deployed using a decentralized, mobile infrastructure with small antennas. Because of this, MBC systems are particularly suited for remote, nonintensive data collection. In the United States, the Department of Agriculture currently operates a MBC system called SNOTEL ("snow telemetry") with over 500 sites for collection of snow depth and other climate data in remote unattended locations. In another application, the Alaskan MBC system is utilized by five U.S. federal agencies to collect data from remote areas in Alaska.

[1] This variability refers to the so-called *sporadic* meteors as opposed to *shower* meteors. The latter do not occur frequently enough to be of use in MBC links. However, shower meteors are at times used by amateur radio operators to establish links in the VHF band.

MBC systems also have interest in some military applications because the ground footprint of the meteor scatter signals is smaller than that of satellite or skywave systems. As a result, MBC links provide a higher level of covertness and good resistance to ground-based interference and jamming. Mobile MBC systems also hold strategic interest as a "high survivability" means of long-distance data communications: in the event of a nuclear conflict, communication systems based upon centralized or fixed stations (satellites, fixed troposcatter, etc.) would be primary targets, and the fallout present in the D layer of the atmosphere would likely disrupt ionospheric skywave links.

ITU-R Recommendation P.843-1 [6] provides useful information for the design and planning of MBC systems, including the estimation of useful burst rates, choice of operation frequencies, and antenna considerations.

12.4 TROPOSPHERIC DELAY IN GLOBAL SATELLITE NAVIGATION SYSTEMS

Global navigation systems such as GPS are based on time delay measurements of signals broadcast by satellites and received by ground or aircraft users. The measured time delays are affected by variations in the refractive index n of the troposphere. Because $n > 1$, there is an excess time delay compared to propagation in a vacuum that can be quite significant, ranging from 6 to 80 ns (corresponding to 2–25 m range differences for GPS systems).[2] Errors due to these delays must be compensated if high-precision positional information is to be achieved [8]. In Chapter 11, we have examined an analogous excess delay in GPS signals caused by the ionosphere. Because the ionospheric delay is dispersive at L band, it can be compensated using dual-frequency receivers: GPS systems use a primary signal (L1) at 1575.42 MHz and a secondary signal (L2) at 1227.60 MHz. Unfortunately, this solution is of no use to compensate the tropospheric delay because the troposphere is not dispersive for frequencies below 15 GHz.

The tropospheric delay is written in terms of the refractive index n as

$$\Delta t^{\mathrm{a}} = \frac{1}{c} \int_S n(s)\, ds, \tag{12.4}$$

where c is the speed of light, and ds is the infinitesimal length along the propagation path S from the satellite to the receiver; c and ds should be in compatible units in this equation (typically m/s and m, with the delay in seconds). Because of refraction, the actual propagation path is curved and hence longer than the geometrical (straight) path length from the satellite to the receiver. The delay along the geometrical path

[2]There is also an excess time delay caused by relativistic effects, but this is considerably smaller and will be ignored here.

when no tropospheric effects are present is given by

$$\Delta t^{\mathrm{s}} = \frac{1}{c}\int_{S_{\mathrm{s}}} ds, \tag{12.5}$$

where S_{s} is a straight line from the satellite to the receiver. The excess delay time is $\Delta t^{\mathrm{e}} = \Delta t^{\mathrm{a}} - \Delta t^{\mathrm{s}}$, that is,

$$\Delta t^{\mathrm{e}} = \frac{1}{c}\int_{S} n(s)\, ds - \frac{1}{c}\int_{S_{\mathrm{s}}} ds. \tag{12.6}$$

This last equation can be rewritten as

$$\begin{aligned}\Delta t^{\mathrm{e}} &= \frac{1}{c}\int_{S} [n(s) - 1]\, ds + \frac{1}{c}\left[\int_{S} ds - \int_{S_{\mathrm{s}}} ds\right] \\ &= \frac{1}{c}\int_{S} [n(s) - 1]\, ds + \frac{\Delta S}{c}\end{aligned} \tag{12.7}$$

or

$$\Delta t^{\mathrm{e}} = \frac{10^{-6}}{c}\int_{S} N(s)\, ds + \frac{\Delta S}{c} \tag{12.8}$$

with the refractive index $N(s)$ expressed in "N-units," as seen in Chapter 6, and where ΔS is the difference in length between actual and straight paths. Sometimes, the term "tropospheric delay" is used to refer to the time delay above multiplied by the speed of light, that is, the length $c\,\Delta t^{\mathrm{e}}$. The tropospheric delay is more pronounced for lower elevation angles (i.e., with the satellite near the horizon) than for large elevation angles (satellite near zenith) because of the longer path length through the troposphere in the former case.

It is convenient to separate two contributions to the tropospheric delay at L band: the *dry delay*, caused by dry gases (mostly nitrogen and oxygen), and the *wet delay*, caused by water vapor. The dry delay typically accounts for 90% or more of the total tropospheric delay. Because the dry atmosphere has a very uniform composition (with the exception of CO_2, which makes a negligible contribution), the dry delay can be modeled with a higher degree of predictability as a function of the local ground temperature and atmospheric pressure. In contrast, the water vapor content is highly variable in altitude, geographical location, and time, making it more difficult to model even when local ground humidity data are available.

The tropospheric delay can be written in terms of a sum of dry delay and wet delay contributions as

$$c\,\Delta t^{\mathrm{e}} = \Delta S_{\mathrm{zd}}\, m_{\mathrm{d}}(\theta) + \Delta S_{\mathrm{zw}}\, m_{\mathrm{w}}(\theta), \tag{12.9}$$

where ΔS_{zd} and ΔS_{zw} are the dry and wet delay, respectively, at zenith, while $m_{\mathrm{d}}(\theta)$ and $m_{\mathrm{w}}(\theta)$ are dry and wet "mapping functions" (obliquity factors), respectively, that model increases in the tropospheric delay as the elevation angle θ decreases. The mapping functions depend on both θ and the atmospheric profile of the refractive index, but for simplicity many approximate predictive models assume dependence only on θ.

Predictive models for the tropospheric delay can be obtained in two ways. One way is to determine the atmospheric profile from measurements and calculate the refractive index of each layer of the troposphere. Once the refractive index profile is known, the delay can be found using ray tracing. The atmospheric profile consists of measurements of temperature, pressure, and water vapor density for varying altitudes. The most sophisticated models are based on numerical weather model data [9]. This approach gives more precise results but it has two drawbacks: first, a typical user rarely has access to such measurements, and second, it is computationally intensive [10]. Another way to find the tropospheric delay is to use empirical models for the combined zenith delay due to both dry and wet components, which rely on either the seasonal average of weather parameters or actual delay measurement data at the receiver location, augmented by mapping functions to model the effect of longer propagation paths for satellites at lower elevation angles [9,10].

One of the simplest approximate models [8] is based on a fit of annual average delay data obtained using U.S. Standard Atmosphere parameters (Chapter 10) at 30°N, 45°N, 60°N latitude and ray tracing. It predicts a combined zenith delay

$$\Delta S_{\mathrm{zd}} + \Delta S_{\mathrm{zw}} = 2.4405 \tag{12.10}$$

in meters and mapping functions of the form

$$m_{\mathrm{d}}(\theta) = m_{\mathrm{w}}(\theta) = \frac{1.0121}{\sin\theta + 0.0121}, \tag{12.11}$$

so that

$$c\,\Delta t^{\mathrm{e}}(0) = \frac{2.47}{\sin\theta + 0.0121} \tag{12.12}$$

in units of meters for a receiver at sea level ($h = 0$). This approximate model gives a value of $c\Delta t^{\mathrm{e}}(0) = 2.44$ m at zenith ($\theta = 90°$) and $c\Delta t^{\mathrm{e}}(0) = 24.9$ m for a 5° elevation angle. By further assuming that the refractivity varies with altitude approximately as $(1 - h/h_{\mathrm{d}})^4$ with h in km and $h_{\mathrm{d}} = 43$ km, the following approximation [8] can be obtained for the delay in meters for a receiver at altitude h:

$$c\,\Delta t^{\mathrm{e}}(h) = c\,\Delta t^{e}(0)\, e^{-0.133h}. \tag{12.13}$$

This simplified model gives only a rough estimate of the tropospheric delay. Also, neither short-term variations affecting the wet delay component nor seasonal/ geographical variations are included.

The term "tropospheric delay" is somewhat of a misnomer because about a quarter of the nonionized delay actually occurs due to atmospheric gases above the troposphere, that is, in the tropopause and stratosphere. Note that the troposphere also produces attenuation of GPS signals at L band, mainly due to oxygen absorption. However, this attenuation is generally small at GPS frequencies: no greater than 0.5 dB for satellites at low elevation angles and 0.05 dB for satellites near zenith [3].

12.5 PROPAGATION EFFECTS ON RADAR SYSTEMS

Radar (radio detection and ranging) systems attempt to determine information about remote targets by transmitting electromagnetic waves and measuring properties of the signals that are returned. Radars are routinely used for tracking of aircraft and ships; for collision avoidance, aircraft guidance, and landing-assistance systems; for determination of weather patterns, terrain altitude, and snow cover thickness; for underground exploration of minerals, gas, and oil; and in many other applications. Radar systems can be classified as either monostatic or bistatic. In monostatic radar, the transmitting and receiving antennas are colocated, whereas in bistatic radar, the transmitting and receiving antennas are in different locations.

In many applications, information about the remote region or target is sought from the power level of the returned signal, that is, the "target echo". To quantify this process, we first consider monostatic radar and propagation in free space for simplicity. Assuming that a discrete target is present in the far field at distance R and look angle (θ, ϕ) from the radar, we can use equation (5.2) to write the power density (Poynting vector magnitude) of the field incident on the target as

$$S_{\mathrm{T}}(R, \theta, \phi) = \frac{P_{\mathrm{T}}\, G_{\mathrm{T}}(\theta, \phi)}{4\pi R^2}, \tag{12.14}$$

where P_{T} is the total input power to the transmitter antenna and G_{T} is the transmitter antenna gain. The basic parameter that describes the reflectivity of the target is the *radar cross section* (RCS), denoted as σ_{RCS}. The RCS is an area such that the product $S_{\mathrm{T}}(R, \theta, \phi)\, \sigma_{\mathrm{RCS}}$ (watts) is equal to the radiated power of an equivalent isotropic radiator at the target location that would produce the same power density S_{R} as observed at the radar receiver antenna. In other words,

$$S_{\mathrm{R}} = \frac{1}{4\pi R^2}\, S_{\mathrm{T}}(R, \theta, \phi)\, \sigma_{\mathrm{RCS}}\,. \tag{12.15}$$

Mathematically, the RCS of a target can also be expressed as the (far-field) limit

$$\sigma_{\mathrm{RCS}} = \lim_{R \to \infty} 4\pi R^2\, \frac{|\underline{\overline{E}}_{\mathrm{s}}|^2}{|\underline{\overline{E}}_{\mathrm{i}}|^2}, \tag{12.16}$$

where $\underline{\overline{E}}_{\mathrm{i}}$ is the electric field incident *on* the target and $\underline{\overline{E}}_{\mathrm{s}}$ is the far-field scattered electric field *from* the target. Among the factors that may influence the RCS are the target's size, shape, and material (constitutive) properties. In general, the RCS also depends on the relative orientation of the target with respect to the direction of propagation of the incident wave, as well as on the polarization of the incident wave. Since the target is assumed to be in the far field of the radar, the RCS does not depend on the range R.

Using equation (12.15), the received power for a receiving antenna with effective aperture area A_{R} can be found as

$$P_{\mathrm{R}} = A_{\mathrm{R}}\, S_{\mathrm{R}} = \frac{A_R}{4\pi R^2}\, S_{\mathrm{T}}(R, \theta, \phi)\, \sigma_{\mathrm{RCS}}\,. \tag{12.17}$$

Combining equations (12.14) and (12.17), and using the fact that $A_R = \lambda^2 G_R/(4\pi)$, we obtain the monostatic *radar equation*

$$P_R = \frac{P_T\, G_T(\theta, \phi)\, G_R(\theta, \phi)\, \lambda^2\, \sigma_{RCS}}{(4\pi)^3 R^4}, \tag{12.18}$$

where G_R is the receiver antenna gain and λ is the wavelength of operation. Note that when the same antenna is used for transmitting and receiving in monostatic radar, $G_T(\theta, \phi) = G_R(\theta, \phi)$. In contrast to the Friis transmission formula, which applies to one-way transmissions,—equation (5.5)—the received power in the radar equation decays as $1/R^4$ instead of $1/R^2$.

Rearranging some factors, we can also rewrite equation (12.18) as

$$P_R = P_T\, G_T(\theta, \phi)\, G_R(\theta, \phi)\, \frac{4\pi\sigma_{RCS}}{\lambda^2} \left(\frac{\lambda}{4\pi R}\right)^4, \tag{12.19}$$

which, when expressed in decibels, becomes

$$\begin{aligned} P_{R,dBw} = {} & P_{T,dBW} + G_{T,dBi}(\theta, \phi) + G_{R,dBi}(\theta, \phi) \\ & + 10 \log_{10}\left(\frac{4\pi\sigma_{RCS}}{\lambda^2}\right) - 2\, L_p^{free}, \end{aligned} \tag{12.20}$$

where L_p^{free} is the free-space path loss in decibels, given by

$$L_p^{free} = 20 \log_{10}\left(\frac{4\pi R}{\lambda}\right). \tag{12.21}$$

For propagation scenarios where the use of the free-space path loss is not justified, L_p^{free} should be replaced by the appropriate path loss model L_p, that is,

$$\begin{aligned} P_{R,dBw} = {} & P_{T,dBW} + G_{T,dBi}(\theta, \phi) + G_{R,dBi}(\theta, \phi) \\ & + 10 \log_{10}\left(\frac{4\pi\sigma_{RCS}}{\lambda^2}\right) - 2\, L_p. \end{aligned} \tag{12.22}$$

Free-space propagation is often assumed in many radar analyses; it is important to recognize when this is appropriate and when it is not. It is also common to see equation (12.22) written as

$$\begin{aligned} P_{R,dBw} = {} & P_{T,dBW} + G_{T,dBi}(\theta, \phi) + G_{R,dBi}(\theta, \phi) \\ & + 10 \log_{10}\left(\frac{4\pi\sigma_{RCS}}{\lambda^2}\right) - 2\, L_p^{free} + 2F, \end{aligned}$$

where F is a factor that describes any propagation gain relative to free space (note that F can be either positive or negative).

When bistatic radar is considered, equation (12.22) is modified to

$$\begin{aligned} & P_{R,dBw} = P_{T,dBW} + \\ & G_{T,dBi}(\theta_T, \phi_T) + G_{R,dBi}(\theta_R, \phi_R) + 10 \log_{10}\left(\frac{4\pi\sigma_{RCS}^{B}}{\lambda^2}\right) - L_{p,T} - L_{p,R}, \end{aligned} \tag{12.23}$$

where (θ_T, ϕ_T), (θ_R, ϕ_R) are the look angles from the transmitter and receiver antennas to the target, respectively, while $L_{p,T}$ and $L_{p,R}$ are the path losses from the target to the transmitter and receiver antennas, respectively. Here, $\sigma_{RCS}{}^{B}$ represents the bistatic RCS of the target, which depends on the bistatic angle (the angle between the transmitter and the receiver, as seen from the target location) and the orientation of the target.

Radar engineering is a vast and fascinating topic, and further discussion on this subject is beyond the objectives of this book. A good introduction to this topic can be found in Ref. [11].

REFERENCES

1. M. P. M. Hall, *Effects of the Troposphere on Radio Communication,* Peter Peregrinus Ltd., 1979.
2. ITU-R Recommendation P.617-1, "Propagation prediction techniques and data required for the design of trans-horizon radio-relay systems," International Telecommunication Union, 1992.
3. Bothias, L., *Radiowave Propagation*, Section 8.1.1.3, McGraw-Hill, 1987.
4. Sugar, G. R., "Radio propagation by reflection from meteor trails," *Proc. IEEE,* vol. 52, pp. 116–136, 1964.
5. Freeman, R. L., *Radio System Design for Telecommunications*, Chapter 13, third edition, Wiley–IEEE Press, 2007.
6. ITU-R Recommendation P.843-1, "Communication by meteor burst propagation," International Telecommunication Union, 1997.
7. "Meteor burst communications: An ignored phenomenon?", *Cryptologic Quarterly*, vol. 9, no. 3, pp. 47–63, 1990.
8. Spiker Jr., J. J., "Tropospheric effects on GPS," in *Global Positioning System: Theory and Applications*, Volume 1, pp. 517–546, (Progress in Astronautics and Aeronautics vol. 163), AIAA, Washington DC, 1996.
9. Boehm, J., A. E. Niell, P. Tregoning, and H. Schuh, "Global mapping function (GMF): a new empirical mapping function based on numerical weather model data," *Geophys. Res. Lett.*, vol. 33, no. 7, p. L07304, 2006.
10. Hobiger, T., R. Ichikawa, Y. Koyama, and T. Kondo, "Computation of troposphere slant delays on a GPU," *IEEE Trans. Geosci. Remote Sens.*, vol. 47, no. 10, 2009.
11. Skolnik, M., *Introduction to Radar Systems*, third edition, McGraw-Hill, 2002.

INDEX

Radiowave Propagation: Physics and Applications. By Curt A. Levis, Joel T. Johnson, and Fernando L. Teixeira